W0036315

ADHESION PROBLEMS IN THE RECYCLING OF CONCRETE

NATO CONFERENCE SERIES

I   Ecology
II   Systems Science
III   Human Factors
IV   Marine Sciences
V   Air-Sea Interactions
VI   Materials Science

VI   MATERIALS SCIENCE

ADHESION PROBLEMS IN THE RECYCLING OF CONCRETE

Edited by

Pieter C. Kreijger

Eindhoven University of Technology
Eindhoven, The Netherlands

Published in cooperation with NATO Scientific Affairs Division

PLENUM PRESS · NEW YORK AND LONDON

Library of Congress Cataloging in Publication Data

NATO Advanced Research Institute on Adhesion Problems in the Recycling of
  Concrete (1980 : Saint-Rémy-les-Chevreuse, France)
  Adhesion problems in the recycling of concrete [proceedings]
  (NATO conference series. VI, Materials science ; v. 4)
  Bibliography: p.
  Includes index.
  1. Concrete—Recycling—Congresses. I. Kreijger, Pieter C. II. Title. III. Series.
TA440.N334 1980   666′.893   81-15411
                                                                AACR2

ISBN-13: 978-1-4615-8314-1   e-ISBN-13: 978-1-4615-8312-7
DOI: 10.1007/978-1-4615-8312-7

Proceedings of a NATO Advanced Research Institute on Adhesion
Problems in the Recycling of Concrete, held November 25-28, 1980,
in Saint-Rémy-les-Chevreuse, France

© 1981 Plenum Press, New York
A Division of Plenum Publishing Corporation
233 Spring Street, New York, N.Y. 10013

Softcover reprint of the hardcover 1st edition 1981

All rights reserved

No part of this book may be reproduced, stored in a retrieval system, or transmitted
in any form or by any means, electronic, mechanical, photocopying, microfilming,
recording, or otherwise, without written permission from the Publisher

PREFACE

The building explosion during the years 1945-1960 will
inevitably lead to increased demolition in the next decades since
the lifetime distribution of structures no longer fulfills its
functional social requirements in an acceptable way. In the
building period mentioned there was a great increase in reinforced
and prestressed concrete construction. Consequently there is now
more and more concrete to be demolished. Increasingly severe
demands will be made upon demolition technology, including the
demand for human- and environment-friendly techniques. On the
other hand, the possibility of disposing of debris by dumping is
steadily diminishing, especially close to major cities and
generally in countries with a high population density. At the
same time in such countries and in such urban areas a shortage
of aggregates for making concrete will develop as a result of
restrictions on aggregate working because of its effect on the
environment and because of the unavailability of aggregate
deposits due to urban development.

From the foregoing it follows that recycling and re-use of
environment- and human-friendly demolished and fragmented building
rubble should be considered. The translation of this general
problem into terms of materials science is possible by forming
clear ideas of adhesion and cohesion: the whole process of
demolition, fragmentation, and recycling or re-use of concrete
is to break the bonding forces between atoms and molecules and
to form new bonds across the interfaces of various particles of
either the same nature or a different nature. Studies of these
processes are especially important since it is now known that
there is a good chance that recycling and re-use of demolished
buildings will be economically possible. Consequently an
evaluation of existing knowledge applied to the adhesion problems
in the recycling of concrete will be of great importance and
could be realized by a (first) conference on this topic, bringing together experts who are specialists in physical, chemical,
and/or fracture mechanical aspects of materials science, with
the aim of discussing and applying their particular knowledge
to the adhesion and cohesion aspects mentioned.

It was a fortunate coincidence that the formulation of the problem as given in the preceding paragraph just fitted in with the program of the Special Program Panel on Materials Science of the Scientific Affairs Division of NATO. Their acceptance of our proposal to hold such a Conference made it possible to realize this Advanced Research Institute on "Adhesion Problems in the Recycling of Concrete" at Saint-Rémy-les-Chevreuse, 25-28 November 1980. Without their financial help this A.R.I. would not have been possible.

The preparation of the A.R.I. was set up by a Programming and Organizing Committee originally consisting of Dr. L.H. Everett (UK), Dr. G. Frohnsdorff (USA), Prof. Dr. Ing. H.K. Hilsdorf (W. Germany), Prof. Ir. P.C. Kreijger (The Netherlands), Dr. J.J. Mills (USA), and Prof. Dr. F.H. Wittmann (at that time, The Netherlands). Later on Dr. Everett and Prof. Wittmann had to resign owing to special circumstances. NATO appointed Prof. Kreijger Director of the A.R.I., and he took care of and is responsible for the realization of the organization.

We are grateful to the "Union Technique Interprofessionelle" of the "Fédération Nationale du Bâtiment" in Paris for their consent to make use of all facilities of their research center "Domaine de Saint Paul" at Saint-Rémy-les-Chevreuse. Our special gratitude goes to Mrs. M. Geudelin of the "Direction de la Recherche," "Relations Internationales," who not only made all organizational preparations for a four-day stay of about thirty persons at the research center, but also looked after technical facilities (Mr. Millet) and typing and copying capacity (Mrs. Vallot) during the meetings, and kept exellent minutes and prepared reviews of the various discussions at the A.R.I., partly assisted by Mrs. M.J. Rubens as a freelance second secretary.

Looking back to this first scientific conference on the recycling of concrete, consisting of a series of invited lectures, giving the state of the art of related background fields, and (eight) workshops under such headings as "Fragmentation," "Contamination," "Recycling," "Re-use," and "Future demolition-friendly materials," with invited introductory lectures, I wish to thank all participants for the cooperative, cheerful, and positive manner in which they approached the assigned work, especially the difficult task of preparing workshop reports. We realize that the success of our discussions was stimulated to a great extent by the great effort that was put into papers and introductory lectures, and our gratitude goes to the authors; we are grateful for all special contributions given during the workshop sessions or later on, and which have been incorporated into these Proceedings.

In line with the foregoing, the Proceedings of this A.R.I.
are composed of four parts:
    Part 1 - Organization of the A.R.I.
    Part 2 - Lectures, discussions, and general discussion.
    Part 3 - Introductory lectures to workshops, workshop reports,
           and contributions to workshops.
    Part 4 - Concluding remarks.

The information that has become available reveals numerous
opportunities for new and demanding research in this field and
should provide helpful directions to those who are already involved
with this research, to those who wish to enter the field, and to
those who are responsible for making administrative decisions.

Finally I want to thank my secretary, Mrs. D.M. Vermeltfoort-
Danen, for her organizational assistance and the care taken with
all correspondence.

Eindhoven   Pieter C. Kreijger
December 1980

CONTENTS

PART 1 – ORGANIZATION OF THE ADVANCED RESEARCH INSTITUTE ON

"ADHESION PROBLEMS IN THE RECYCLING OF CONCRETE"

ORGANIZATION OF THE ADVANCED RESEARCH INSTITUTE ON

"ADHESION PROBLEMS IN THE RECYCLING OF CONCRETE"

Pieter C. Kreijger

Technological University of Eindhoven

Den Dolech 2, 5600 MB Eindhoven, The Netherlands

1. Introduction

The idea for a scientific conference on the recycling and re-use of concrete originally was put forward by Prof. Kreijger to the Special Program Panel on Materials Science of the Scientific Affairs Division of NATO, of which he was a member at that time. This Panel on Materials Science had just been formed by the Science Committee of NATO next to the already existing program panels for ecosciences, systems science, human factors, marine sciences, and air-sea interaction. The panel decided that its special program should be concerned with

- the conservation of scarce resources.
- the more effective use of abundant resources.

Based on these goals, the panel decided to concentrate on

- more effective processing and/or use of bulk materials
- development of materials that may affect future usage patterns

The materials concerned fell in the following categories:

- industrial materials
- novel materials
- materials for energy systems.

Thus, in principle the proposed subject seemed to fit in well with the stated program of the panel. After discussion it was decided that emphasis should be given to the scientific rather than the engineering side of the subject, which restricted the subject to "Adhesion Problems in the Recycling of Concrete," and a proposal for

3

such a conference was invited by the panel. Prof. Kreijger contacted
Dr. L.H. Everett (Building Research Establishment, U.K.), who at that
time was president of RILEM Technical Committee 37 DRC: Demolition and
recycling of concrete; Kreijger himself was a member of this committee.
Together they formulated a motivated proposal for a four-day conference of invited experts, complete with a budget and a proposal for a
Programming Committee. Both were accepted by the Special Program
Panel on Materials Science, and the necessary means were made available to form the Programming and Organizing Committee with the assign-
ment of drawing up the scientific program and proposing a list of experts to be invited. The Committee consisted of Dr. L.E. Everett (Build-
ing Research Establishment, Garston, U.K.), Dr. G. Frohnsdorff (leader,
Building Composites Group, Structures and Materials Division, Center
for Building Technology, NEL, Washington, D.C., U.S.A.), Prof. Dr.-Ing.
H.K. Hilsdorf (Institut für Baustofftechnologie, University of Karls-
ruhe, W. Germany), Prof. Ir. P.C. Kreijger (Group Materials Science,
Technological University Eindhoven, The Netherlands), Dr. J.J. Mills
(Martin Marietta Laboratories, Baltimore, Md., U.S.A.), and Prof. Dr.
F.H. Wittmann (Group Materials Science, Technological University Delft,
The Netherlands). The Committee met at the Building Research Station
(Garston, U.K.) on February 26-27, 1979, and based on Dr. Everett and
Prof. Kreijgers' proposal, the scientific working program was developed, consisting of one-and-a-half days of invited lectures, two days
of (eight) workshop sessions, each preceded by an invited introductory
lecture, and a final session. The subjects and contents of all lectures and workshops were outlined, and lecturers as well as conference
participants were proposed. After the Committee reported to the
Special Program Panel on Materials Science, this Panel agreed to the
proposal and made available the necessary means, while Dr. Everett
and Prof. Kreijger were appointed Directors of the Advanced Research
Institute on "Adhesion Problems in the Recycling of Concrete."

Soon thereafter, however, Dr. Everett had to resign for reasons
of health; some time later Prof. Wittmann had to resign as a member
of the Programming and Organizing Committee because of his transfer
to the Technological University of Lausanne. Consequently, Prof.
Kreijger together with the remaining members of the Committee arranged
the organization of the A.R.I, which finally was held November 25-28,
1980, at Domaine de Saint Paul, the research center of the "Union
Technique Interprofessionelle" of the "Fédération Nationale du Bâtiment" at Saint-Rémy-les-Chevreuse, France.

2. Program of the A.R.I. on "Adhesion Problems in the Recycling of Concrete"

The Programming and Organizing Committee considered materials
science and materials technology as inseparable parts of knowledge
in the problem of the recycling of concrete. The basic scheme of
the program of the A.R.I. therefore was based on Fig. 1, as was the
choice for the invitation of participants.

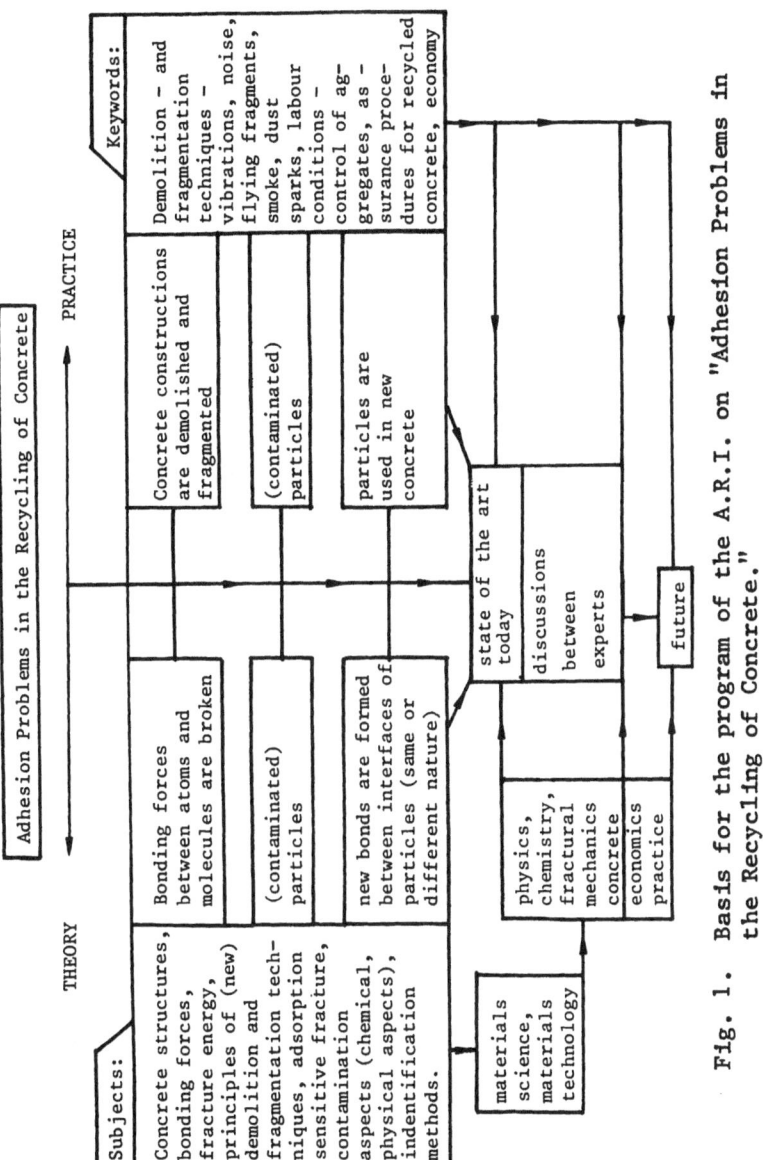

Fig. 1. Basis for the program of the A.R.I. on "Adhesion Problems in the Recycling of Concrete."

The scheme of Fig. 1 was then translated into a desired program according to Table 1.

Table 1. Desired program for the A.R.I. on "Adhesion Problems in the Recycling of Concrete"

THEORY	STATE OF THE ART A. Introduction and stating the problem	PRACTICE

B. Lectures on concrete:
 	
1. Structure of concrete
2. Principles of adhesion-bonding
   in cement and concrete
3. Chemical and physical aspects
   of contamination in concrete
4. Fracture of concrete
 	
On demolition/fragmentation
of concrete:
 	
5. Principles of demolition and
   and fragmentation techniques
   of concrete
6. Absorption sensitive fracture
   and fragmentation
	7. Economics of demolition, re-
	   cycling and re-use of concrete
	
	C. Workshops with introductory
	   lectures:
	
	- Demolition and fragmentation
	  of plain, reinforced and
	  prestressed concrete (3)
	- Contamination effects on frag-
	  mentation and on recycling
	  and re-use of concrete (2)
	- Recycling of concrete (1)
	- Re-use of concrete (1)
	
	DISCUSSIONS	

- future demolition - friendly materials (1) entries		

The principal contents of lectures and workshops were described, and suggestions for invited lectures and participants listed. The director of the Institute contacted the lecturers in Europe, while Dr. Frohnsdorff did so for the U.S.A., and both succeeded in obtaining acceptances from the suggested persons for nearly all parts of the program. It was a pity, however, that Dr. L.H. Everett (Building Research Station, England), who despite his resignation from the Organizing Committee agreed to be responsible for lecture B-5 (see Table 1), was unable to prepare this lecture and did not attend the meeting. Faced with this problem, we gratefully acknowledge the (unprepared) lecture of Mr. P. Lindsell (University of Surrey, England) on "Demolition Techniques for Concrete Structures" (with many slides), a clarifying lecture which is reproduced in Part 3 of these Proceedings as a contribution to Workshop 1. Also, there was a last-minute excuse of Mr. P.A. Griffiths (Charles Griffiths Ltd., Demolition and Excavation Concretors, London), who originally agreed to give an introductory lecture on demolition and fragmentation of plain and reinforced concrete (see Table 1) and whose task was voluntarily taken over by Prof. Dr. S. Mindess (University of British Columbia, Canada), which again was appreciated very much. Still, one can say that the originally planned program was realized to a good approximation.

Table 2a. Program of the A.R.I. on "Adhesion Problems in the Recycling of Concrete" on November 25, 1980

Lectures: (chairman of the day P.C. Kreijger)

- Introduction and stating the problem - P.C. Kreijger (University of Technology, Eindhoven)
- The structure of concrete: a macroscopic view - C.D. Pomeroy (Materials Research Department, Cement and Concrete Association, Wexham Springs)
- Principles of adhesion- bonding in cement and concrete- D. Tabor (Cavendish Laboratory, University of Cambridge)
- Contamination problems in the recycling of concrete - J.F. Young (University of Illinois)
- Fracture energy of concrete - H.K. Hilsdorf, S. Ziegeldorf (Institut für Baustofftechnologie, University of Karlsruhe)
- Absorption sensitive fracture and fragmentation - J.J. Mills (Martin Marietta Laboratories, Baltimore)

A Chairman was invited for each workshop, and we gratefully acknowledge their willingness to fulfill this difficult task, for they too promised - together with the introductory lecturers and the secretaries, Mrs. M. Geudelin and Mrs. M.J. Rubens - to write the workshop reports. In Tables 2a-d the real program of the A.R.I. on "Adhesion Problems in the Recycling of Concrete" is given. The sequence of the various workshops was established based on the participants' preferences for attending the workshops. Nearly all lectures were sent to the participants two to four weeks in advance of the A.R.I. Table 3 gives an overall view of the introductory lectures.

Table 2b. Program of the A.R.I. on "Adhesion Problems in the
          Recycling of Concrete" on November 26, 1980

Morning session: Lecture and general discussion

- Economics of concrete recycling in the United States -
  S. Frondistou-Yannas (Management and Technology Associa-
  tes, Inc., Newton, U.S.A.
- General discussion (chairman P.C. Kreijger)

Afternoon session: Workshop 1 and 2

Workshop 1 Fragmentation of plain and reinforced concrete	Workshop 2 Recycling of concrete (Aggre- gates for use in concrete)
chairman and introduction: J.J. Mills secretary: Mrs. M. Geudelin discussion report to plenary meeting	chairman: Mrs. S. Frondistou- Yannas secretary: Mrs. M.J. Rubens introductory lecture: B. Mather discussion report to plenary meeting

Table 2c. Program of the A.R.I. on "Adhesion Problems in the
          Recycling of Concrete" on November 27, 1980

Morning session: Workshop 3 and 4

Workshop 3 Re-use of concrete (other than as aggregates for concrete)	Workshop 4 Fragmentation of prestressed concrete
chairman: C.D. Pomeroy secretary: Mrs. M.J. Rubens introductory lecture: S.H. Carpenter discussion report to plenary meeting	chairman: H.K. Hilsdorf secretary: Mrs. M. Geudelin introductory lecture: P. Lind- sell discussion report to plenary meeting

continued on next page

Afternoon session: Workshop 5 and 6

Workshop 5 Contamination (effects on fragmentation, fibre - and polymer concrete)	Workshop 6 Future demolition - friendly materials
chairman : S.P. Shah	chairman and introduction: Mrs. D.M Roy
secretary: Mrs. M. Geudelin introductory lecture: S.P. Shah discussion report to plenary meeting	secretary: Mrs. M.J. Rubens discussion report to plenary meeting

Table 2d.   Program of the A.R.I. on "Adhesion Problems in the
Recycling of Concrete" on November 28, 1980

Extended morning session: Workshops 7 and 8, final session

Workshop 7 Contamination (effects on re- cycling and re-use)	Workshop 8 Fragmentation, all types of con- crete
chairman and introduction G. Frohnsdorff secretary: Mrs. M.J. Rubens discussion report to plenary meeting	chairman: S.P. Shah secretary: Mrs. M. Geudelin no introductory lecture discussion report to plenary meeting

final session: proceedings of the A.R.I., concluding remarks

Table 3.   Introductory lectures to the workshops

Workshop 2 - Recycling of Concrete (aggregates for use in Concrete)
Crushed Concrete as Concrete Aggregate - B. Mather
Workshop 3 - Re-use of concrete (other than as aggregates for concrete)
Recycling of Concrete into New Applications - R.L. Berger, S.H. Carpenter
Workshop 4 - Fragmentation of prestressed concrete
Demolition of Prestressed Concrete Structures - P. Lindsell
Workshop 5 - Contamination effects on fragmentation, fibre - and polymer concrete
Material characterization for fragmentation - S.P. Shah

3. Participants in the A.R.I. on "Adhesion Problems in the Recycling of Concrete"

In all, forty participants were invited while finally thirty-one participants from eleven countries attended the meetings of the A.R.I. Lists of the participants in alphabetical order and by country are given at the end of these Proceedings.

4. Editing the Proceedings

The input of the A.R.I. consists of lectures, discussions, and special contributions; these Proceedings can be considered to be the output. The contents list of Proceedings was proposed by the Organizing Committee (Frohnsdorff, Hilsdorf, Kreijger, Mills) and approved by the participants, but the final format is the responsibility of the editor.

Part 1 (Organization) and Part 4 (Concluding remarks) were written by the editor. In Part 2 (Lectures and discussions), after each lecture the direct discussion is given, and at the end a report of the general discussion. The reproduction of all discussions in Part 2 is based on the minutes of all discussions taken by Mrs. Geudelin, on the questions and answers drafted by those participants who asked questions and sent their reproduction to the editor, and finally on the notes of the editor himself. In Part 3, for each workshop first the introductory lecture (if any) is given followed by the minutes of the workshop discussions, kept by Mrs. Geudelin, the (summary) reports to the plenary meeting given by the chairman (if received), and finally the contributions to the session.

PART 2 - LECTURES, DISCUSSIONS AND GENERAL DISCUSSION

1.1 INTRODUCTION AND STATING THE PROBLEM(S)

Pieter C. Kreijger

Eindhoven University of Technology
Department of Architecture, Building and Planning
Group Science of Materials
P.O. Box 513 - 5600 MB Eindhoven - Netherlands

1. THE GENERAL PROBLEM

In the design of buildings and other structures no attention is paid to the question how they are to be demolished
when its economic, social or technical service life has expir-
ed. This is indeed understandable, bearing in mind that most
if not all designers are at the design stage hardly likely to
entertain the idea of their structural creations subsequently
being demolished. Yet it can be presumed that the majority of
structures and the parts thereof will in course of time no longer
fulfill the functional requirements applicable to them. Having
regard to the rapid evolution of our society, this is indeed
likely to arise within an ever decreasing length of time.
This means that, except for buildings designated for preservation as monuments, the greater number of structures will
sooner or later have to be demolished in order to be replaced by
others which are better suited to the needs of the time. Until
fairly recently, the structures to be demolished dated from a
time when building was for the most part done in masonry or
brickwork with timber and steel beams.
Concrete, a difficult material to demolish, was a comparatively
rare occurence. Finding suitable sites for dumping the rubble
presented no difficulties either.
    In recent times however, a number of changes have been
taking place at a relatively rapid rate, which will indeed give
rise to problems if they are not forestalled in good time:
- The great increase in reinforced and prestressed concrete
   construction after the Second World War must inevitably result in more and more concrete having to be demolished in

comparison with former times. Partly because reinforced concrete is a difficult material to demolish, increasingly severe
demands will be made upon demolition technology.
- The strenuous physical effort demanded of men employed in
demolition work will become less and less acceptable as labour
suitable for human beings: for example, an eight-hour working
day spent operating a hand-held hydraulic or pneumatic demolition hammer.
- Partly, an account of increased environmental awareness,
nuisance due to noise and dust will become increasingly unacceptable, especially in densely populated areas.
- The need for achieving the highest possible degree of adaptability of buildings to enable them to adjust to the rapid
developments that are taking place is becoming increasingly
manifest. This adaptability may necessitate the partial or
total demolition of structures, with, in extreme cases, the
rebuilding of entire urban districts.
- The possibility of disposing of rubble by dumping is steadily
dimishing especially close to major cities, whereas the quantities of rubble to be disposed of are increasing.
- Within the foreseeable future a shortage of aggregates for
making concrete will develop in a number of countries as a
result of restrictions on aggregate working, because of their
effect on the environment and because of the unavailability
of aggregate deposits due to urban development.
- From the foregoing it follows that re-use of environment- and
human-friendly demolished and fragmented building rubble should
be considered.
In general recycling of materials depends on two factors, the
collection of the material and the processing of it. In this
connection reference is possible to the Department of Chemical Engineering at the University of Delaware where Russell
and Swartzlander [1] developed a Recycle Index (R.I.) for
chemical waste, defined as the product of the two operations:
RI = (CI).(PI) where CI = Collection Index and PI = Processing Index = ranging from 0 (bad prospects) to 1 (good
prospects) and both indexes well defined according to given
scales, for the re-use of building waste the RI $\geq$ 0.35
means an economical possibility for recycling. According to
their scales, for the re-use of building waste the RI could be
calculated as about 0.4 which means that there might be a
good change that the recycling of demolished buildings
economically could be possible.
- The general problem holds social, economical and technological
al problems. It seems useful to describe somewhat more in
detail the engineering problem which can be derived from
this general problem and seems to be related to the keywords:
concrete, demolition and re-use.

2. THE ENGINEERING PROBLEM

The engineering questions that have to be answered are:
- How should demolition (of concrete) be done? What is the most economical method of demolition and what restrictions must be laid down with regard to noise nuisance, danger to surroundings and dust nuisance?
- What is to be done with the (concrete) rubble? Is dumping it on a tip the most economical solution in the long run or should the structure in view of increasing shortance of raw materials (and weighed against increasing energy prices) and increasing environmental awareness of the public, be endeavoured to be demolished in such a way that the building materials can be wholly or partly retrieved for re-use?
- Should not be investigated too, whether it can be ensured in the design state that the component parts of a (concrete) structure can be subsequently dismantled and if possible, re-used?
  The advantage of this so-called demountable building construction is moreover that those (concrete) parts which are not re-usable need not be demolished on the spot, but can instead be removed to "demolition and processing plants" which specialize in such activities.

While the last mentioned question is more a building construction problem the first mentioned ones are more technological (or material) engineering problems and therefore more related to material science.

These questions can be united further to one subject called "the recycling of concrete". It seems worthwhile to look further to the technological problems of the recycling of concrete.

3. RECYCLING OF CONCRETE AS A TECHNOLOGICAL PROBLEM

Recycling of concrete presupposes a certain amount of concrete available for recycling, the size and place of which at a time may have considerable influence on further actions which include:
- Application of demolition techniques to separated large concrete-units from reinforced or prestressed concrete-structures and its transport;
- Application of fragmentation techniques to separate large concrete-units or to concrete rubble (including separation of reinforcement) and its (mostly internal) transport;
- Recovery and/or re-use of demolished, fragmented and separated materials taking into consideration possible contamination and the properties of the fragmented materials.

3.1. Estimate of amount of demolished concrete

Although estimates of quantities of demolition-materials are
very approximate, they do show that these quantities are
significant [2,3], see table 1.

Table 1 Production of demolition waste in 10^6t/year:

U.S.A.	25	(18 concrete)
Canada	3	
Japan	7-12	(concrete only)
U.K.	21-23	(7.5-10.5 concrete)
Sweden	2-5	

An estimate of the amount of concrete to be recycled may be
based on the amount of (cement or) concrete used as function
of time [4] starting at about 1920.
Table 2 and fig.1 give such an estimate for the trend of the
amount of concrete to be demolished in the EEC during the
years 1970-2020 [5a,b].

Table 2   Trend of concrete to be demolished in EEC
           during 1970-2020 in 10^6ton/year concerned:

	average life of an object 50 years (curve B, fig.1)	average life of an object 50 years, however 20% stands 30 - and 30% 70 years [1] (curve C, fig.1)
1970	50	70
1980	130	130
1990	210	220
2000	220	310
2010	460	470
2020	800	680
1) Industrial buildings and flats 30 years, public buildings and dwellings 70 years.		

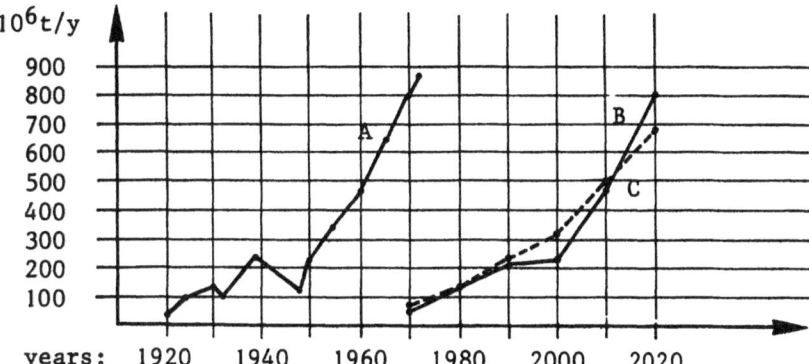

Curve A: concrete production
Curve B,C : amount of concrete rubble (see table 2)

Fig.1. Estimate of concrete production and amount of
        concrete rubble to be expected from demolition
        in EEC [5b]:

Table 3 Estimates of demolition waste from dwellings
         in the Netherlands (10^6t/year)

Demolition waste from:	1977 min.	1977 max.	1980 min.	1980 max.	1990 min.	1990 max.	2000 min.	2000 max.
demolition	1.1	1.1	1.2	1.3	1.2	1.5	1.2	1.4
renovation	1.2	1.2	1.0	1.4	0.6	0.8	0.4	0.4
new dwellings	0.8	0.8	0.7	0.7	0.4	0.4	0.4	0.4
Total	3.1	3.1	2.9	3.4	2.2	2.7	2.0	2.2

Botman and Kreiter [6] calculated a minimum and maximum
scenario for demolition from dwellings in the Netherlands on
the basis of now-a-days values of demolition-probabilities per
class of building year and expectations for futural development of these probabilities, see table 3.

From further investigations on the composition of demolition
waste it was found that during demolition in the Netherlands
already 20% of the wood, 90% of the steel and 100% of zinc,
lead and copper are removed, leaving an average building
waste composition of:

brickwork	62 %
concrete	24 %
brickrubble	6.1%
tiles	2.3%
bituminous mat.	0.2%
wood	4.7%
steel	0.1%
rest	0.6%
	100 %

This means that in 1980, 1990 and 2000 the amount of
demolished concrete to be expected from dwellings are:
resp. $0·67.10^6$ $0·55.10^6$ and $0·56.10^6$ ton/year.
Totally one expects in the Netherlands (population 14.10^6
inhabitants, concrete consumption about $36.10^6 t$/year)
$1-1.5 \times 10^6 t$/year demolished concrete and $3-3.5 \times 10^6 t$/year
demolished brickwork on an amount of about $6.10^6 t$/year
demolition waste.
A substantial growth in future is to be forseen.
Bekker [7] found that the cumulative (relative) distribution
of the life time variable of both demolished and withdrawn
dwellings could be represented best by a Weibull distribution
which has the additional advantage of being the basis for
a hazard-rate-function h(t) that turns out to be a parabola:
$h(t) = 5.645 \times 10^{-6} t^2$ in which t=lifetime in years and
5.645×10^{-6} a constant factor (based on statistical material
1968, Netherlands). Maybe this approach can be used for
improved calculations on the amount of demolished concrete
to be expected in the future, better than a sometimes heard
rule of thumb of $\frac{1}{2}-\frac{1}{4}$ ton per capita per year for major cities.
Although all estimates are very approximate they do show that
the quantities already now available are significant and
probably will show a substantial growth.
Slabs, walls, beams and columns may be partly or wholly
reinforced, prestressed or posttensioned. Especially the
last two cases mentioned are difficult to demolish.
Here it is a priority to have knowledge of the erection-
procedure and a complete set of "as-built" record drawings,
when any demolition sequence is being planned including the
relieve of the tension [9].

3.2. Demolition and fragmentation techniques

It may be useful to realize that most concrete buildings have
some system in their construction [8] as given in table 4.

Table 4 Construction systems in concrete buildings

type of building	type of construction
industrial and commercial buildings (< 4 floors)	mostly restricted to 3-4 floors: the skeleton consists of columns (as a whole) restrained in the foundation; beams are supported by cantilevers and slabs by beams; for parking garages only the span in one or two directions will be greater.
office buildings, school- and university buildings, banks (> 4 floors)	a) a skeleton of columns, beams and slabs, columns consist of more parts, horizontal rigidity/stability offered by cores (with staircases and elevators) or transverse walls, the connection column-beam is more complex since cantilevers are not wished; b) skeleton consists of bearing facade-components and slabs, horizontal forces are taken up by cores and transverse walls.
apartment buildings, elderly people-homes, etc.	a) transverse walls placed perpendicular to facades with slabs parallel to facades, lengthwise rigidity is provided by cores; b) joined slabs and walls of small dimensions (3-4m).

If not explosive blasting methods for the structure as a
whole are used, the first stage is to separate large concrete
'units' from the structure (demolition) while, mostly after
transport, elsewhere the 'units' are fragmented and
reinforcement separated from concrete.
Also in this stage explosive blasting can be used as well as
a variety of other methods. More expensive 'selective demolition'(different materials are separated during the demolition
on the site) is studied in various objects and still has to prove
its importance for recycling of demolition waste.

Table 5 gives an overall view of demolition techniques used
nowadays as well as in preparation for the future. Most of
last mentioned techniques still are in the laboratory stage.

Table 5   Demolition techniques

Nowadays techniques [9, 10, 11]	based on
- hydraulic and pneumatic hammers - ball and crane - cylinder drop weight - hydraulic crane with ripping tooth and shovel - "Arrow" drop ram - explosives	breaking and impact loading (or explosion)

- "Darda" and "Nibbler"
  hydraulic splitting
  and breaking equipment
	splitting and
	breaking
	(bending)

- diamond or carborundum
  based sawing and drilling
  equipment (+ hydr. bursting)
- thermic boring
- flame cutting
	cutting,making
	holes and slots
	by abrasion,
	pression and/or
	temperature

New techniques [10, 11] (mostly in laboratory stage)	based on

- high pressure water jet
- carbon dioxide expansion bursting
- ultrasonic vibrations
- laser
- electromagnetic waves or fields
- micro waves
- artificial frost attack
- artificial corroding of reinforcement
- artificial solution
- low frequency and high frequency
  currents
- plasma cutting torch
- disruption of micro structure by
  small explosives
- miniaturisation of nuclear explosives
	pression and/or
	temperature
	(sometimes locally)
	and/or electro-
	chemical action.

The choice of technique or combination of techniques will depend on the requirements or the restrictions imposed in connection with the demolition work to be carried out as well as on the nature and strength of the structure to be demolished. The aspects which are of importance in assessing any particular demolition technique are:

    technical feasibility (plain, reinforced or post-tensioned / pre-tensioned concrete and thickness of concrete, maintenance of apparatus)
    skill required,
    physical effort required of the demolition worker,
    suitability of application in built-up areas,
    precision of the method,
    noise,
    vibrations,
    flying fragments,
    danger,
    smoke, dust and/or sparks,
    energy consumption and cost.

After the separation of concrete "units" from the structure as occurs for most of the demolition techniques, these "units" have to be fragmented.
Fragmentation techniques are mostly developped in the mining industry in the form of crushers and which also are used for the fragmentation of building waste, inclusive concrete eg. jaw crushers (single or double toggle) with reduction rates of particle size of 9:1/7:1, cone and gyratory crushers with reduction rates of 6:1/10:1, incidently up to 15:1.
These types seem to be based mainly on compression and shear forces while the "prall"- and hammer-types function more by tensile and shear force. The material here is crushed by jetting it with high velocity to a wall or plates and the particle size reduction here is about 20:1 and even 40:1/100:1 respectively while the shape of the crushed particles is much more cubic than for jaw- and cone crushers. Also reinforcement in concrete needs not to be a problem if this jetting principle is used. The physics of comminution is a complex scientific field, see fig. 2, in which fracture mechanics is becoming an important tool. The symposium "Fragmentation" held in Paris, Dec. 1978, has tried to combine much knowledge in the field of comminution [12], while Méric [13] recently gave a review on the grinding of cement clinker.
A relative new idea in this field of comminution is the possibility to decrease the zetapotential and consequently, as has been shown by Westwood [14] experimentally, the fracture energy. So crushing in an appropriate electrolyte-solution in principle could save energy if the crack propagation is not too high. Neerhoff [15a] showed for both portland cementstone and blast furnace cementstone the relation between electrolyte concentration, zêtapotential and fracture energy (fig. 3).

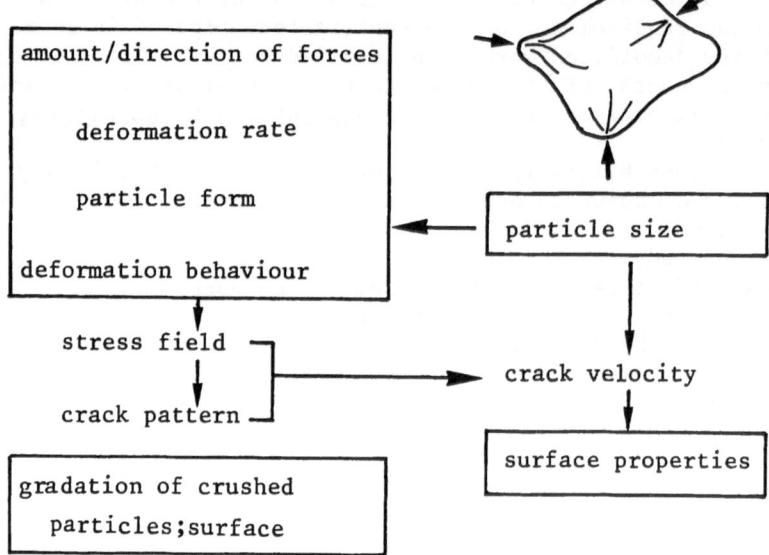

Fig. 2 - Aspects of fragmentation [see 12, Schönert]

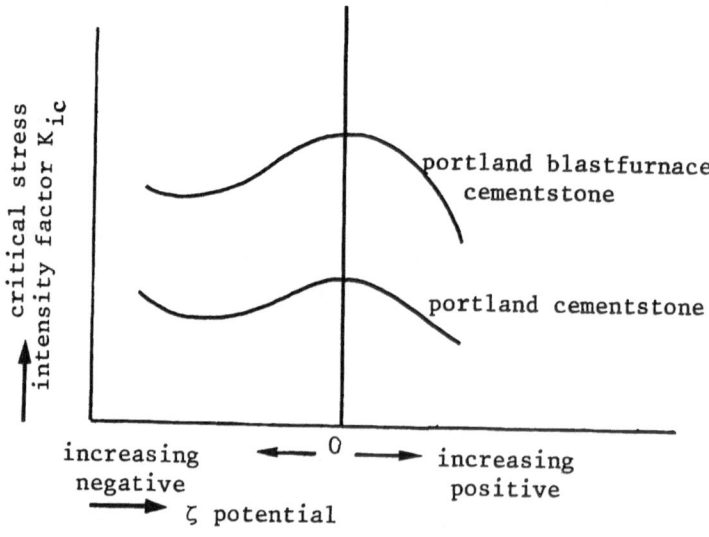

Fig. 3 - Correlation between critical stress intensity factor
         and zeta potential for cement stone   [15b]

For a zero zeta potential the kic-factor proved to be maximum while eg for increasing negative value of the zeta-potential a minimum in fracture energy occured at a certain value. In this connection, D.M.Roy and Daimon [16] recently also showed the effect of the concentration superplasticisers on the zeta potential of cement.

It will be hoped that this line of research will be able to improve the poor efficiency of crushing techniques.

In Belgium [17] research is carried out on the use of explosives (solid explosives in cartridge form, powdered or highly viscous liquid explosives, special explosives, hollow charges and external charges) for fragmentation of concrete particularly for the separation of the reinforcement.

Good results have been obtained so far already: by a good choice of the charge of explosive, the reinforcement is separated totally from the concrete while at the same time the gradation of the fragmented concrete could be predetermined.

3.3. Re-use of demolished, fragmented aggregates and possible contamination

Concrete structures which are to be demolished may have been used in sea, river or domestic waste water, in the ground, for nuclear reactors etc. and therefore may contain carbonates, sulphates, shell, chlorides, reactive aggregates (alkali-aggregate reaction), radio activity, ferrous and non-ferrous metals, heavy metals etc. "Normal" structures (dwellings, office buildings, etc.) may also have become contaminated by the use of insulation materials, foamed plastics etc. The cement matrix can also be of different types (alumina-, portland-, blast-furnace-, pozzolanic cement) Moreover the concrete may be bio-degraded or contain organic coatings, mortars and plasters either containing sulphate or contaminated through usage.

If selected demolition economically would be possible, a lot of possible trouble could be overcome in a recycling plant. Any case its quality control of aggregates and quality assurance of concrete should distinguish these things which means that those who manage such plants have to call in the help of people who know the principles of concrete making and the judgement of aggregates for such a purpose. Even then there are difficulties in judging the tolerable amounts of con-taminations of the fragmented aggregates. Is there agreement about the maximum permissible value of impurities in aggregates and concrete?

Some data in this respect are given in table 6.

Table 6 - Some data found for maximum permissible values
of contaminations in aggregates for concrete
(% of aggregate)

plaster : - SO_3 (1% SO_3 ≡ 2.1% $CaSO_4$)	- 0.5 - 2.5 (m/m) [18,19] and depending on particle size
water soluble chloride	- 0.55 - 0.15 (m/m) [various standards]
organic matter	- < Gardner colour scale nr.11
vinyl acetate paint	- 0,2 (vol) [20]
staining components	- staining index 60 [ASTM - C331 - 44]
sulfur calculated as SO_3	- 1.0 (m/m) [standards for cinders]
asphalt	- 2 (vol) [20]
wood	- 4 (vol) [20]
soil	- 5 (vol) [20]
sugars, phosphates or zinc	- 10 mg/l in mixing water of concrete (?)
fines smaller than 63 μ	- 2 m/m [various standards]
glass content depending on fineness	- according to ASTM alkali-aggregate reaction tests
hardness	- crushing value [BS, ASTM]

And what to do with the great amounts of fines smaller than
63 μ resulting from the fragmentation process? Partly the
fines consist of partly hydrated cement (mostly only grains
smaller than 20-30 μ are fully hydrated). Would it be possible
(theoretical or practical) to re-use these for making a recycled (masonry)cement?
Regarding the quality of concrete made of recycled fine and
coarse aggregates there seems to be a more or less general
agreement that compression strength, tensile strength and
modulus of elasticity are decreased by about 30 - 40% while
shrinkage is increased with some ∿30 to 60 % [eg 20].
If only recycled coarse aggregate is used in combination with
natural fine aggregates like sand, then the reduction in
properties mentioned would be about 15%, 20% and 25% respectively [eg 20]. About the cause of these reductions not much
is known. It could be the increased porosity of the recycled
aggregates and consequently increased water adsorbtion
(∿ 4-8% for coarse and ∿ 8 - 12% for fine aggregates) leading
to increased w/c ratio. Other causes however could be thought
of like: more microcracks due to demolition/fragmentation
techniques, minor contaminations, fine dust layers on aggregates leading to decreased adhesion etc.

As far as is known no data are available on the sustained
loading strength of recycled concrete which is related to the
amount and type of microcracks formed during loading (at about
70 - 90% of ultimate load mortar cracks are formed bridging
the adhesion cracks which on their turn have been increased
in amount, length and width in the range between $\sim$ 30 and 70%
of the ultimate load) . This micro-cracking could be different
for recycled concrete if more micro cracks in aggregate could
lead to more adhesion cracks, formed during the first days due
to sedimentation, plastic shrinkage and cooling of concrete
(after heat of hydration of cement has increased temperature)
And/or could it be possible that the greater moisture content
of the recycled aggregate (and more soluted ions available)
causes a decreased zeta potential of the existing cement skin
(and other micro layers) around the aggregates leading to
smaller fracture energy (see 3.2) and so causing decreased
properties of the concrete? And how many times can recycled
concrete be recycled?
Finally the behaviour of recycled concrete in the long run
(durability) has been judged mainly by results of frost-
resistance tests which proved not to differ much from those
of the reference concretes. Carbonation of recycled concrete
however has been found considerably greater [20] as was the
case for creep (for steam cured recycled concrete about 1.5
- 2.5 times according to [20]).
Given a lower modulus of elasticity and a greater creep,
a lower adhesion to reinforcement (about 15% - 20% [20])the
deflection of recycled concrete beams would be much greater
than that made of reference concrete-beams for which specifications in standards are representative, while corrosion of
reinforcement would start earlier. Perhaps it would be wise,
until more data are known and agreed upon, to use recycled
concrete in the first place for concrete products and not for
constructional concrete.
From the technological point of view points of interest are
- Demolishing techniques, environment-friendly and economic.
- Characterising the demolished waste and the fragmentated
  aggregates and its quality control.
- Fragmentation of demolished waste on an industrial scale.
- Permissible amounts of contamination in recycled aggregates.
- Possible re-use of fragmentated brickwork as recycled
  aggregate for concrete .
- Effect of different proportions of recycled and natural
  aggregates on the quality of recycled concrete.
- Effect of successive recycling on the quality of recycled
  concrete.
- Codes for recycled concrete.
- Economic aspects of recycled concrete.
Last mentioned point is not easy either for reason of the
variability of cost-factors. (see table 7)

Table 7 - Costfactors for recycled concrete

phase	costfactor	dependent on
demolition phase	1. extra adsorption demolition place 2. dump cost 3. transport cost to landfill-place 4. transport cost to recycling plant	- quality of demolition waste - requirements for recycled aggregate - government - place of landfill - place of recycling plant
recycled aggregate production phase	1. fragmentation cost and cost of quality controll 2. transport cost to concreting place	- quality and quantity of demolished waste - requirements for recycled aggregate - distance place of fragmentation to place of user (or recycling plant)
concrete production phase	1. cost of concrete production	- quality of recycled aggregates - extra provisions in production and quality assurance

4. RECYCLING OF CONCRETE AS A MATERIAL SCIENCE PROBLEM

In writing the word "material science" one could question what are the distinctions between "material- and constructional engineering", "material technology" and "material science", particularly regarding the recycling of concrete as a part of the building industry. In table 8 it is tried to give the coherence of the different parts of knowledge necessary in building. So the material engineering part of the recycling of concrete is the more coarse materiality directed on performance of the product in use, the problems in this field have been discussed under 3.3.

Now material science and material technology in fact ought to be integrated somewhat more since properties are strongly determined by deviations of the ideal structure, so the fundamental properties (primary, secondary forces) of ideal materials can be changed much by introduction of micro structural parameters (eg. see table 8). The microstructure mostly is a situation of imbalance and is determined strongly by the choice of the process. So there is a strong interaction between material science and material technology.

Table 8 - Build-up of knowledge necessary for building
         constructions

1. atoms molecules	fundamental properties	primary secondary	bonding forces

to obtain from this:
specific effects and technical
interesting properties
= material science

2. materials	micro-structural parameters	pores, particle sizes and boundaries, inter- faces, phases, precipi- tation, texture, pure-
	technical properties	ness, additives, mol. weight distributions, defects, clusters

- possibilities to realize
  materials with specific
  micro structure and most-

3. material parts, products	technical properties coupled to per- formance	directed upon well or not: bearing , protecting against moisture, temperature, frost, fire, sound (indoors/outdoors), esthetical, transport (installations)

ly in a distinct form by
way of proper choice of
process =
material technology

4. building products, building components constructions = joining of material parts	behaviour in use	design (inclusive material choice), calculations (strength, rigidity, stability, building physics)

material engineering, constructional and architec-
tural engineering

not always included
in practice

5. lifetime (durability)	behaviour in use as function of time, specific harmful influences in practical situations	loss of function by mechanical, physical, chemical, biological attack

maintenance engineering,
material engineering and
material science

6. recycling	microstructural parameters and contaminations	collection and process technology

material science,
material technology,
material engineering

To make such a distinction in complicated practical materials like recycled concrete is hardly possible and therefore
already in 3.2 and 3.3 some problems have been discussed which
as well could have been dealt with here. In fact we should combine the two distinctions in a new word like: product science.
If we look from the point of view of materials science or
product science to concrete recycling we may cover this whole
subject by one concept: adhesion.

The whole process of demolition , fragmentation and recycling
of concrete is to break the bonding forces between atoms/
molecules and to form new bonds between the interfaces of the
various particles either of the same nature or of different
nature.

To meet the demands of demolition/fragmentation technology the
molecular forces which hold the concrete together must be assessed and evaluated in terms of their extent and form so that
the most appropriate way of fracturing the bonds may be considered. The basic principles involved in this process, the
adhesion/cohesion of materials, commencing at the level of
atomic structure and building up from here to polycrystalline
solids and heterogenous bodies need to be known if significant
strides are to be made in concrete demolition, fragmentation
and re-use.

Concrete,consisting of fibrous strands of hydrated material,
incorporated in an essentially silica matrix presents a range
of unique bonding types, each having a different bond strength.
For example, the Si-O bond, with a strength of some 88. kcals/mol,
differs essentially from the bonding found in the hydroxyl ion
in water (O-H, with a bond strength of 110 kcals/mol), which in
turn is different in character from the van der Waals forces,
with bond energies of about 10 kcals/mol, which exist between
the whiskers of hydration products providing the basic strength
of the concrete matrix.

Concrete is inherently weaker than calculation of bond energies
would suggest so that, fracture can be nucleated from regions of
high stress concentration. Such stress concentrations may be
dislocations in crystalline bodies, or microcracks in a brittle
material. When such a stable micro-crack of a given depth is
formed, in principle energy released must be equal to the increase in free energy of the two new surfaces formed.

The condition for failure than is that the crack depth exceeds
a critical value above which this relationship no longer holds.
Fracture energy however is much greater than surface energy because of the non-elastic deformations in a sub-microscopic
volume around the crack tip. This zone increases much in
temperature (up till 4000 K) and may be considered as a moving
source of heat.[12, Schönert).

Such a fracture mechanism offers the basis on which demolition
techniques can be considered and emphasises the importance of
the forces of adhesion between molecules in evaluating the

problems in the recycling of concrete. It also illustrates the approach of the material   scientist or engineer in considering how the nature of these binding forces may be changed to his advantage in demolition or in rebinding the constituent parts of the concrete, for example the unhydrated cement grains, to form a useable material.

Concrete may alternatively be regarded on the macro scale as pores in a solid body. The theoretical treatment of such a system is derived as a special case from the general equations for the theory of mixtures. These equations all depend on a model such as rigid spherical particles evenly distributed in an elastic matrix and become too complicated to handle for real bodies with varying elasticities, irregular particle shape and graded particles.   Fair agreement between theory and experiment is possible by simplification so that the mixture is designed to resemble the model. The presence of a second phase introduces an important feature in such a model by the change in surface energy and the extent of the new surface. The problem then becomes one of adhesion - adhesion being defined as a surface property while cohesion is a bulk property.

Not only the mean value but the range of materials properties must be known if the forces of rupture and the bonding forces in a given circumstance are to be given proper statistical significance. For this reason it is always necessary to consider the concrete at all levels of structure and size.
Chemical structure which is relevant to strength, elasticity,
                            creep, thermal expansion;
Physical micro-structure which is relevant to absorption of
                            liquids, gases, permeability, dimensional
                            stability, thermal conductivity, visco-elastic properties, strength, elasticity;
Macro-structure which is relevant to strength, dimensional
                            stability, thermal conductivity, thermal
                            stresses, acoustic properties.

Some of these factors are known and understood but others are at best incompletely considered in the context of concrete demolition and reuse.
On this basis we have to look to the subject of our Symposium:
a. Fundamental mechanisms of bonding in cementitious materials
   and their influence on recycling of concrete
b. Application of demolition and fragmentation techniques to
   separated large concrete units or to concrete rubble (including separation of reinforcement)
   Demolition and fragmentation implies destruction of the adhesion between aggregates and cement matrix (or between
   cement natrix and reinforcement) and the question raised is:
   what are the basic mechanisms of the demolition and fragmentation processes (traditional, current, future) with
   regard to the scales mentioned earlier.

With current knowledge of the properties of materials on the macro to atomic scale, can the basic principles for demolition and fragmentation be more clearly defined?

c. Recovery and/or re-use of demolished, fragmented and separated materials taking into consideration possible contamination. The questions raised are how such contamination can affect the forces of cohesion/adhesion and what limits and in which form from the material point of view can be tolerated for effective re-use.

d. Since cement in concrete constructions is not normally fully hydrated recovery of cement from the crushed concrete may be considered.

The opportunities which exist here from the material-science point of view in developing a binder which has low bond strength so can be separated at any time in a convenient way need be considered.

It is accepted that a binder so produced would have very variable properties which would impose serious problems in regard to the properties developed and also to effective quality control.

5. PROCEDURES TO MEET THE NEEDS OF BROBLEM(s) SOLVING (purpose and goals)

In the foregoing the problems have been stated, ranging from general to specific. It is the aim of this Advanced Research Institute to bring together approximately 30 experts who are specialists in physical, chemical and/or fractural-mechanical aspects of material science with the aim of discussing and applying their particular knowledge to adhesion aspects of the recycling of concrete.

As procedure to reach this goal, the Organizing and Programming Committee proposed a four-day meeting. On the fist one and a half days a series of presentations would acquaint all participants with the state of the art within the subjects. Also here is followed the line going from general to specific topics, while the economical considerations of the total subject are included. Indeed it is interesting to know the economical implications of the recycling of concrete.

During further two days the subjects mentioned under 4 will be discussed in Workshops which have been headed under the names "Fragmentation", "Contamination", "Recycling" and "Future demolition -friendly materials", which last one too might be considered as a kind of brainstorming session. For each of the Workshops an expert is found to present an introductory lecture. The summary reports of the Workshops will be discussed at the last afternoon and may be modified as required.

It is hoped that the outcome of the Advanced Research Institute is worthwile the time of the participants and the cost.

ACKNOWLEDGEMENTS

    I wish to thank the other members of the Organizing Committee
(Dr. Everett, Dr. Frohnsdorff, Prof. Hilsdorf, Dr. Mills and
Prof. Wittmann) for their stimulating help, thoughtful comments
and suggestions and the lecturers and speakers for their
lectures and papers and last but not least the other participants
for their interest and contributions. I also wish to acknowledge
the Special Programme Panel on Materials Science of the Scientific
Affairs Division of NATO for the organizational and financial
support which made this Institute possible.

SELECTED LITERATURE

1. T.W.F. Russell, M.W. Swartzlander – The recycling index –
   32 Chemtech, Jan. 1976, p 32 – 37
2. D.W. Wilson, etal. – Demolition debris: quantities, composition
   and possibility for recycling – Proc. of the 5th Mineral
   Waste Utilization Symposium, Bureau of Mines Chicago, 1976
   pp 8 – 15.
3. P.J. Nixon – The use of materials from demolition in construction – Resources Policy, Dec. 1976 pp. 276 – 283
4. P.C. Kreijger – De betekenis van schaarste-, milieu-, en ener-
   gieoverwegingen voor het gebruik van bouwmaterialen (The
   impact of shortages, the environment and energy in the use
   of building materials) Cement XXIV nr. 2, Febr. 1977 pp 46
   – 52.
5a. Letter BC 594-B/CH of '76-12-16 of Secretary of European
   Demolition Association.
5b. C. de Pauw – Kringloopbeton (Recycled Concrete) – WTCB Tijd-
   schrift – nr.2 June 1980, pp 2 – 15.
6. J.J. Botman, B.G. Kreiter – Bouwafvalprognose met splitsing
   naar materialen en provincies (Demolition wastes divided
   into materials and provincies) – Bouw nr. 16 – '79-08-04,
   pp. 29-31.
7. P.C.F. Bekker – A theory on material consumption and the
   "Probability of survival" of dwellings – Materials and
   Society Vol. 3, 1979, pp. 175 – 190.
8. G. Huyge – Demontabel bouwen (demountable building) – Lezing
   Studiedag "Slopen, demonteren en kringloopgebruik van beton"
   (Lecture on Study-day "Demolition, demountable construction
   and recycling of concrete") – Technologisch Instituut KVIV,
   Antwerpen, 29 Nov. 1979.
9. P. Lindsell – Demolition of post-tensioned concrete – Concrete
   Jan. 1975, pp 22 – 25.
10. Y. Malier – Le découpage thermique des bétons et précontraints-
   Ann. de l'Inst. Techn. du Bat. et des Tr. Publ. nr. 353,
   Sept. 1977 pp. 94 – 111

11. J.C. Cubaud etal. - Procédés de déstruction des ouvrages en
    béton armé - Matériaux et Constructions vol 10 - nr. 57
    pp. 127 - 138.
12. UTI-CISCO Séminaire La Fragmentation - Dec. 12-14, 1978,
    St.-Rémy-Les-Chevreuse (Proceedings still in the press)
13. J.P. Meric - Influence of grinding and storage conditions of
    clinker - 7th International Congres on the Chemistry of
    Cement, Paris 1980, Vol I, pp. 4/1 - 4/16.
14. A.R.C. Westwood - Tewksbury lecture: Control and application
    of environment - sensitive fracture processes - Journal of
    Material Science 9 (1974) p. 1871 -1894
15a.A.T.F. Neerhoff ¬ Energetisch voordeliger fragmentatie van
    beton voor recycling doeleinden (energy efficient fragmen-
    tation of concrete for recycling purposes),Report ML 79-4,
    Laboratory of Material Science, Eindhoven University of
    Technology, Aug. 1979, 46 pp.
15b.P.C. Kreijger - Beton uit beton en nog wat (Concrete from
    Concrete and other matter) Cement XXXII (1980) nr.4, pp. 214
    - 218.
16. D.M. Roy, M. Daimon - Effects of admixtures upon electro-
    kinetic phenomena during hydration of C_3S, C_3A and cement -
    7th International Congres on the Chemistry of Cement, Paris
    1980, Vol II, pp. II-242/246.
17. C. Fossé - Utilisation d'explosifs pour la fragmentation du
    béton de demolition - Lezing studiedag "Slopen, demonteren
    en kringloop gebruik van beton" (Lecture on studyday:
    "Demolition, demountable construction and recycling of con-
    crete") - Technologisch Instituut KVIV, Antwerpen, 29 Nov.
    1979.
18. A.D. Buck - Recycled concrete as a source of aggregate -
    ACI Journal, May 1977, pp 212 - 219.
19. P.J. Nixon - Demolition and re-use of concrete - Materiaux et
    Constructions 1978 Vol 11, no. 65.
20. Takeshi etal - Study on re-use of waste concrete for aggregate
    of concrete - Seminar "Energy and resources conservation
    in concrete technology"- San Fransisco Sept. 10 - 13th, 1979.
21. 7th International Congress on the Chemistry of Cement,
    Paris 1980, vol I, II, III.

1.2 CASE STUDY OF THE NETHERLANDS

(ILLUSTRATING THE GENERAL PROBLEM)

Pieter C. Kreijger

Technological University of Eindhoven

5600 MB Eindhoven, the Netherlands

Additional information given during the presentation of the introduction. The Netherlands (see fig. 1) are a small country with a high population density: width about 150 km.., length about 350 km., surface 41.2×10^3 km^2, population 14.3×10^6 and population density 435/km^2 land. The Western part of the Netherlands (called Holland) is highly industrialized and situated on ground consisting mainly of (sea) clay and peat while the Eastern part has relative small, more isolated industrialized areas and contains raw materials like (some) marl, sand and gravel (much) and (river) clay (much) which are used for cement, concrete and bricks. The dividing line (going from S.W. to N.E. - see dotted line in fig. 1) between these two areas has about the level of the sea, the lowest part of the Western part being about 7m. under sea level; if dunes or dikes, which protect the Netherlands against the North Sea, should break, about 50% of the country would be flooded.

Now concentrating on concrete, the consumption of this and of composing parts cement, sand and gravel are given in fig. 2 as function of time [1,2]. In the same figure the (dutch) production is given too. It follows from this graph that most concessions for excavation end between 1990 and 1995. There is great resistance against conferring new concessions on base of the general idea that one can not make again water from our country after we have made land of water with so much effort. And this is understandable if it is realized that for each ton of building material produced one needs e.g. for sandlime bricks 0.2 m^2, for clay bricks 0.8 m^2, for concrete 0.4 m^2 and for steel 5 m^2 ground (3), producing holes in the ground of this surface and depths varying from 6 m (clay) to 30 m (sand, gravel). So for the yearly consumption of the main

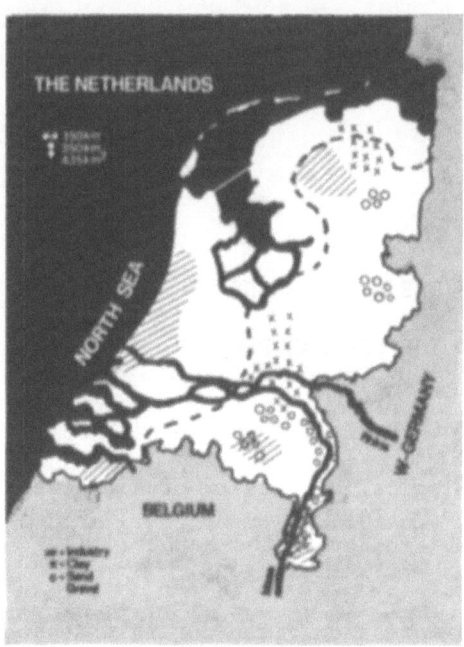

Figure 1

building materials (concrete, bricks, steel) one produces about 40 km^2 of holes from which in the Netherlands about 15 km2/ year (due to some import of cement and agfregates - see fig. 2 - and the import of iron ore which is not found in Dutch ground). For the next 20 years this means a surface of 300 km^2 (a lake of 50 x 6 km^2). Moreover for debris one needs (mostly in the Western part) dumping tips; per year about 1 km^2 surface (depth 20 m) so in the next 20 years about 250 isolated holes of 0.1 km^2 because of the increasing amounts of concrete to be demolished. Consequently one now studies intensively not only the possibilities of recycling of concrete but reuse fo all waste materials like brick- and concrete rubble, fly ash, fosfogypsum, incinerator residues, dredging soil and sewage sludge and colliery spoil left from earlier coal mining (S.E. of the Netherlands). Also sketched and a possible scenario is given for the use of these materials which indeed will decrease substantially the need for raw materials for the building industry like marl, clay, sand and gravel.

    It is a good thing to know that the Goverment is deeply convinced of the necessity for the re-use of materials, plays a coordinating role and stimulates amongst others research in this area.

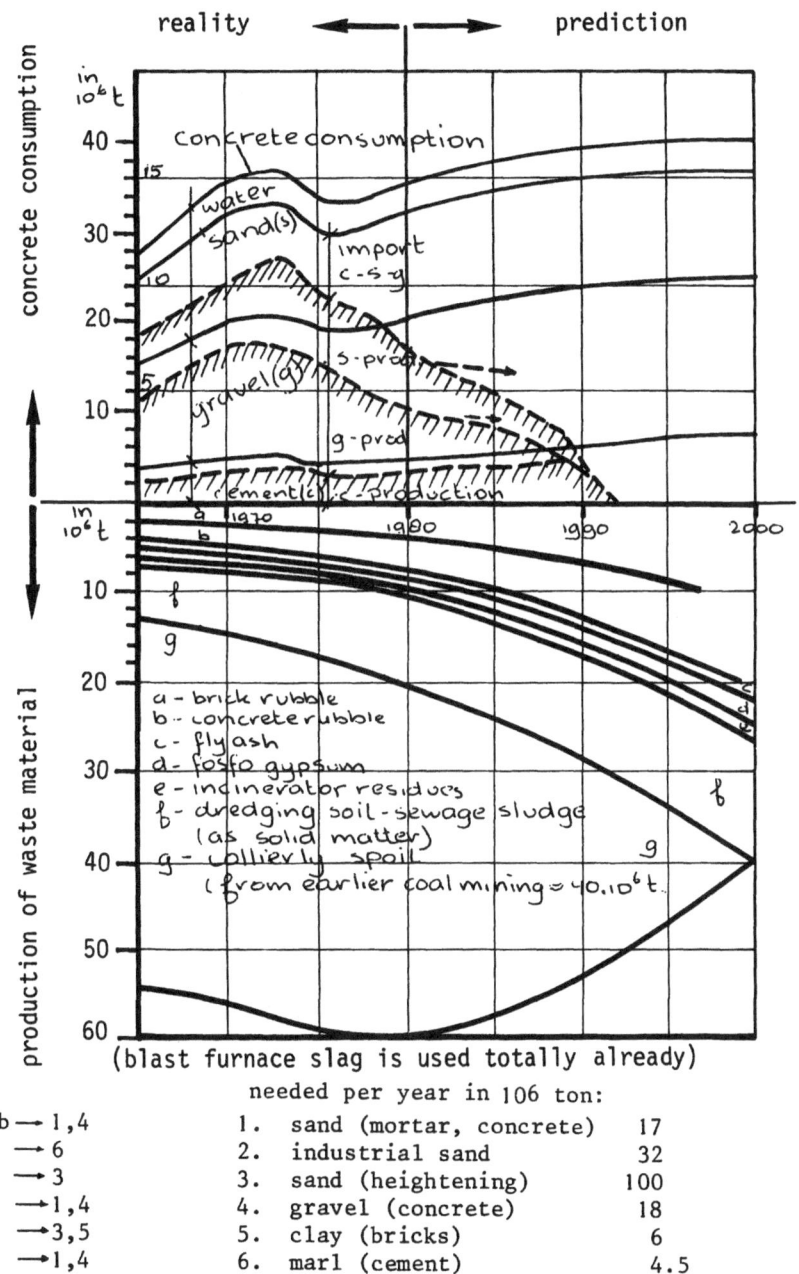

(blast furnace slag is used totally already)

needed per year in 106 ton:

a.b → 1,4	1. sand (mortar, concrete)	17
c → 6	2. industrial sand	32
d → 3	3. sand (heightening)	100
e → 1,4	4. gravel (concrete)	18
f → 3,5	5. clay (bricks)	6
g → 1,4	6. marl (cement)	4.5

Fig. 2 – Consumption and production of concrete; production of waste materials and possible scenario for using these.

Literature: 1,2 P.C. Kreijger – De betekenis van schaarste-,
                 milieu- en energie-overwegingen voor het gebruik
                 van bouwmaterialen (the impact of shortages, the
                 environment and energy on the use of building
                 materials)
                 1. De Ingenieur nr. 4, jrg. 89, 27 January 1977,
                    p 81–83
                 2. Cement nr. 2 jrg. 29, February 1977, p 46–52
             3. P.C. Kreijger – Bouwmaterialen en hedendaagse pro-
                 blemen in de samenleving – Matériaux de construc-
                 tion et problèmes actuels d'environnement – Tijd-
                 schrift der Openbare Werken van België nr. 1-1980,
                 p. 1-17.

2.1 THE STRUCTURE OF CONCRETE : A MACROSCOPIC VIEW

C. D. Pomeroy

Materials Research Department
Cement and Concrete Association
Wexham Springs, Springs, Slough SL3 6PL, U.K.

SUMMARY

Some of the factors that affect the strength, elastic modulus and deformational stability of concrete are discussed and examples are given of their effects. The importance of the cement paste matrix is emphasised as the mechanical characteristics of concrete are controlled largely by this constituent.

1. INTRODUCTION

Concrete consists of natural or synthetic aggregates bound together with an hydraulic cement. The properties of the concrete depend upon many factors that include the strength and porosity of the cement-paste matrix and the properties and grading of the aggregate.

There are many cements and aggregates that may be used in a wide range of combinations in the manufacture of concretes of very different properties and some of the factors that may be varied to produce specific properties are discussed in this paper. Concretes are frequently compared on the basis of compressive strength which can be as low as a few MPa or as high as 100 MPa. Concrete may be used plain (i.e. unreinforced) or reinforced with steel bars or with steel, glass or polymer fibres. It is important to consider the effects of the reinforcement upon the response of concrete to loads and to the environment if concrete is to be properly assessed.

Many books have been written about concrete and it is not intended to reproduce these here. Instead an attempt will be

made to highlight the parameters that control some of the
properties of the concrete and which may be controlled to produce
concretes with specific properties.

    It is first necessary to introduce the materials that are
commonly used in concrete and to indicate their principal
characteristics. This is followed by a discussion of some of the
specific properties of concrete. In view of the breadth of the
subject it is possible to consider only the basic principles,
but some recommendations are made for further reading at the end
of the paper.

2. Materials used in the manufacture of concrete

    The principal constituents of concrete are cement, aggregate
and water. In addition small quantities of chemical admixture
might be used to alter the rheological and setting properties of
concrete and the quantities of air entrained. Concretes are
strong in compression, but weak in tension and hence the need for
reinforcement. For many years asbestos fibres have been included
in cement matrices and in mortars, particularly for thin sheet
fabrications and more recently alternative fibres have been used.
There are thus large numbers of variables that can be exploited
in the design of a whole family of concretes for use in a wide
variety of applications and it is the way in which these variables
can be manipulated that makes a study of concrete so fascinating.

2.1 Cements

    There are several types of Portland cement that are in use,
but of these ordinary Portland cement is most widely used.
However the specification for this basic material differs from one
country to another, particularly with regard to the permissible
inclusion of small quantities of inert material. Portland cement
is hydraulic i.e. it reacts with water to form a porous solid
that has a strength and stiffness that increase as the porosity
decreases progressively as the hydration process continues. The
mechanism of strength development will not be discussed here. By
making changes in the raw feed to the cement kiln, in the firing
process and in the degree of grinding it is possible to change
the properties of Portland cement, particularly with regard to its
rate of reaction and its resistance to sulphate attack.

    Portland cement may also be combined with blast-furnace slags,
with natural pozzolanas and with pulverised fuel ash. The
natural pozzolanas and pfa react with the lime that is released
when Portland cements hydrate to form a solid hydrate, while the
slags are themselves hydraulic cements, although slow reacting.

2.2 Aggregates

Most concrete contains dense aggregates, such as natural gravels or crushed rocks, the grading of the material being carefully selected to give close packing in the compacted concrete. The aggregate shape and grading affect the workability of the fresh concrete.

For special purposes the fines may be omitted or reduced to give a weaker and lower density material, with a high porosity.

Lightweight aggregates are also used. These include natural materials like pumice and manufactured aggregate formed by the expansion of shales and slate, for example or by sintering pulverised fuel ash pellets. Expanded polymer beads have also been used for concretes of low thermal conductivity. Lightweight aggregates may be used with dense fines (sands or crushed rocks) or with crushed lightweight fines.

The stiffness, porosity, strength, surface texture and water absorption of the aggregates affect the behaviour of the concrete and some of these factors are discussed later.

2.3 Reinforcement

Reinforcement is discussed only in so far as it affects the behaviour of the concrete under load. Conventionally steel bars and meshes are embedded in concrete to carry the tensile stresses and to resist shear. The steel may be prestressed or post-tensioned so that the hardened concrete is put in compression, thereby effectively increasing the tensile stress the concrete may carry before it cracks.

Stirrups and secondary reinforcement can constrain the lateral expansion that occurs when concrete is compressed and help to enhance its apparent ductility.

Fibres (asbestos, steel, polymer and glass) are used to reinforce concrete, and more importantly, cement paste or mortar to facilitate the manufacture of thin sheets that can be corrugated for roofing or folded into complicated shapes for many different purposes.

2.4 Admixture

Chemicals are added to a concrete mix for a variety of reasons, the more important being

a) to reduce the water required for a specified workability

b)   to accelerate the setting, particularly in cold weather

c)   to retard the setting, when there is likely to be a delay
     in the placement of the concrete or in hot weather

d)   to entrain a controlled quantity and size of air bubbles
     for enhanced resistance to freeze/thaw damage

e)   to impart a super plasticising or free flowing
     characteristic to the concrete.

With so many potential variables available it is not
surprising to find it possible to design concretes with properties
suitable for very diverse purposes. In the next part of this
paper some of the factors that control the properties of plain
concrete are discussed. Emphasis is given to strength and
stiffness since these are two parameters that are usually
specified in the design of concrete structures.

3.   The properties of concrete

Before consideration is given to the factors that control the
properties of concrete a list of properties of interest to the
designer or user is presented.

a)   Strength. This is defined as the maximum stress that may
     be carried and different values apply for compression,
     tension, shear or for combinations of such stresses

b)   Elastic modulus and Poisson's ratio

c)   Crack resistance and fracture toughness

d)   Shrinkage

e)   Creep

f)   Thermal expansion and conductivity

g)   Porosity and permeability, factors relevant not only to
     the mechanical properties of the concrete but also to the
     resistance to chemical attack consequent to the ingress
     of aggressive gases or liquids.

In addition the constituents of concrete, the nature of the
cement, the type and severity of the mixing, the temperature,
the manner the concrete is placed and other factors have a large
effect upon the early age properties of the concrete. The
rheological behaviour, the rate of setting, the generation of heat
and temperature rise as the exothermic cement hydration occurs,

the entrapment of air and so on, depend upon these factors.
These important topics are not discussed in detail in this paper,
although some of the effects will be mentioned when particular
properties of hardened concrete are being considered.

3.1 Concrete strength and stiffness (Uniaxial compression)

It has already been inferred that "strength" is not a unique
parameter but as is customary practice the compressive strength
is considered first. If we consider concrete to be comprised
of an assembly of aggregate particles in a matrix of hardened
cement paste it is pertinent to seek relationships between the
properties of the aggregate and of the matrix and those of the
composite - the concrete. A complication of such an enquiry
concerns the changing properties of the matrix with time as the
cement hydrates, and studies on hardened cement paste itself,
unencumbered by aggregate, show that the crushing strength
increases as the porosity decreases as shown in Figure 1. Within
normal practical limits the strength-porosity relationship holds
even when the hardened paste is autoclaved or subjected to other
forms of accelerated curing.

A similar relationship holds between the stiffness, or
elastic modulus of the paste and the porosity.

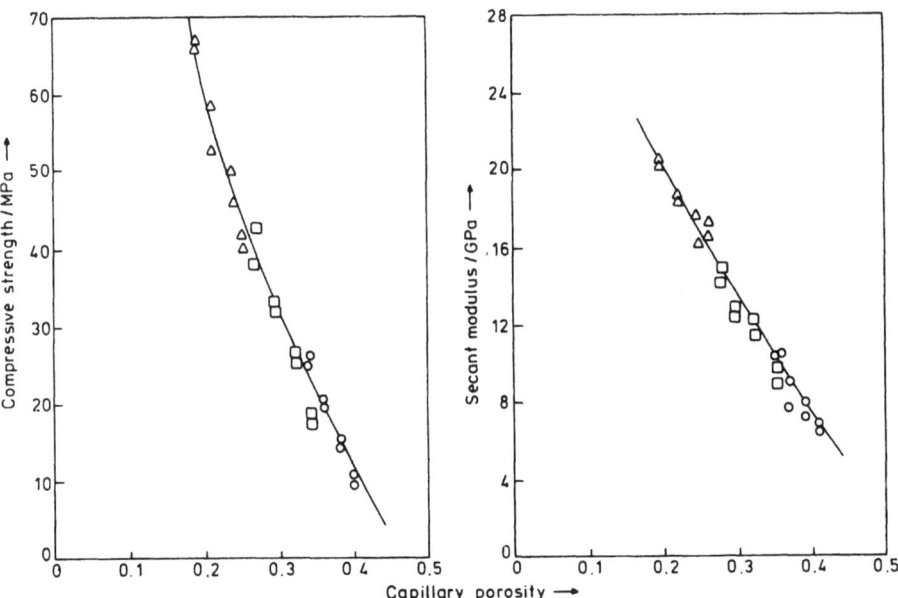

Fig. 1   Relationships between compressive strength, secant modulus
         and capillary porosity for water cured cement pastes, aged 7
         - 84 days. Water-cement ratios 0.59(o), 0.47 (□), and 0.35(△).

Equations have been fitted to these curves and attempts made to account for the observed relationships on a theoretical basis. Typically Hansen (1965) proposed the semi-empirical equation

$$E_p = E_s \; \frac{(1-p)}{(1+2p)} \quad \cdot \quad \cdot \quad \cdot \quad \cdot \quad \cdot \quad (1)$$

for the relationship between the elastic modulus E_p of the material of porosity p, where E_s is the modulus of the pore-free solid. Powers (1961) had also fitted the equation

$$E_p = E_s \; (1-p)^3 \quad \cdot \quad \cdot \quad \cdot \quad \cdot \quad \cdot \quad (2)$$

to observed measurements.

Helmuth and Turk (1966) found that of these and other theoretical and empirical relationships only the two equations given satisfactorily fitted their experimental measurements on hardened cement paste.

For strength Nurse (1966) used the empirical equation $\sigma_p = \sigma_s \; e^{-kp}$ to represent the observed relationship between the crushing strength σ_p of a hardened cement paste of porosity p and the interpolated strength σ_s of the pore-free solid. Lawrence (1970) found this equation applied over a wide porosity range.

It follows, therefore, that both the crushing strength and the elastic modulus of hardened cement paste relate closely to the porosity, which depends in part on the initial water/cement ratio of the paste, partly upon the cement hydration at the time it is tested and partly upon the quantity of air that is entrapped in the paste matrix. The last component will be affected by the way the material is cast and by the inclusion of air-entraining admixtures.

Normal dense aggregates have elastic moduli that exceed the moduli of the hardened cement paste (HCP) by a factor of five or more (80 GPa compared with 10 - 20 GPa for the HCP). It follows therefore, that the stiffness of most dense aggregate concretes will be greater than that of the matrix. Many attempts have been made to combine the properties of the constituents in predictive equations.

It is generally assumed in theoretical analyses that concrete is a two phase material, the aggregate and the hardened cement paste matrix. It is assumed that both phases are linearly elastic over the working stress range, and while this is true for most dense aggregate it is not for HCP. However from the basic assumptions in which E_a and E_p are the elastic moduli of aggregate and paste it is possible to derive mathematical models that predict the concrete modulus E from the volumetric proportion of aggregate V_a.

The most simple geometrical representations are those attributed to Reuss and to Voigt (Figure 2). In the former the paste and the aggregate are subject to equal compressive stresses when a compressive load is applied and

$$E = (V_a/E_a + {}^{(1- V_a)}/E_p)^{-1}$$

In the Voigt model both the paste and the aggregate are subjected to equal strain when the compressive load is applied and

$$E = V_a E_a + (1 - V_a) E_p$$

Neither model is realistic but they provide an upper and lower bound for the moduli of two phase materials.

A number of subtle refinements in which the aggregate component is distributed within the matrix were devised and some of these led to equations that predicted the modulus of concretes reasonably well. The first major development from the strict geometrical portrayals was made by Hashin and Shtrikman (1963). Variational principles were used to bound the strain energy of a multiphase material and thus also the elastic modulus. Hobbs (1969a, 1969b) also derived a relationship from considerations of the bulk modulus of a composite in which the aggregate and matrix retain contiguity. In the situation where the aggregate and paste matrix have the same Poisson's ratio of 0.2, the Hashin and Shtrikman and the Hobbs equation for the elastic modulus of a two-

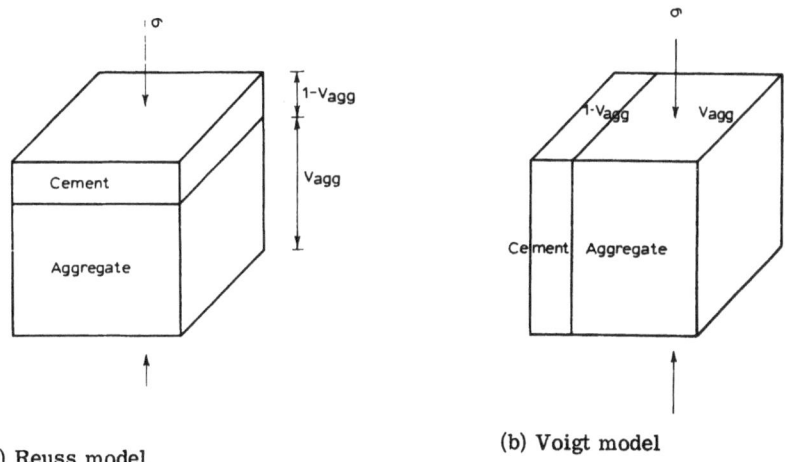

(a) Reuss model

(b) Voigt model

Fig. 2 Reuss and Voigt models of a two phase material

phase composite are identical, the modulus being given by

$$E = \frac{(E_a - E_p)\ V_a + (E_p + E_a)}{(E_p + E_a) + (E_p - E_a)\ V_a}\ E_p \quad . \quad . \quad . \quad . \quad (3)$$

and this equation represents experimental observations both for high modulus and low modulus aggregates very satisfactorily (Figure 3).

Extensions of the models to include a third component, the voids, have been made by a number of workers, in general confirmity with the porosity/modulus relationship previously discussed for hardened cement paste, and these have been summarized by Hobbs (1973).

A difficulty in the application of these equations in practice is that the moduli of the aggregate are seldom known. In consequence attempts have been made to relate the modulus of concretes to their compressive strengths which are more readily determined and are, in fact, usually specified for the concrete to be used.

The predictive equations are empirical and debate continues regarding the most acceptable one. CEB/FIP recommendations (1978) expressed the modulus E of a dense aggregate concrete to the cylinder compressive strength f_{cy} by the equation

$$E = 6.6\ f_{cy}^{0.5} \quad . \quad . \quad . \quad . \quad . \quad (4)$$

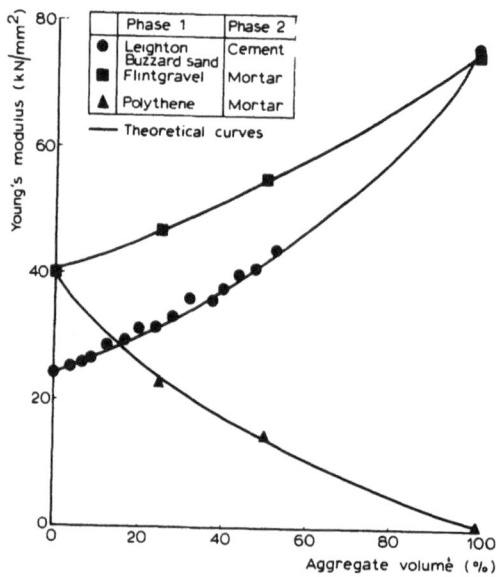

Fig. 3 Comparison of Hashin and Shtrikman equation with observed moduli for high and low modulus aggregates.

where f_{cy} is in MPa and E in GPa. The equation is modified for lightweight aggregate concrete of density ρ to

$$E = 1.8 \, (\rho^3 . f_{cy})^{0.5} \qquad . \qquad . \qquad . \qquad . \qquad . \tag{5}$$

More recently the CEB (1976) have suggested the alternative equations

$$E = 9.5 \, (f_{cy}^1 + 8)^{0.33} \text{ for dense aggregate concrete.} \qquad . \tag{6}$$

and $\quad E = 9.5 \, (w/_{2400})^3 \, (f_{cy}^1 + 8)^{0.33} \text{ for lightweight}$

$$\text{aggregate concretes} \tag{7}$$

the number 8 being introduced in an attempt to allow for the differences between the specified characteristic cylinder crushing strength f_{cy}^1 and the average strength of the concrete in a structure but such an approach seems liable to error.

Teychenné (1978) reported tests made on a large number of concretes of widely different mix proportions made from 24 crushed rock aggregates. Teychenné, Parrott and Pomeroy (1978) used these and other results to see whether there would be a significantly better prediction of the elastic modulus of concrete if the aggregate modulus was known in addition to the compressive strength of the concrete. They concluded that the modulus of concretes at any age t could be estimated from the equation

$$E_t = C_o + 0.2 \, f_t \, . \qquad . \qquad . \qquad . \qquad . \qquad . \qquad . \tag{8}$$

where f_t is the cube strength in MPa at the relevant age of the concrete. C_o was a constant that is linearly related to the modulus of the aggregate. For most dense concretes a value for C_o of 20 will give an acceptable value of modulus for design, but where greater precision is required from the estimate the appropriate value for the aggregate is needed. This may be estimated from equation 8 using a strength and modulus test for a concrete made with the particular aggregate. Values of the moduli of other concretes made with the aggregate can then be estimated.

Thus the way in which the aggregate, the hardened cement paste and the air voids interplay to provide a composite material of a certain elastic modulus can be explained on a mathematical basis, but from the practical standpoint empirical relationships between crushing strength and modulus are, at present, more useful. If a measure of the stiffness of the aggregate is available, though, the empirical relationship (Eqn. 8) between modulus and compressive strength is more accurate.

All the above analyses assume linear elasticity for the aggregate and the HCP and this assumption is not true especially for the HCP. Hence concrete is not linearly elastic.

The cause of the non-linearity has been discussed by many workers. Certainly HCP creeps under a sustained load and water movement contributes significantly to the creep mechanism. A discussion of this complex subject is beyond the scope of this paper but it is important to realise that movements within concrete can accommodate and redistribute stresses without cracks being formed over a greater stress range than might be inferred from short term tests.

The non-linearity is explained by a combination of creep and microcracks which form within HCP and concrete when it is loaded.

Some people think that microcracks are created only above some threshold value of stress. When concrete is loaded in compression there is at first a reduction in the volumetric strain as the load is increased. The rate of reduction in volumetric strain decreases to zero and there is subsequently an increase that is attributed to the development of a macro-crack pattern (Figure 4).

However this view has been challenged by Spooner et al (1976) who showed that microcracks can form in concrete the very first time that a load is applied to the material, even at low stresses. This is not unexpected since the irregular shapes of the cement grains or the aggregate particles and the differences in modulus between the HCP and the aggregate will induce high stress concentrations from which cracks will develop. Moreover there will be strength variations within the composite, especially close to the interface between the aggregate and the matrix.

If sufficient energy is available the cracks may propagate rapidly through the solid and brittle failure occurs. It is more

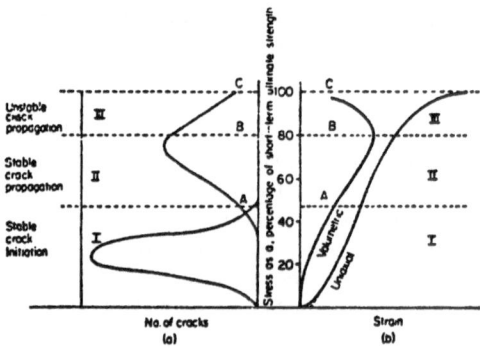

Fig. 4   Relationship between crack development and volumetric
strain (Newman and Newman 1971)

likely that when the initial cracks are formed they will not
propagate but will, in effect, redistribute the stresses in the
concrete and reduce the peak stress concentrations to a lower
value. Further where the strength of a composite varies from point
to point a crack will initiate at a weak zone and stop when it runs
into stronger material (e.g. when its path intersects an aggregate
particle).

The arrest of cracks has been demonstrated by flexural tests
on notched mortar and concrete beams subjected to cyclic deflections
to progressively greater values (Brown 1972). A measure of the
effective driving stress at the crack tip (pseudo K_{ic}) shows that
an initial single crack requires a higher driving force as it
advances and this can be attributed to crack branching around
aggregate particles and to crack arrest (Figure 5).

The propagation and arrest of cracks will depend upon the
quality of the concrete and the properties of the aggregate used.
If the cement paste matrix is much weaker than the aggregate,
cracks formed in the paste or at the aggregate/matrix boundary will
skirt around the aggregate particles. If the aggregate is of
strength comparable to or weaker than the matrix, cracks will
probably continue through the aggregate so that there is small
likelihood for crack arrest or branching to occur and failure is
much more abrupt.

The change from a situation in which cracks are not easily
propagated to one where cracks advance without hindrance affects

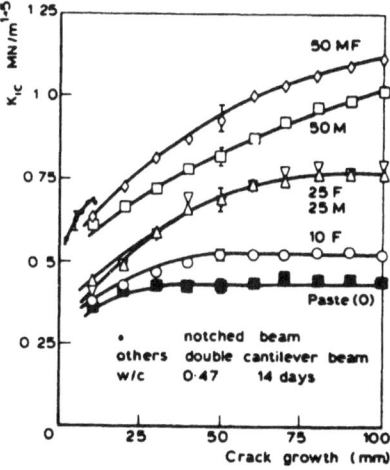

Fig. 5   Relationship between K_{IC} and crack growth for mortars
          containing 10-50% of medium (M) or fine (F) aggregate
          Brown, 1972)

the falling part of the complete stress/strain curve for concrete
in compression (Figure 6). The ductile failure applies to concretes
of moderate strength made with dense aggregates. High strength
concretes have a much more rapid drop in load carrying capability
after the peak stress has been reached.

Wittmann (1980) has summarized work on fracture on a microscale
and discusses the conditions for crack propagation to take place,
both through and around aggregate particles.

At this point it must be emphasised that only plain concrete,
subjected to simple compressive loading has been considered and
that the deformation and failure of concretes subjected to more
complicated loading can show different characteristics.

3.2 Concrete strength and stiffness (Multi-axial stresses)

It has been realised for many years that the load bearing
capacity of a material depends upon the constraints to the
deformation as the loads are applied. For example a pile of loose
sand cannot carry any load but if it is constrained within a
bucket it is able to do so. In about 1900 Considere (1906)
undertook some experiments to study the effect of lateral
constraint upon the load carrying capability of concrete and found
that if a cylinder was compressed by a stress of σ_3 on the curved
surface the axial load that could be carried increased from the
unconstrained uniaxial compressive strength f_{cy} by about $3\sigma_3$

$$\sigma_1 \quad = \quad f_{cy} \quad + \quad 3\sigma_3$$

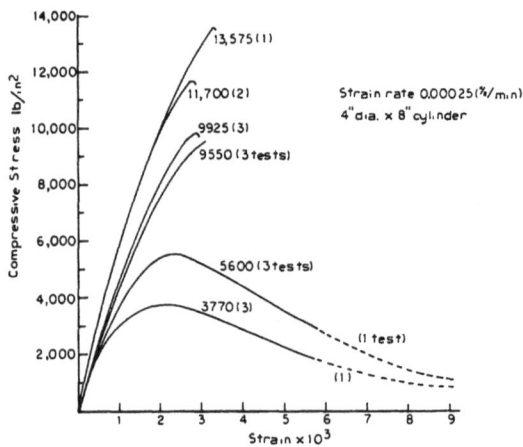

Fig. 6   The effect of concrete strength on the ductility of
concrete in compression

During the succeeding 80 years many more experiments have been
undertaken and while slightly more precise numbers have been
proposed the basic law still holds. Further the constraint also
affects the failure process and a brittle material that breaks
explosively in uniaxial compression when the peak stress is applied
can become ductile with a much more non-linear stress-strain curve
(in the direction of the applied major principal stress) and a
gradual fall off in the load bearing capacity after the peak
is reached. The provision of ductility to a material that would
fail catastrophically in the uni-axial mode is of great importance
to the Civil Engineer, since a truly brittle material that is
notch sensitive can only be used at very low stresses, well below
the theoretical strength, since a single crack formed at a stress
- magnifying discontinuity can initiate sudden collapse at a low
average stress. The fact that the stronger concretes are more
susceptible to the influence of stress concentrations is
demonstrated by experiments that relate the size of a concrete
cube to its compressive strength (Pomeroy 1972). The higher
strength concretes show a significant decrease in strength with
cube size (Figure 7) whereas no such fall is observed with the
weaker concrete. This is attributed to the fact that large
cubes are more likely to contain a crack initiating flaw of a
given severity than a small one so that failure is more likely at
a given average stress in a large cube than in a small one,
provided that the crack so initiated can propagate and lead to
total failure. If the crack can be interrupted (as discussed
earlier) the influence of the extremely high stress concentrations
will be small and hence the size of the specimen will be of little
consequence.

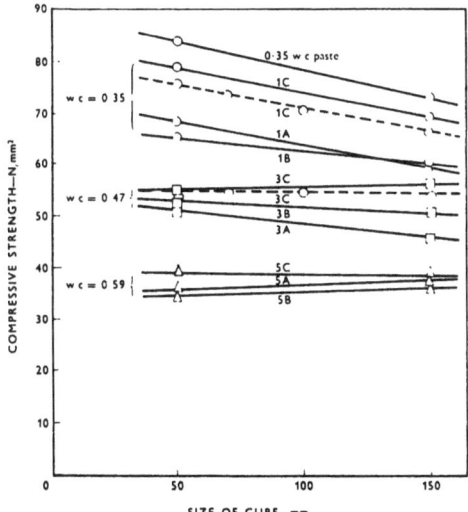

Fig. 7   The effect of specimen size on the compressive strength
of weak and strong concretes (Pomeroy, 1972)

3.3 The effect of reinforcement

    Although the primary use of steel reinforcement is to carry
the tensile loads in a structure the presence of steel also may
impose constraints on the concrete that is in compression and by
so doing the bearing capacity of the concrete may be higher than
that of the concrete in uniaxial compression or, of even greater
significance, it prevents sudden collapse and gives ductility to
the structure. Such ductility may seldom be necessary as the
design loads are usually well below the stresses likely to result
in severe disruption to the concrete. It is important to impact
situations and where progressive or sequential collapse could
otherwise occur. In the first case the containment of an explosion
or the absorption of the energy from mechanical impact, such as
the collision between a lorry and a bridge column, or a ship and
an oil platform benefit from the capacity of the concrete to
deform without losing its full load carrying capability. In the
second case accidental damage to one part of a structure should not
result in the step-wise build up in overloads in adjacent
structural elements so that total collapse continues like the
collapse of a pack of cards. Again a ductile behaviour enables
local disruption to be contained.

    The provision of ducility in this way has been discussed by
(Bazant, 1980, Pomeroy 1973 and Shah and Rangan, 1971 amongst
others). Figure 8 shows a typical effect on the compressive
stress/strain curve for concrete without and with spiral or
stirrup reinforcement.

3.4 Fibre reinforcement

    An alternative way to inhibit crack growth in concrete is to
include a proportion of steel, glass or polypropylene fibres in
the composition. Fibre reinforcement has been widely studied
during recent years and several publications (Concrete Soc., 1973,

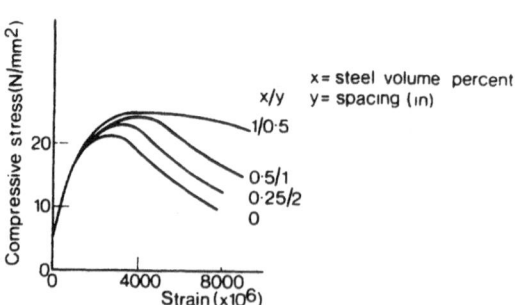

Fig. 8 Ductility provided to concrete by provision of stirrup
        (after Shah and Rangan)

Hannant 1978, GRCA 1980) describe the technology and performance
characteristics of fibre composites. It is universally agreed that
one of the greatest benefits imparted by the fibres is the energy
adsorbing capability. However to utilise this potential the
concrete must be cracked, since it is the crack arrest mechanisms
and the pull-out of the fibres from the matrix that absorbs the
energy. Within normal load ranges the fibres are not as efficient
as conventional reinforcement in providing stiffness to the
concrete member, nor do they increase significantly the load at
which cracking is initiated. However where overloads are to be
expected (explosive or impact) the manyfold increase in energy
absorption can be of great benefit. Figure 9 shows how for small
deflections the inclusion of fibres has no practical effect, but
at large deflections the benefits are immediately apparent.

   The figure also includes data for polymer Portland cement
concretes since claims are also made that polymer impregnation or
addition to the mix (ACI, 1978) provide a means of raising the
strength and improving the properties of the concrete. It can be
seen that in terms of the energy to cause a given beam deflection
there is no large benefit attributable to the polymer treatment,
but if both polymers and steel fibres are included (albeit at
considerable cost) a very energy absorbing composite is formed.
Probably the polymers increase the fibre matrix bond, and hence
the energy used to pull out the fibres.

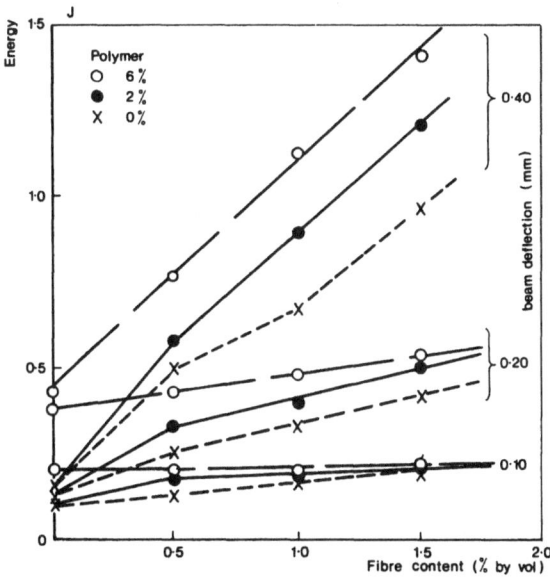

Fig. 9 The effect of steel fibres and polymer emulsions on the
          energy expended in beam deflection

3.5 Strength and Stiffness - Discussion

An attempt has been made to discuss some of the factors that affect the mechanical behaviour of concretes under load. At the beginning mention was made of the many ways the properties of concrete can be varied. Not only is the choice of cements, aggregates and admixtures wide, but the proportions of each component and the grading of the aggregate all affect the ultimate product, either directly or in the way they affect the water needed for the fresh concrete to be usable. The importance of the cement paste matrix has been emphasised, in relation to the stiffness and strength of the composite and in the modes of failure of brittle and ductile types. These concepts are relevant to the theoretical potential of concrete. In practice other factors affect its performance. These include the placing and compaction of the concrete into the structure, the curing and early age temperature history of the concrete, the subsequent loading and environmental history of the concrete and the long-term stability of the constituents. These topics will be discussed briefly in the remaining part of this paper.

4.0 The influence of the environment upon concrete behaviour

It is easy to believe that experience gained with concretes in one part of the world automatically applies elsewhere, yet in Canada or Norway concrete could be placed at sub-zero temperatures while in the Middle East not only are the ambient temperatures high but the prevailing humidity can be very low. The hydration of Portland cement and the products of reaction depend upon the temperature to a considerable extent, the early age strength development generally increasing with the temperature but this may be offset by the subsequent gain in strength which takes place (Figure 10 Dalziel 1980). In fact the long term strength (180 days or more) is frequently lower for concretes cast at a high temperature than for those cast at 10°C or less which gain strength slowly yet ultimately attain higher strengths. The temperature in the concrete rises as hydration proceeds as a consequence of the exothermic reaction, the rate of temperature rise increasing as the casting temperature rises, so that concrete properties in a structure may differ from those of a concrete specimen cured and tested at 20°C, say.

It is easy to see that, apart from the effect that temperature has on the hydration process, thermal gradients will be induced and that the coefficients of thermal expansion of aggregate and matrix may well be incompatible. As the concrete cools stresses will be set up, both as a stress gradient in the bulk concrete and locally close to the aggregate/matrix interface. The concrete is thus liable to crack as the surfaces cool first and tensile stresses are induced. Concrete technology

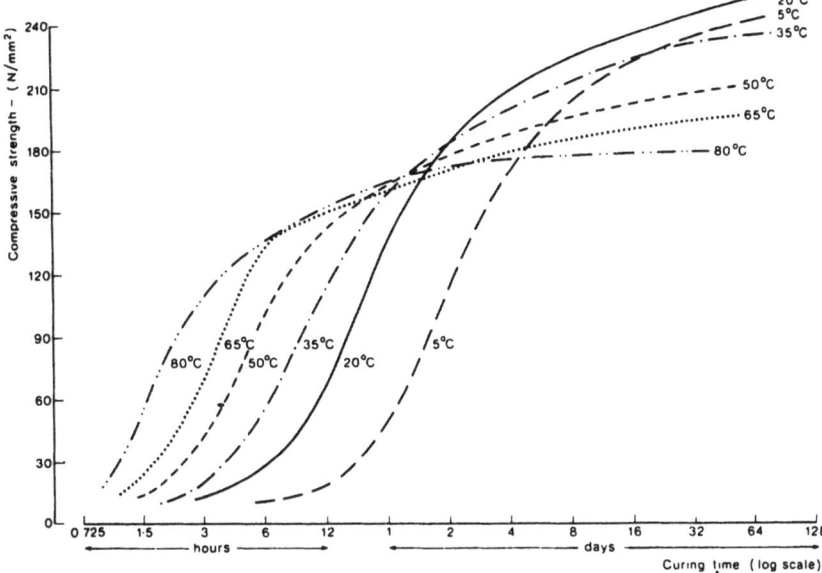

Fig. 10   The effect of curing temperature on strength development
of ordinary Portland cement (after Dalziel 1980)

has been developed that minimizes these effects.   These include
a limitation to the size of the structural units so that thermal
movements can be accommodated without cracks forming;   the use of
insulating mats to delay and control the rate of surface cooling
that can occur and hence to reduce the strain gradients in the
concrete;   the use of diluents such as pulverised-fuel ash or
pozzolana that may act as water reducers that permit concrete
mixes to be formulated that meet strength requirements at lower
Portland cement levels, although at casting higher temperatures pfa or
pozzolana reactions may lead to high temperature rises;   the use
of low heat Portland cements or blastfurnace slag.

A simple measurement of the temperature history of the
concrete is an inadequate measure of the likelihood for cracks to
occur if composite cements (e.g. pfa + opc) are used since the
properties of the concrete also depend upon the constituents and
there is some evidence (Bamforth 1980) that the tensile strain
that a Portland/pfa concrete can carry without cracks forming is
lower than that of an equivalent OPC concrete.

4.1   Drying shrinkage

The moisture history is also of considerable importance.   If
fresh concrete is permitted to dry before the cement has properly
hydrated not only will the strength be impaired and cracking may

be caused but the concrete will be highly porous and have poor
durability. It is therefore imperative that sufficient moisture be
retained in the concrete while hydration is particularly active
(for a few days at high temperatures and for many days as 0°C is
approached).

When the hardened concrete dries it will contract and
dependent upon the properties of the aggregate and the HCP matrix,
cracks may form. These would "weaken" the concrete and give it
a reduced effective modulus. Some aggregates shrink and these can
seriously disrupt concrete if excessive drying is permitted.

Carbonation of the calcium hydroxide formed during the
hydration process also causes some local shrinkage to occur and if
cracks are formed the depth of penetration of the atmospheric CO_2
can increase and ultimately reach the reinforcing steel. If this
happens the steel is likely to rust and expand until the concrete
surface spalls. It is thus obvious that if proper care is not
taken to use a suitable concrete mix design and to cure it
properly, a porous layer of concrete close to the surface is liable
to be carbonated and that this can lead to rusting of the
reinforcement and subsequent disruption to the concrete.

From the structural point of view knowledge of the bulk
shrinkage of concrete needs to be available. Several parameters
affect the shrinkage and this subject has been studied widely by
Bazant (1975). Hobbs (1979) has used simplifying assumptions that
lead to equations and graphical representation of sufficient
accuracy for most practical applications. The information
available to a structural concrete designer is usually sparse and
apart from the quality of the concrete (specified usually by
strength), the geometry of the structure and the prevailing
environment little else is known. From these parameters the
shrinkage is given in Figure 11, for dense aggregate concretes.

The stiffness of the aggregate provides resistance to the
shrinkage which occurs as the HCP loses water, but it is the
proportion of the aggregate in the concrete that is of greater
importance. With low density or lightweight aggregate concretes
there is a greater potential shrinkage than when dense aggregates
are used.

4.2 Creep and Dimensional Stability

Mention has already been made of the time dependent
deformation of concrete under a sustained load. This creep is
affected not only by the "quality" of the concrete and the loads
applied but also by the temperature and moisture history. Creep
and shrinkage are not independent and if creep occurs at the same
time as drying takes place the observed movements exceed those if

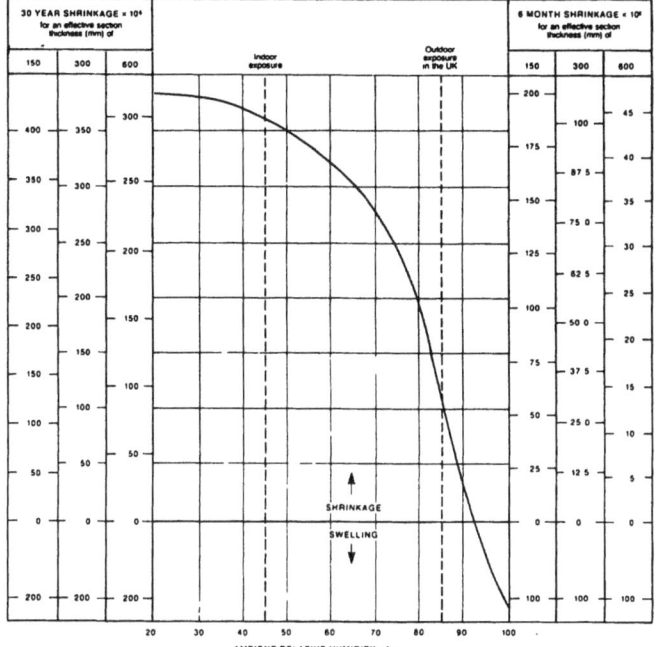

Fig. 11   Shrinkage prediction for concrete members of different
          size, stored in different environments (Parrott, 1980)

both processes occur separately – namely creep at an equilibrium
moisture state plus shrinkage of unloaded concrete. This is just
one example of the complex interactions that must be taken into
account when the performance of structural concrete is being
predicted. Other factors include the size of the structural
member (and hence the rate of drying), the age of the concrete
when it is loaded, the temperature sequence and the ambient
humidity.

     The creep phenomena have provided a fruitful basis for much
research as exemplified by the recent International Symposium held
at Lausanne (Wittmann 1980). From the practical point of view the
engineer wants simple methods that he can use with confidence and
at the present time the available methods include those proposed
by the CEB/FIP (1978), by Bazant (1975) and by Parrott (1980).

4.3 Concrete Durability

     Before leaving the subject of the dimensional stability of
concrete mention must be made of two causes of damage to concrete.
These are sulphate attack and expansive reactions between siliceous
and carbonaceous aggregates and the alkalis in the concrete.
Sulphate attack normally arises in foundations placed in water
bearing ground that contains a high proportion of sulphate.

Special sulphate resisting cements with a low C_3A content are highly
resistant to this form of attack. The use of pozzolanas or
pulverised fuel ash in sufficient proportions (30% or so) will also
provide a better resistance than an OPC with a high C_3A level,
although there is some doubt about their long term efficacy and
sulphate resisting Portland cements are preferred. Good quality,
dense and well compacted concrete is probably adequate for most
foundations and bad workmanship cannot be remedied by better
materials.

In the case of alkali silica attack, it is generally found that
an opaline constituent is present in a critical proportion and this
component reacts with the alkalis in the concrete to form silica
gel. This gel may imbibe water and swell to such an extent that
the concrete cracks. Several factors must be present
simultaneously for disruption to occur: 1. A critical proportion
of the reactive component. 2. Sufficient alkali and 3. An
excess of water.

The cause of distress from the alkali-carbonate reaction is
different. Typically an orgillaceous dolomitic limestone
aggregate may deteriorate by surface or rim reactions with alkali
that forms calcite and brucite and this is akin to a weathering
action, which leads to a loss in strength by the concrete.

5. CONCLUDING REMARKS

A number of factors affecting the properties of concrete have
been discussed, particularly with regard to strength and deformation.
These properties are largely controlled by the strength of the
hardened cement paste matrix, the bond between the matrix and the
aggregate particles, the strength and stiffness of the aggregate
and the volume concentration and grading of the aggregate.

It is normal to find that lower strength concretes, with
their more porous matrices, fail in a more ductile manner than
those in which the matrix is strong, especially when the matrix
and the aggregates have comparable strengths. However lateral
confinement of a concrete that is loaded in compression will
change the failure mode very considerably and even mild constraint
can change the failure from brittle to ductile. Thus it is seen
that the behaviour of concrete under load is not unique and a
proper appreciation of its performance can be complex, especially
if long term creep, shrinkage, carbonation and durability effects
are also considered. In fact when the performance of concrete in
a structure is being appraised reliance is often put upon National
Codes of Practice and these often allow for the differences in
performance between a cubic or cylindrical specimen by suitable
empirical equations, so that the effects mentioned are taken into
account, albeit without full realisation of the causes of the

differences between laboratory tests and practice.

Gradually the methods used to select concretes for specific purposes are being improved. For example concrete can be used to replace cast iron or timber (Pomeroy 1978) provided a performance specification is available. The ideas propounded here help the achievement of better and novel uses for concrete.

References

American Concrete Institute (1978) "Polymer in Concrete" Publication SP 58.

Bamforth, P.B. (1980) "In-situ measurements of the effect of partial Portland cement replacement using either fly ash or ground granulated slag on the performance of mass concrete". Proc. Instn. Civ. Engnrs. Part 2 69 : 777-800.

Bazant, Z.P. (1975) "Theory of creep and shrinkage in concrete structures. A precis of recent developments". Mechanics Today (Ed. S. Nemat-Nasser, Pergamon Press) Vol 2 : 1-93.

Bazant, Z.P. (1979) "Material behaviour under various types of loading". High Strength Concrete (Ed. S.P. Shah) University of Illinois at Chicago Circle, U.S.A.

Brown, J.H. (1971) "Measuring the fracture toughness of cement paste and mortar" Magazine of Concrete Research 24 : 185-196.

CEP-FIP Model Code for Concrete Structures, 3rd Edition 1978. Published by Commité Euro-International du Béton (CEB).

Concrete Society (1973) "Fibre-reinforced cement composites". London.

Considere, A. (1906) "Experimental researches on reinforced concrete". Translation and Introduction by L.L. Moisseiff, McGraw-Hill Book Co., New York, 2nd Edition.

Dalziel, J.A. (1980) "The effect of curing temperature on the development of strength of mortar containing fly ash". Proceedings of 7th International Congress on the Chemistry of Cement Paris. Volume 3 : IV : 93-97.

Hannant, D.J. (1978) "Fibre-cements and fibre concrete". John Wiley and Sons, Ltd.

Glass-fibre Reinforced Cement Association (1980) "The Developing Success of GRC". Proceedings of the Second International Congress, U.K.

Hansen, T.C. (1965) "Influence of aggregate and voids on the modulus of elasticity of concrete". J. Am. Concrete Inst. Proceedings Vol 62 : 193-216.

Hashin, Z. and Shtrikman, S. (1963) "A variational approach to the theory of the elastic behaviour of multiphase materials" J. Mech. Phys. Solids. II : 127-140.

Helmuth, R.A. and Turk, D.H. (1967) "The reversible and irreversible shrinkage of hardened Portland cement and tricalcium silicate pastes". Journal of Portland Cement Association, Skokie, U.S.A. : 9 : 8-21.

Hobbs, D.W. (1969 a) "Bulk modulus, shrinkage and thermal expansion of a two-phase material" Nature 222 : 849-851.

Hobbs, D.W. (1969 b) "The dependence of the bulk modulus, Young's modulus, shrinkage and thermal expansion of concrete upon aggregate volume concentration". Cement and Concrete Association, London Technical Report 42.437.

Hobbs, D.W. (1973) "The strength and deformation of concrete under short term loading : a review" Cement and Concrete Association, London Technical Report 42.484.

Hobbs, D.W. and Parrott, L.J. (1979) "Prediction of drying shrinkage" Concrete, 13 : 19-24.

Lawrence, C.D. (1970) "The properties of cement paste compacted under high pressure". Cement and Concrete Association, London Research Report 19.

Newman, K. and Newman, J.B. (1971) "Failure theories and design criteria for plain concrete". Structures, Solid Mechanics and Engineering Design, Ed Te'eni, M. Southampton, U.K. Part 2 : 963-996.

Nurse, R.W. (1968) "Cohesion and adhesion in solids". The Structure of Concrete. Proceedings of an International Conference, London 1965. Cement and Concrete Association, London : 49-58.

Parrott, L.J. (1980) "Simplified methods of predicting the deformation of structural concrete". Cement and Concrete Association, London. Development Report 3.

Pomeroy, C.D. (1972) "The effect of curing conditions and cube size on the crushing strength of concrete". Cement and Concrete Association, London. Technical Report 42.470.

Pomeroy, C.D. (1973) "Fibre-reinforced concrete - its properties and applications". Cement and Concrete Association, London DN 4023.

Pomeroy, C.D. (1978) "Concrete, an alternative material". 14th John Player Lecture, Proceedings Inst. Mech. Engnrs. 192 : 135-144.

Powers, T.C. (1961) "Fundamental aspects of shrinkage of concrete". Revue des Materiaux, No 545 : 79-85.

Shah, S.P. and Rangan, B.V. (1971) "Fiber reinforced with fibrous wire". J. Am. Concrete Inst. Proceeding Vol 68 : 126-135.

Spooner, D.C., Pomeroy, C.D. and Dougill, J.W. (1976) "Damage and energy dissipation in cement pastes in compression". Magazine of Concrete Research 28 : 21-29.

Teychenné, D.C. (1978) "Concrete made with crushed rock aggregates" Building Research Establishment, Watford, U.K. Current Paper CP 63/78.

Teychenné, D.C., Parrott, L.J. and Pomeroy, C.D. (1978) "The estimation of the elastic modulus of concrete for the design of structures". Building Research Establishment, Watford, U.K. Current Paper CP 23/78.

Wittmann, F.H. (1979) "Micromechanics of achieving high strength and other superior properties". High Strength Concrete (Ed. S.P. Shah). University of Illinois at Chicago Circle, U.S.A.

Wittmann, F.H. (ed.) (1980) "Fundamental research on creep and shrinkage of concrete". École Polytechnique Fédérale de Lausanne, Switzerland.

Recommendations for further reading

Neville, A.M. (1972) Properties of Concrete, Pitman Publishing Corp. New York.

Orchard, D.F. (1979) Concrete technology, Vol. 1 Properties of materials 4th edition. Applied Science Publishers Ltd.

Illston, J.M., Dinwoodie, J.M. and Smith, A.A. (1979) Concrete, Timber and Metals. Van Nostrand Runhold, Wokingham, U.K.

Short, A. and Kinniburgh, W. (1963) Lightweight Concrete. John Wiley and Sons Inc. New York.

Hydraulic cement pastes : their structure and properties (1976) Proceedings of a Conference, University of Sheffield, Cement and Concrete Association, London.

Brooks, A.E. and Newman, K. (Editors) (1968) The Structure of Concrete. Cement and Concrete Association, London.

2.2 DISCUSSION AFTER THE LECTURE OF C.D. POMEROY

Ramachandran asked if he was correct that in one of the figures shown the creep strain increases as the condensation of C-S-H increases. In his laboratory it was observed that shrinkage of cement pastes made with a number of admixtures increases as the surface area of the paste increases. It is thus not easy to differentiate between the relative influences of degree of condensation and surface area on creep or shrinkage.

Pomeroy explained that the test was undertaken on parallel samples of unloaded material and the graph showed time dependence. Load would only have a small effect whereas when temperature changed, condensation increased with creep modification. There was evidence that the two phenomena were linked. Further work should be done to delineate the relative influence of the factors.

Ishai asked for the quantitative measure of ductility and Pomeroy answered it was the area under the stress-strain curve up to failure, or the area under the complete load-deformation curve. Energy is best based on the "work done" principle but with creep there are great deflections. He closed the debate by emphasizing the importance of using stiff testing machines.

3.1 PRINCIPLES OF ADHESION - BONDING IN CEMENT AND CONCRETE

David Tabor

Cavendish Laboratory
University of Cambridge
Cambridge UK

SUMMARY

This paper provides a simplified account of the types of bonds
formed at interfaces and the role of such bonds in determining the
mechanical strength of the interface. The final section shows how
these ideas may be applied to the adhesion of cement and concrete
if it is assumed that the major participant in interfacial bonding
is portlandite and/or calcium silicate hydrate.

I INTRODUCTION

Adhesion between solids involves two fundamental concepts.
The first concerns the strength of the bond formed at the interface when one solid is brought into contact with the other. This
is essentially a problem in the physics and chemistry of surfaces
and is sometimes referred to as the thermodynamic factor. It also
demands a knowledge of the true area of atomic contact. The second
concept concerns the processes involved in pulling the joined solid
apart in order to measure the mechanical strength of the interface.
This is essentially a problem in mechanics and is often understood
most clearly by applying the principles of fracture mechanics.
This again demands a knowledge of the true area of contact and, in
many cases, the detailed geometry of the contacting surfaces. These
two concepts, the thermodynamic and mechanical are, of course,
interrelated though not always in an obvious way. For example weak
bonds generally imply low adhesion but in some situations the
adhesive strength can be high. Again strong bonds usually imply
high adhesion but in some cases, particularly with brittle solids,
the adhesive strength may be small. We shall first deal with these

concepts separately and then discuss their relevance to the
adhesion of cement and concrete.

II   THERMODYNAMIC OF ADHESION

We consider the simplest, most idealized case of two atomically
flat surfaces that are brought into atomic contact over a specified
interfacial area. The first idea we have to recognise is that of
surface energy. A solid exists as a solid because of the attractive
forces between its constitutent atoms. If we think of the atoms
being brought together from infinity the attractive forces lower
the energy of the interacting atoms, so that the solid has a lower
energy than the isolated individual atoms. The solid is therefore
the stable form.

The bond energy of the solid depends on the strength of the
individual bonds and the number of bonds formed between each atom
and its neighbour. At the free surface of a solid the surface
atoms have lost, approximately, one half of their nearest neighbours. Their energy is therefore lowered less than the atoms in
the bulk. Consequently, the surface atoms have a greater energy
than those in the bulk. This excess is the surface energy and it
is expressed as energy per unity area. We are familiar with this
in liquids where the surface energy tends to draw drops into
spherical shape since the sphere is the shape that has the smallest
surface-area/volume ratio. This is not usually observed with solid
particles at normal temperatures because they are too rigid but
shape changes may occur at temperatures approaching the melting
point. We may, parenthetically, remark that with liquids the
surface energy is exactly equivalent to the surface stress* whereas with solids this equivalence does not hold. We shall not pursue
this matter further. Typical values for γ are given in Table 1.
It is clear that γ covers a very wide range: this is primarily
associated with the type of bond formed between neighbours. We
shall discuss this in greater detail below.

Suppose we have two solids of surface energies γ_1 and γ_2
respectively. If they are brought into contact the interfacial
energy γ_{12} will not be equal to $\gamma_1 + \gamma_2$ since the surface atoms
will acquire new neighbours. Consequently their energy will fall
and γ_{12} will be less than $\gamma_1 + \gamma_2$. In the limiting case if two
identical solids are brought into contact the interface will become
identical to the bulk material and γ_{12} will be zero. More generally

---

*   In the older English literature the surface stress was referred
    to as surface tension. In American usage surface tension is used
    to mean surface energy.

Table 1. Typical values of surface energy γ of various solids

Solid	γ mJ m^{-2} (erg cm^{-2})	Type of bond
Argon	10	van der Waals
Paraffin wax	30	van der Waals
Ice	100	Hydrogen
Rock salt	350	Ionic
Gold	2,000	Metallic
Diamond	5,000	Covalent

we can say that γ_{12} will always be less than $\gamma_1 + \gamma_2$. We may therefore write

$$\gamma_{12} = \gamma_1 + \gamma_2 - \Delta \tag{1}$$

where Δ is yet to be determined.

We are now in a position to describe the thermodynamic energy of adhesion when the two surfaces 1 and 2 are brought into atomic contact. For each unit area of contact we destroy unit area of surface 1 (energy γ_1), unit area of surface 2 (energy γ_2) and create unit area of interface (energy γ_{12}). The system has thus lowered its energy (per unit area) by

$$\gamma = \gamma_1 + \gamma_2 - \gamma_{12} \tag{2}$$

For a reversible thermodynamic process this would therefore be the energy required to separate the surfaces back to their original condition. Substituting (1) into (2) gives for the energy of adhesion

$$\gamma = \Delta \tag{3}$$

In order to derive an estimate for Δ we introduce a different way of regarding surface energy which links this parameter with the type of bonding existing between constituent atoms and their neighbours. We consider the following "thought experiment" for a solid consisting of a single species of atom. We take a block of the solid of unit cross section and pull on it so that it separates across some specified atomic plane so creating two new unit areas of surface each of which has surface energy γ. If the experiment is carried out slowly and reversibly the work done W is equal to 2γ. In this model W is equal to the work involved in breaking atom-atom bonds across the interface. If we know the number of atoms n per unit area and ε is the bonding energy between each atom and its neighbour we have

$$W = 2\gamma = n\varepsilon \tag{4}$$

This model of course, ignores the role of more distant neighbours, the possible changes in atomic spacing or in atomic structure of the outermost atomic layers, the changes in entropy* etc. Its main virtue is that it emphasises the importance of the number of atoms per unit area and the bond strength ε. Since atoms do not vary greatly in size, n does not cover a wide range. The main factor determining γ is the bond energy ε.

Consider now the interaction between two atoms for each of which the bond energy is ε_1 and ε_2. If the bonds are of similar type, we expect the bond energy ε_{12} to lie between ε_1 and ε_2. A reasonable average which is fairly widely used is of the form $(\varepsilon_1\varepsilon_2)^{\frac{1}{2}}$ and there are certain theoretical models which justify this relation.

On this model the energy per unit area to break the bonds at the interface is

$$n(\varepsilon_1\varepsilon_2)^{\frac{1}{2}} = n[\frac{2\gamma_1}{n} . \frac{2\gamma_2}{n}]^{\frac{1}{2}} = 2(\gamma_1\gamma_2)^{\frac{1}{2}} \tag{5}$$

This is the quantity γ or Δ in equation (3). More generally we may write (following Good)

$$\gamma = \Delta = 2\phi(\gamma_1\gamma_2)^{\frac{1}{2}} \tag{6}$$

where ϕ is near unity if the two solids have bonds of similar type and something less than unity when bonding types are substantially different.

We now consider the commonest types of bonding which determine ε and therefore γ.

van der Waals Interactions

The weakest type of interaction is that which occurs between

---

* The work done is really the total surface energy, h, which is related to the free surface energy γ by the thermodynamic relation $h = \gamma - T(\delta\gamma/\delta\tau)$. For simplicity we may ignore the entropy term $(-T\delta\gamma/\delta\tau)$ or alternatively assume that all our experiments are carried out at absolute zero.

non-polar atoms (Ar, Ne) or non-polar molecules (N_2, O_2). These
materials are non-polar only on a time average. If we could study
the electron distribution over very short intervals of time we
would find that the density of the electron cloud fluctuates, at
some instant it will be more on "one side" of the atoms (or molecule) than the other. This produces an instantaneous dipole which
interacts with a neighbouring atom, polarises it, and attracts it.
Since the frequency of these fluctuations is of the order of 10^{15}Hz,
the electrons respond to fields of this frequency: this is in the
optical range of wavelengths and leads to optical dispersion. For
this reason these forces are often called dispersion forces or
London forces after the physicist who first explained their origin.

The expression for the interaction energy between two identical
molecules (or atoms) of polarisability α is given by the London
equation

$$\varepsilon_d = \tfrac{3}{4}h\nu_o \alpha^2/4\pi\varepsilon.x^6 \qquad (7)$$

where x is the separation between them and $h\nu_o$ represents a characteristic energy often identified with the ionization potential of
the atom or molecule. It is seen that the dispersion energy falls
off as $(\text{separation})^{-6}$ so that a doubling of the separation reduces
the interaction energy by a factor of 64. Clearly, any separation
appreciably greater than closest packing greatly reduces the interaction energy. For two argon atoms in contact $\varepsilon_d \simeq 2.10^{-21}$J.

If one molecule has a permanent dipole (e.g. HCl) and the
other is non-polar the polar molecule will induce polarity in the
non-polar molecule and again produce an interactive energy ε_{in}
comparable with, though rather smaller than, the dispersion energy
and it also falls off as $(\text{separation})^{-6}$. In the context of the
present paper this is not an important interaction and we do not
need to quote an equation for it. On the other hand if both
molecules have permanent dipoles the interaction energy depends on
the mutual orientation of the dipoles and if they are free to move,
on the temperature. If they are not free to move (as is approximately the case in solids) the strongest interaction occurs when
they are oriented in line with the positive end of one dipole
pointing towards the negative end of the other. The interactive
energy between two identical molecules of dipole moment μ when
they are all oriented to give the maximum interaction is given by

$$\varepsilon_\mu = 2\mu^2/4\pi\varepsilon.x^3 \qquad (8)$$

where x is the separation between the molecules. It is seen that
the energy falls off as $(\text{separation})^{-3}$. The energy also depends
on the permanent dipole moment. For argon $\mu = 0$, for water $\mu = 1.8$

and for nitrobenzene $\mu = 4$ Debye (0, 6 and 13 Cm respectively). For
a highly polar material such as nitro-benzene the dipole-dipole
energy for the most favourable orientation at molecular contact is
nearly twenty times greater than the dispersion energy. With
many materials the difference between ε_d and ε_μ is not so marked.
For most polar molecules ε_μ is of the order 10^{-20}J. Some typical
comparisons are given in Table 2.

Table 2. van der Waals interactions energies between identical
species at equilibrium

Separation $x_0 \simeq 0.3$ nm			
Molecule	Energy in 10^{-21}J			Total
	Dispersion ε_d	Induction ε_{in}	Dipole-Dipole (max) ε_μ	
Ar	2	0	0	2
Ice (Water)	3	1	20	24
HCl	6	2	8	16
CH_3NO_2	7	6	$\sim$100	120

Of course in all these calculations we have ignored the short range
repulsion which ultimately provides equilibrium. This reduces the
effective interaction energy by a factor of about two (see Figure 1).

Closed Shell Interactions

      The simplest example in this category is the interaction
between say, two argon atoms. We have already described this in
terms of fluctuating dipoles which produce the van der Waals dispersion forces: the potential energy increases as (separation)$^{-6}$.
This relationship holds reasonably well until the atoms are so
close together that the orbitals begin to interact with one another.
When "atomic contact" is achieved this additional interaction is of
the same order of magnitude as the calculated dispersion energy ε_d.
Thus the potential energy of two argon atoms due to dispersion and
closed shell interaction is approximately doubled. Allowing for
the repulsive energy mentioned above the binding energy reverts

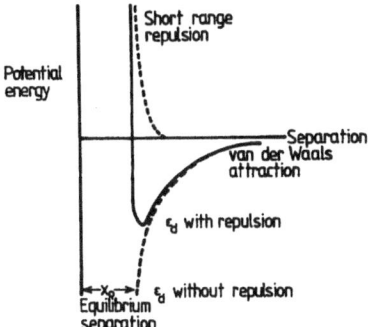

Fig. 1.   Potential energy of two atoms or molecules as a function
          of separation, assuming van der Waals' dispersion forces.
          If repulsion forces are taken into account the energy ε_d
          at equilibrium is approximately halved.

approximately to the value 2×10^{-21}J quoted earlier.

The interaction energy between closed shells in "contact" may
be regarded as a quadropole, octapole type of dispersion interaction
or as an electron-electron interaction. The main point is that it
is a general feature: if, for example, atomic contact occurs
between two non-polar oxides this interaction energy may be at least
as large as the simple London dispersion energy.

Hydrogen bonding

The hydrogen atom consists of a proton and a single electron.
When it combines to form a compound the electron is "drained"
away leaving a rather exposed proton. In the presence of another
molecule which contains a highly electro-negative atom (such as O,
Cl, F, N) with a lone pair of electrons available a relatively
strong bond will be formed. This is known as the hydrogen bond
and it operates over a very short range. Its energy is 20 to 40
$\times 10^{-21}$J so that it is large compared with van der Waals dispersion
energies. The bond occurs with hydrogen and not with other positively charged atoms because the proton has no orbital electrons
to screen it.

By its very nature a compound with an exposed proton will have
a dipole moment which can be deduced from dielectric measurements
in the gaseous phase. If it forms a hydrogen bond with another
molecule this increases the molecular polarisation and the interaction energy is increased. The energy to break the bond is then

partly due to the short range bond and to the residual dipole interaction of the separating molecules. Since the division of the energy
into various parts is theoretically difficult and in practice is of
little interest it is usual to refer to the whole of the interaction
energy as the hydrogen bond energy. For example, the dipole moment
of an isolated water molecule is 1.84 Debye = 6.1 Cm. If two such
molecules were brought into contact (separation ca. 0.3 nm) in the
orientation giving the maximum interaction and if no distortion of
the dipoles occurred the energy would be of order 20×10^{-21}J (see
Table 2). By contrast the total energy to break the hydrogen bond
in ice (deduced from the lattice energy or the enthalpy of vapour-
ization) is over 40×10^{-21}J. It is interesting to note that the
average energy required to break the covalent O-H bond in the water
molecule itself is of order 600×10^{-21}J. The hydrogen bond is
thus weak compared with a true valency bond but, as mentioned above,
large compared with dispersion energies.

The hydrogen bond plays a major part in the bonding of
hydroxides. We shall discuss this in greater detail below in
dealing with the properties of portlandite and calcium silicate
hydrate.

Ionic bond

This is a purely electrostatic bond: for each interacting
charge the energy falls off as (separation)$^{-1}$ so that it appears
to be a long range force. However, since matter is electrically
neutral every positive charge has a neighbouring negative charge
associated with it and these considerably reduce the effective
interaction energy. The ionic bond gives an interaction energy,
ε_i, about 10 times larger than the hydrogen bond. However, the
ionic character of a bond is sometimes deceptive. With Cs-F the
bond is almost 100% ionic: with H-Cl it is only about 20% ionic,
the remainder being covalent.

Metallic bond

In the metallic state the metal atoms lose their valency
electrons and the assembly may be regarded as a regular array of
positive charges in a sea of electrons. The electrons themselves
provide a relatively powerful fluctuating field and contribute to
the bond a dispersion energy of order 10^{-20}J while the metallic
bond itself has an average value of order $\varepsilon_m = 10^{-19}$J. With metals
one may generally assume that $\varepsilon_d \simeq 0.1$ to 0.3 ε_m.

Table 3. Order of magnitude values of bond energies between molecular pairs in contact

Type of bond	Example	Bond energy 10^{-21} J
van der Waals –		
dispersion ε_d	Ar–Ar	3
dipole-dipole (max) ε_μ	HCl–HCl	10
Hydrogen	H_2O–H_2O	40
Ionic	NaCl–NaCl	200
Metallic	Au–Au	400
Covalent	Diamond C–C	600

Covalent bond

The covalent or chemical bond involves electron sharing. Each
bond contains two electrons of opposite spin which pair-up to form
a stable configuration. Not all chemical bonds are 100% covalent.
Quite often there is a certain amount of charge transfer from one
atom to the other so that part of the bond is of an ionic nature.
Thus in H–Cl, 80% of the bond is considered to be covalent, 20% to
be ionic. Covalent bonds are very strong compared with the other
types of bonds we have considered – of order 500 to 1000 x 10^{-21} J.
They are strongly directional and they are saturated. By contrast
ionic bonds and van der Waals dispersion forces are central and
they are not saturated.

It is seen that these order-of-magnitude bond energies are
approximately proportional to the surface energies quoted in Table
1. Of course with solids which contain different atoms it is
important to choose only those bonds which are actually broken in
forming the free surface. For example with ice the relevant bond
is the hydrogen bond, not the covalent O–H bond. With paraffin wax
the bond would be the van der Waals dispersion bonds between the
hydrogen chains and not the C–C bonds within the chain. Similarly,
in a layer material such as MoS_2 the bond that is relevant is not
the chemical bond between Mo and S but the van der Waals bond
between parallel sheets of MoS_2 lamellae. Similarly with $Ca(OH)_2$
the relevant bonding, as we shall see later, is between groups of
protons in neighbouring sheets.

With oxides it is not clear which bonds determine the surface energy. This must depend on the crystal structure and the crystallographic plane under consideration. With hydroxides it is probable as indicated above that the weakest bonds will be the hydrogen bonds between the hydroxyl groups and these will determine the surface energy.

In the above discussion we have dealt primarily with the interaction between a species and itself to derive models for the surface energies of simple solids. Before we go on to consider how we can apply the idea of such surface energies to the adhesion between different species we must refer to two other types of bonding.

Acid-base bonds

Physical chemists are familiar with Lewis acids which are electron acceptors and Lewis bases which are electron donors. In the original form in which it was proposed the Lewis acid accepted two electrons from the donor to form a chemical bond, but the term is now used more loosely to cover electron acceptors and donors in general. A strong Lewis acid in contact with a strong Lewis base will clearly involve a strong bond, the energy being of the orders of 100×10^{-21}J. It is comparable with a chemical bond. With weak acids and bases the bond energy is smaller and the bond may be readily dissociated.

Charge transfer

If two surfaces are brought into contact, one being an insulator and the other a conductor or semiconductor electrons will flow from one to the other to equalise the Fermi levels. In the insulator these charges will remain close to the surface. Coulombic forces may then be more important than all the other interactions so far considered. Indeed Deryagin and his colleagues (Deryagin et al. 1978) hold that all adhesive processes are dominated by charge transfer but this sweeping generalisation is not widely accepted. Nevertheless, the possible role of charge transfer must always be born in mind.

The thermodynamic interfacial bond

We are now in a position to consider the extent to which we can use surface energies as a means of determining the thermodynamic interfacial bond γ between different materials, that is, the extent to which we can use the relation

$$\gamma = 2\phi(\gamma_1\gamma_2)^{\frac{1}{2}}$$

With solids the direct determination of γ_1 or γ_2 is difficult but
not impossible. There is however no satisfactory way of determining γ_{12} directly. Most of the information that does exist is
derived from experiments between a solid and a liquid, or between
two liquids, or in experiments where a vapour is adsorbed on a
solid. The interfacial energy may then be determined from calori-
metric measurements while with liquid-liquid systems a direct
determination of γ_{12} may be made. Such experiments suggest the
following generalizations.

(a)   for two contacting materials for which the type of bonding is
      the same and which are not too dissimilar in "strength" ϕ is
      close to unity. If we consider the way in which dispersion
      forces, dipole-dipole forces, ionic forces and metallic bonds
      arise it is easy to show that this is a good first approx-
      imation.

(b)   for non-polar bodies (e.g. van der Waals dispersive solids)
      in contact with a metal, only the dispersive part of γ metal
      (γ_m) reacts with that of the solid. As mentioned above, with
      metals $\varepsilon_d \simeq 0.1$ to $0.3 \varepsilon_m$ which implies that $\gamma_d \simeq 0.1$ to
      $0.3 \gamma_m$. Thus if we insert the full value of the surface
      energy of the metal in the bracket of equation (2) we must
      reduce ϕ by a factor of the order $(0.1)^{\frac{1}{2}}$ to $(0.3)^{\frac{1}{2}}$, i.e. 0.3
      to 0.5. In fact the available data suggest a lower value of
      ϕ of about 0.5.

(c)   for polar bodies in contact with metals it would seem that in
      many cases only the dispersive part of both bodies interact,
      i.e. dipole-image dipole forces may be ignored. There is no
      theoretical justification for this and as a generalization it
      must be applied with caution.

(d)   for polar bodies in contact with other polar bodies it may be
      preferable to split the surface energy of each body into its
      constituent parts γ_d and γ_μ and assume that the dispersion
      parts and the dipole parts interact independently.

(e)   for two solids which can form a covalent bond across the inter-
      face, theory suggests that a better estimate for the energy
      of adhesion γ (see equations (2) and (6)) is $(\gamma_1 + \gamma_2)$:
      rather than $2(\gamma_1\gamma_2)^{\frac{1}{2}}$ but if γ_1 and γ_2 are of comparable mag-
      nitude this does not differ markedly from $2(\gamma_1\gamma_2)^{\frac{1}{2}}$. A more
      important point is that it is not at all clear that covalent
      bonds will form across a solid-solid interface without some
      type of activation process. In that case one can only rely
      on the non-covalent parts (e.g. dispersion, dipole, closed
      shell interactions) to form interfacial bonds.

(f)   for lamellar solids in which the bonding between the lamellae
      is relatively weak the interaction will again approximate to
      $2(\gamma_1\gamma_2)^{\frac{1}{2}}$. This would apply to the interaction between say
      MoS_2 and graphite where the bonding is primarily of a van der
      Waals dispersion type: it would also probably apply to the
      interaction between $Ca(OH)_2$ and other similar hydroxides.

Thermodynamic criterion of cohesive and adhesive failure

        If we assume that only thermodynamic factors are involved we
are in a position to determine whether the interface will be
"weaker" or "stronger" than either of the two contacting bodies;
this in turn will determine whether the bodies will pull apart
truly at the interface (adhesive failure) or within one of the
bodies (cohesive failure).

        We use the energy criterion. In separating unit area of inter-
face we do net work $2\phi(\gamma_1\gamma_2)^{\frac{1}{2}}$. If this is greater than $2\gamma_1$ or $2\gamma_2$
it will be advantageous for the system to fail in one of the
solids (cohesive failure): is smaller than $2\gamma_1$ or $2\gamma_2$ it will be
advantageous for failure to occur in the interface (adhesive
failure). The comparison is thus between $\phi(\gamma_1\gamma_2)^{\frac{1}{2}}$ and γ_1 or γ_2 .
Figure 2 shows a plot of $\phi(\gamma_1\gamma_2)^{\frac{1}{2}}$ against γ_1 (in units of γ_2). It
is seen that for $\phi = 1$ the interface is always "stronger" than the
"weaker" of the two solids (γ_1) so that cohesive failure always
occurs. There is one important exception to this: it occurs when
$\gamma_1 = \gamma_2$ for then the energy to separate at the interface is exactly

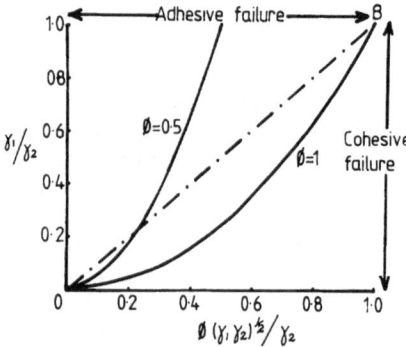

Fig. 2.   Plot of $\phi(\gamma_1\gamma_2)^{\frac{1}{2}}$ against γ_1 in units of γ_2 for two values
          of ϕ.   All points to the left of the OB give adhesive
          failure: all those to the right of OB give cohesive
          failure.

equal to the energy to produce cohesive failure (Point B, Figure 2).
In this situation one expects failure to occur randomly in the
interface, and in the hinterland. For bonds of dissimilar nature
where ϕ may be as small as 0.5 there is an appreciable range where
adhesive failure occurs.

This discussion of course ignores the relative sizes of the
atoms in the surfaces of the bodies and other parameters discussed
in greater detail by Good (1966). But it brings out the fact that
with two materials of similar bond type (where $\phi \approx 1$) the interfacial bond will tend to give cohesive rather than adhesive failure.
By contrast, when the bond types are very different there is a wide
range over which adhesive failure may occur.

There are two additional comments that should be made. In our
simplified treatment we have mixed up atomic models, involving bond
energies ε, with macroscopic models involving surface energies γ.
When atomic contact occurs it is more realistic to revert to the
atomic model since at this stage the interface is "aware" of the
graininess of matter. In this situation atomic mismatch at the
interface may be more important than simply affecting the number
of atoms per unit area. For example with two ionic solids in
contact a lack of register in the spacing may produce repulsive, as
well as attractive forces at various points in the interface.

The second comment concerns the area of true atomic contact at
the interface. When solids are placed together there will be some
regions of intimate atomic contact and other regions where there
are gaps. Since the range of atomic forces is relatively small, a
gap of say 0.5 nm is enough to reduce the force to a very small
value. Consequently arguments concerning interfacial energies or
bonds can only be applied realistically to those areas where the
contact is atomic.

Strength of adhesion

Figure 3a shows the potential energy ε between two atoms or
molecules as a function of the separation x. Clearly the force
between then is $-d\varepsilon/\delta x$. This is shown in Figure 3b. As an
increasing tensile force f is applied the separation will increase
from its equilibrium value OA towards AB. In principle it should
be possible to follow the curve BCD along the decreasing part of
the force curve. In that case the total area under the curve ABCD
will be equal to the interaction energy ε_0 shown in Figure 3a. In
practice the system will be highly unstable at B and f_0 will mark
the stage at which once the separation has been increased by the
distance AE, the molecules will pull apart.

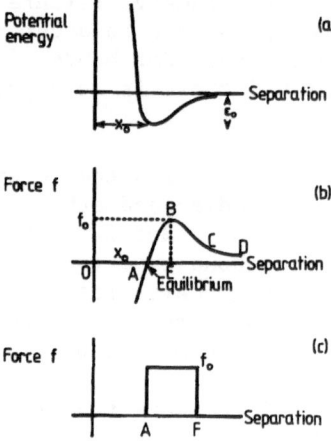

Figure 3.   (a)   Potential energy ε between two atoms or molecules
                  as a function of separation x.
             (b)   Force between the atoms or molecules. Point B
                  represents the maximum force f_0 required to
                  separate the particles. The area under curve ABCD
                  is the work to achieve separation and is equal to
                  the interaction energy ε_0.
             (c)   Equivalent block-like curve in place of Figure 3b.

        It is possible to make detailed calculations of f_0 in terms of
the shape of the f-x curve involving the law of force between the
molecules (attractive and repulsive) but there is little point in
doing so. We replace the curve ABCD of Figure 3b by the equivalent
block-like curve of Figure 3c. We then specify that $f_0.AF = \varepsilon_0$.
The distance over which the forces operate depends on the type of
force. Valence bonds, hydrogen bonds, metallic bonds, closed shell
interactions are effective only over distances of the order of 1 Å
(0.1 nm). Dispersion forces and dipole forces operate over somewhat larger distances. If we assume that, on average, AE is of
order 1-2 Å, the equivalent distance AF in Figure 3c may be taken
as 3-5 Å say 4 Å (0.4 nm).

Then $f_o = \varepsilon_o/AF$ (9)

If this applies to all the atom pairs across a plane of separation
in a solid composed solely of these atoms we may convert f_0 into

the tensile strength per unit area of cross-section (σ_0): we have

$$\sigma_0 = nf_0 = n\varepsilon_0/AF = 2\gamma/AF \qquad (10)$$

Consider a typical example - a hydrocarbon wax for which $\gamma = 30$ mJ m^{-2}. Then $\sigma_0 = 15 \times 10^7$ Nm^{-2} or 1500 kg cm^2 (over 1 tonne cm^{-2}). This is far larger than the observed strength. Similar calculations on other solids show a comparable discrepancy. The real strength is far less than that deduced from arguments based on thermodynamics. The reason for this is the presence of flaws or cracks which act as sources of high stress concentration.

III   FRACTURE MECHANICS

     Consider a block of material, as shown in Figure 4, which is under a uniform tensile stress σ. If there is a crack or flaw in the surface, tensile stress at the tip of the crack will be very much larger than σ. If the length of the crack is ℓ and the radius

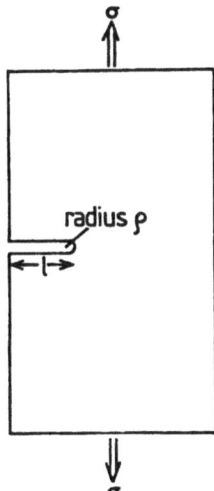

Fig. 4.   A block of material containing a crack of length ℓ and tip-radius ρ. If the applied tensile stress is σ, the tensile stress at the tip of the crack is increased by a factor of the order of $(\ell/\rho)^{\frac{1}{2}}$. This is known as the "stress concentration factor".

of the tip of the crack is ρ, the stress concentration factor is of order $(\ell/\rho)^{\frac{1}{2}}$. Thus the stress at the crack tip will be $\sigma(\ell/\rho)^{\frac{1}{2}}$. If

$$\sigma(\ell/\rho)^{\frac{1}{2}} \geq \sigma_0 \tag{11}$$

the atoms at the crack tip will pull apart, the crack will propagate, and the block will fracture. A fine crack may have a tip radius of say 10 nm. Then if $\ell = 1$ μm the ratio $(\ell/\rho)^{\frac{1}{2}} = 10$ and the block will appear to be 10 times weaker than its intrinsic strength. A surface flaw or crack of length 1 μm is a common occurrence. If the crack had a length of 0.1 mm the stress concentration factor would be 100. Thus, unless such flaws are avoided real solids will always be weaker than expected. Only in recent years has it proved possible to produce fine fibres without surface flaws and these do, indeed, possess their theoretical strength.

What then is the situation if we produce a sharp crack in which $\rho = 0$? Is the stress concentration infinite and does the material fail for a vanishingly small applied stress? At this point a new issue, first discussed by Griffith in 1921, becomes of crucial importance. It is not enough for the stress at the tip of the crack to exceed σ_0 for crack propagation to occur. An energy balance criterion must be observed. If the crack is to extend, this involves the creation of new surface areas. The energy required for this is equal to the increased area of the crack multiplied by 2γ (the crack has two surfaces). The crack can grow only if this energy is supplied and it can be provided only from released elastic strain energy in the hinterland surrounding the crack. This is essentially Griffith's energy criterion. It involves the surface energy γ, the elastic modulus E (as a measure of the stored elastic energy) and the crack length ℓ. To a first approximation the applied stress necessary to propagate a crack is then given by

$$\sigma = \left(\frac{\gamma E}{\ell}\right)^{\frac{1}{2}} \tag{12}$$

It turns out that this relation is equivalent to assuming that even the sharpest crack behaves as though it had a radius of curvature of the order of two or three times an atomic radius.

It is useful to express the Griffith criterion in a different way which provides a means of extending it to various classes of materials. We write

Energy release rate per unit area of crack propagation $= \Gamma$

where $\Gamma = 2\gamma$ if the solid fails in an "ideally" brittle manner.

If the material is viscoelastic or ductile considerable energy may be dissipated by these processes. We then write

$$\Gamma = 2\gamma + \text{viscoelastic or plastic energy dissipation}$$

The viscoelastic term with polymers may be thousands or tens of thousands of times bigger than the surface energy γ so that the effective value of Γ may be as high as 5000,000 mJ m^{-2}. This will increase the effective breaking stress by a factor of 100 or more (see equation 12). Thus the strength which is low because γ is small and because of flaws or cracks is greatly increased by this additional form of energy dissipation. The adhesive strength of adhesive tapes is largely due to the viscoelastic factor. Again with metals the smallest amount of plastic deformation at the crack tip will completely swamp the surface energy term giving effective values of Γ of order 10^7J m^{-2}(10^7mJ m^{-2}). It is for this reason that ductile metals appear so strong - not because of their interatomic forces. Indeed some alloys can be cooled to temperatures where they cease to be ductile: they then fail in a brittle manner at relatively small stresses although there has been no change in the interatomic forces.

    We see that there are three factors involved in the real strength of solids and interfaces

(a) The surface or interfacial energy which is a measure of the interatomic forces.
(b) The presence of flaws which are a source of weakness.
(c) Other dissipative processes which are a source of strength.

We may now mention three other important factors in the strength of contacting solids.

(a) The area of true contact. Anything less than atomic contact plays practically no part in the strength of the junction.
(b) The size and shape of the gaps since these may determine the severity of the stress concentrations.
(c) The topography of the surfaces before they are brought into contact. This is not only because this affects items (a) and (b) above. A few high asperities may tend to forces the surfaces apart and so reduce the effective strength of adhesion. One might say that high asperities provide an additional source of elastic-energy-release in the Griffith sense.

    It is evident from this brief discussion that the thermodynamic factors in adhesion can provide only a partial explanation of the observed adhesive strength.

IV THE ADHESION OF CEMENT AND CONCRETE

Concrete consists of a mixture of cement, sand and aggregate, the cement providing the adhesion with the other constituents. It is important, therefore, to have some idea of the materials that form at the interface. When cement hydrates fully its main constituents are:-

(a) 60 - 70% calcium silicate hydrate (CSH) which appears to be almost entirely amorphous and the main constituent of set cement,

(b) $\sim$ 20% crystalline calcium hydroxide (portlandite) which has a hexagonal lamellar structure (see below). There is also some additional amorphous material present,

(c) $\sim$ 5% calcium aluminium sulphate hydrate (CA$\bar{S}$H). This occurs as the trisulphate (ettringite) and as the monosulphate.

Pinchin and Tabor (1978a) showed, in pull-out experiments of stainless steel wires embedded in cement, that the failure was partly adhesive, partly cohesive. The pull-out force to produce "dehesion" did not depend on the extent to which the failure was adhesive or cohesive. They found that the material near the wires was rich in calcium hydroxide. More recently Al-Khalaf and Page (1979) studied the adhesion between mild steel and mortar. They found that the failure is predominantly, though not entirely, adhesive whereas with copper the adhesion is somewhat greater while failure is largely cohesive. Of greater significance is their observation that there is a marked segregation of portlandite at the interface. Similar results have been quoted by Professor Diamond and his colleagues (1978a, 1978b, 1980). They have shown that many solid surfaces, when placed in cement paste, become covered with a thin film of $Ca(OH)_2$. On the outer surface of the $Ca(OH)_2$ film a thin intimately bonded layer of CSH develops. This "duplex film" appears to play a basic part in cement adhesion. It may therefore be relevant to describe briefly what is known of the structure and bonding of these materials.

Portlandite [$Ca(OH)_2$] forms as hexagonal crystals with the CdI$_2$ structure, i.e. it has a lamellar structure. The Ca^{++} ions lie in a plane, each being surrounded by 6 OH ions, 3 above and 3 below the plane. A sketch is given in Figure 5a. The bonding within each lamella is covalent and strong. On the other hand the bonding between one lamella and its neighbour is relatively weak. It is for this reason that the crystal cleaves so easily. The type of bonding between the lamellae (see Figure 5b) is the subject of some dispute. Busing and Levy (1957), basing their conclusions partly on the OH bond length (0.958 Å), suggest that the group resembles the OH$^-$ radical (0.96 to 0.98 Å) rather than the OH$^+$ ion (1.03 Å). Consequently hydrogen bonding is unlikely since, as pointed out above, this type of bonding is favoured by an unshielded proton.

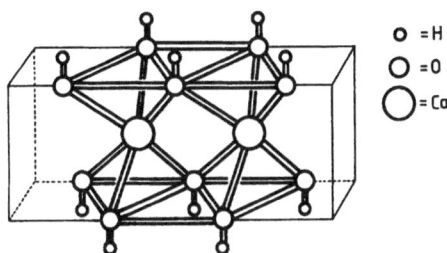

Fig. 5a.   Sketch showing structure of part of a single layer of
           calcium hydroxide (not to scale).

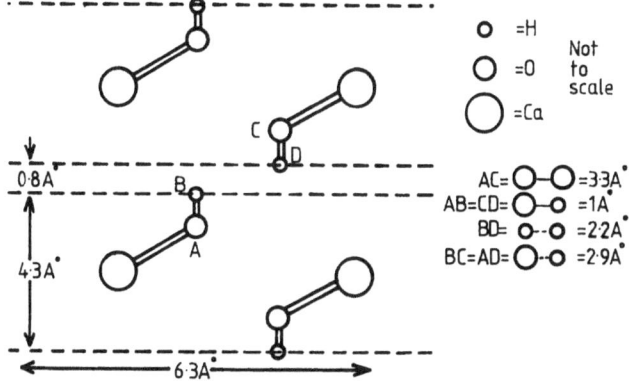

Fig. 5b.   Scale drawing of a plane through two neighbouring
           lamellae of Ca(OH)$_2$. The figure, in particular,
           indicates the distances between the oxygens and
           hydrogens in the neighbouring layers.

Again the distance between the hydrogen atom in one lamella and
the nearest oxygen in the neighbouring lamella is of order 2.9 Å:
this is rather a large distance for effective hydrogen bonding (in
water it is about 1.8 Å). Finally each oxygen is linked to three
calcium ions and one hydrogen ion which suggests that all its
orbitals are occupied. If hydrogen bonding is thus excluded one
might ask what bonding is available to hold the lattice together.
Busing suggests van der Waals forces (a convenient hold-all when
no other type of bonding is clearly present). This is a relatively
weak bond (say 2 x 10^{-21}J). On the other hand the outermost hydrogens of the OH$^-$ ions are in close contact across the lamellae
(separation about 2.2 Å). This could well provide proton tunnelling
across the layers involving a binding energy considerably greater
than that due to van der Waals forces. Again the presence of
some localised electron concentration between the hydrogens might
lead to the formation of a "weak" covalent bond across the lamellae.
The disparity between a van der Waals bond (say 2 x 10^{-21}J) and a
full covalent bond (500 x 10^{-21}) is enormous. Even a "weak"
covalent-bond could well have a strength of order 50 x 10^{-21}J. If
we assume such a value for the bonding across the lamellae we
shall be invoking a value intermediate between the weakest and the
strongest possible bonds. It is comparable in strength with a
hydrogen bond.

As a starting point then, let us assume a value of 50 x 10^{-21}J
for the bonding across the lamellae. There will be approximately
10^{19} OH groups per m^2. We thus obtain a surface energy γ of order
$\gamma = \frac{1}{2}$ x 50 x 10^{-21} x 10^{19} = 250 mJm^{-2}. (If we assumed only van der
Waals forces the value of γ would be about 10 mJm^{-2}. This is an
exceptionally low value. Since bonding and adhesive strength are
related to the surface energy it would clearly be very useful to
have a more direct measurement of the surface energy of Ca(OH)$_2$.
We shall discuss this below.) Meanwhile assuming a value of
γ = 250 mJm^{-2} we obtain a theoretical crystal strength of about
10^9 Nm^{-2}. If we extended this to the adhesion between cement and
metals we may reasonably assume that the metals are oxidised and
hydrolised to some extent. The oxygens might then be free to
form hydrogen bonds with the hydrogen end groups of portlandite: on
the other hand ferrous or aluminium surfaces may favour acid-base
bonding. In either case the interface would be comparable in
strength with the cohesive strength of the portlandite. We
should thus expect both adhesive and cohesive failure; the theoretical strength again being of order 10^9 Nm^{-2}. The observed
strength is 100 times smaller according to Page's results. In
this connection it is interesting to note that in friction experiments carried out be Pinchin on cement paste, he deduced a shear
strength of order 10^7 Nm^{-2}. Thus the relevant strengths are very
much less than the theoretical. On the other hand, if we consider
only van der Waals' interactions the observed strengths are very

close to the theoretical. In view of the discontinuous nature of the cement and the imperfections necessarily present in the crystals and the interface such agreement is most unlikely.

This discussion so far has dealt only with crystalline portlandite and its bonding to metal oxides (and presumably silica particles in the case of mortar). Bonding with the aggregate is less easy to describe.

We now turn to calcium silicate hydrate (CSH) which is probably the most important phase in set and hardened cement. According to Taylor "such order as exists (in CSH) is based on fragments of pseudo-hexagonal Ca-O sheets, not wholly unlike those known to exist in $Ca(OH)_2$ but having the OH groups replaced in varying degrees and in a largely random way with oxygen atoms of silicate anions and in the case of cement pastes, sulphate anions and Al- and Fe- containing complexes." This account clearly matches the description of the "duplex film" referred to by Diamond. Evidently the silicate and/or sulphate anions are linked to the CaO framework by strong covalent bonds. Consequently the question which concerns us here is the bonding between (i) the CSH and itself, (ii) the CSH and other materials in the composite, since it is this weaker bonding which will determine the strength of the whole. The most likely bond for such hydrated materials will be hydrogen bonding. Since in our previous calculation for portlandite we deliberately chose a bond strength of the same order as the hydrogen bond it is evident that the CSH bonding and the theoretical adhesive strength will be of the same order as that quoted above, i.e. 10^9 Nm^{-2}. Again this is about 100 times larger than the experimental value. If these assumptions are valid we must attribute the adhesion of concrete to the formation of bonds which have a strength comparable with hydrogen bonding. Further we are led to assume that the materials are "ideally" brittle and because of the presence of imperfections are appreciably weaker than their theoretical strength. Since the CSH is amorphous it is probably "stronger" than the portlandite and it is therefore probable that most of the failures are cohesive failures within the portlandite.

Before summarizing our conclusions we may remark that in a somewhat earlier paper on the adhesion of cement to steel Page, Khalaf and Ritchie (1978) showed that controlled polarisation of the steel influenced the observed adhesive strength. The relation between the adhesive strength and the potential resembled a classical electro-capillary curve in which the interfacial energy γ_{12} is influenced by the electrode potential. But whether it is sensible to explain this in terms of changes in γ_{12} or whether it is more realistic to view it in terms of polarisation of the metal oxide remains an open question. Polarization may also affect acid-base bonding.

Finally this discussion makes no useful contribution to the

possible role of interlocking fibrils in providing adhesion in
cement.

V CONCLUSIONS

The main conclusion is that the adhesive bonds in mortar and
concrete are provided by portlandite and calcium silicate hydrate
(CSH). The bonds are either hydrogen bonds or a composite of
van der Waals and weak covalent bonds and possibly some proton
tunnelling which together have a bond strength comparable with that
of the hydrogen bond. The adhesive bonds and cohesive bonds are of
comparable strength so that both adhesive and cohesive failure are
equally likely. However, since portlandite is crystalline and has
a well defined easy plane of cleavage, failure is more likely to
occur within the portlandite crystals rather than within the
amorphous CSH. If then the crucial "weak link" is the portlandite
crystal we may make the following four comments:

(i)   Little can be done to increase the adhesive strength by
        improving the bonding. It is conceivable that the aggregate
        could be chemically treated to provide acid—base bonding
        with the portlandite. In that case the interfacial bond
        would be stronger than the bonding within the portlandite
        and cohesive failure would occur within the portlandite
        itself. Consequently, this would not achieve any overall
        increase in the adhesive strength.

(ii)   We need more direct information concerning the interlamellar
        bonding in portlandite, the surface energy and the fracture
        properties. A study of the bonding could prove to be a very
        valuable theoretical exercise. The surface energy could
        best be determined by the method of Gilman (1960) if large
        single crystals can be prepared. In this method a wedge is
        pressed into a preformed crack in the crystal and the force
        to propagate the crack is determined. The whole experiment
        is carried out in liquid nitrogen so as to "freeze out" any
        possible plastic deformation. It would be interesting to
        know if the resulting value is of order $10 - 20$ mJm^{-2}
        (implying only van der Waals interactions) or a value some
        ten times bigger. If the experiment is then carried out at
        room temperature and the deduced value of the surface energy
        increases this implies that some plastic deformation is
        taking place.

(iii)   Since portlandite is a lamellar solid like CdI_2 it may be
        capable of undergoing intercalation, i.e. metals such as
        sodium or potassium which readily give up an electron might
        diffuse into the crystal and provide a bond between the
        lamellae. The effect of intercalation on the electronic

properties of lamellar solids has been widely studied but little attention has been paid to the effect of intercalates on mechanical properties. It is possible that they can strengthen the interlayer bonding and hence increase the strength - or they might provide a measure of ductility otherwise lacking.

(iv)   In the work of Pinchin and Tabor (1978b) on adhesion between stainless steel wires and cement it was found that near the steel surface the density of the cement was less than in the bulk. Apparently the presence of a solid surface restricts the packing of the cement particles. In their pull-out experiments a hydrostatic pressure increased the pull-out force presumably by increasing the area of intimate atomic contact. However, as soon as some displacement of the wire relative to the cement occured the pull-out force decreased markedly. Subsidiary experiments showed that this was due to compaction of the cement which reduced the "misfit" between the wire and the cement. This suggests that shear under hydrostatic pressure may be a practical means of increasing the sensity and the true area of atomic contact at the cement-aggregate interface. Without changing the nature of the adhesional bond this could conceivably increase the resultant adhesive strength of the composite.

(v)   The whole of the above discussion is based on the assumption that the source of **weakness** is the low cleavage energy of portlandite. However portlandite is extremely anisotropic and it may be that in the polycrystalline state it is much more resistant to fracture. (It is known, for example, that sintered polycristalline diamond cutting tools are far more resistant to fracture than cutting tools composed of single crystals). In that case there may be benefit in improving adhesion between aggregate and portlandite. Chemical treatments might improve acid-base bonding. Roughening of the aggregate might increase the surface area and hence the amount of bonding. Again surface active materials in the mix might favour better wetting. But such modifications will almost certainly change the mechanism and/or kinetics of hydration so that the final outcome is rather unpredictable. I am not competent to discuss this further. Finally if surface treatments of the aggregate are envisaged a coating of a slightly ductile material might well prove advantageous.

(vi)   It remains to consider briefly the possibility of reducing the adhesive strength of concrete so that the work of fragmentation might be reduced. In principle there are two approaches (a) thermodynamics (b) facture mechanics.

(a) The possibility of reducing the adhesion in existing concrete is very small. There may be penetrating fluids of extremely low surface tension that can penetrate the concrete and weaken the interface but the diffusion rates would probably be very long. Again ultraviolet irradiation can break chemical bonds but this does not seem a very practical approach. Does prolonged exposure to X-rays have any effect on strength? (This is rather unrealistic since even if it were effective the health hazards would rule it out).

(b) There might be more chance of reducing the energy consumption involved in the fracture process itself. This could be achieved if plastic flow of the cement could be reduced. With most solids including metals, alloys, organic and inorganic crystals and polymers, flow can be reduced and fracture faciliated by working at very low temperatures. This does not seem to occur with concrete. Even if it did occur it would not be economic. However a basic study of why low temperatures have so little effect on the fracture energy of concrete might, in itself, shed further light on the fracture process. Another approach would be to consider ways of inhibiting multiple-crack formation or some other way of reducing the fracture energy. Engineers might give some thought to practical tricks that might achieve this. For example, in the presence of a liquid at high hydrostatic pressure, fracture is sometimes facilitated if one can make use of the released elastic strain energy in the liquid.

Acknowledgements

I wish to express my sincere thanks to Dr. A.J.Majumdar for critical comments and for providing me with background information on adhesion in cement and concrete: and to Dr. A.R.Beal and Dr. J.V. Acrivos for constructive discussions on the nature of the bonding in lamellar solids.

References

Al-Khalaf, M.N. and Page, C.L. (1979), Cem. Concrete Res., 9, 197.
Barnes, B., Diamond, S. and Dolch, W.L. (1978a) Cem. Concrete Res., 8, 233.
Barnes, B., Diamond, S. and Dolch. W.L. (1978b) Cem. Concrete Res., 8, 263.
Busing, W.R. and Levy, H.A. (1957) J. Chem. Phys., 26, 563.
Deryagin, B.V., Krotova, N.A. and Smilga, V.P. (1978) "Adhesion of Solids", Consultants Bureau, New York and London.
Diamond, S., Ravina, D. and Lovell, J. (1980) Cem. Concrete Res., 10, 297.
Gilman, J.J. (1960) J. Appl. Phys., 31, 2208.
Page, C.L., Al-Khalaf, M.N. and Ritchie, A.G.B. (1978) Cem. Concrete Res., 8, 481.

Pinchin, D.J. and Tabor, D. (1978a) Cem. Concrete Res., 8, 15.
Pinchin, D.J. and Tabor, D. (1978b) Cem. Concrete Res., 8, 139.

As I have little first hand knowledge of the literature on cement and concrete my references in this field are very limited. Concerning the general approach to adhesion and adhesive strength I have found the first volume of "Treatise on Adhesion and Adhesives" edited by R.L. Patrick (Marcel Dekker, N.Y., 1966) extremely useful particularly the chapters by R.J. Good, J.R. Huntsberger, G.R. Irwin and F.M. Fowkes. But I have not necessarily agreed with all the points of view expressed by these authors.

3.2 DISCUSSION AFTER THE LECTURE OF D. TABOR

Participants were eager to discuss the points underlined by
Tabor in his conclusions as possible areas of research:
- If Portlandite determines strength, surface treatment will not
  improve bond, because Portlandite is always present in the inter-
  laminar areas whereas if CSH is important, surface treatment may
  be effective
- Densification as investigated in an experiment placing a wire in
  cement paste, revealing that the paste density in the vicinity
  of the wire is lower than in the bulk Applying hydrostatic pressure caused little effect, but adding shear to this pressure increased the densification between wire and cement paste because
  the true area of atomic contact is increased, thereby improving
  strength.
- Regarding the problem of facilitating fragmentation, suggestions
  were made of using thermodynamics, incorporating a surface agent
  to decrease bond, fracture mechanics with lowering temperature.

The use of negative temperature first was commented by
Lambotte who has found increased compressive and tensile strength
at −60°C. Kreijger reported work on freeze-thaw cycling in liquid
nitrogen, −190°C, revealing the effect of the water phase, see
fig. 1 (strength) and fig. 2 (temperature)

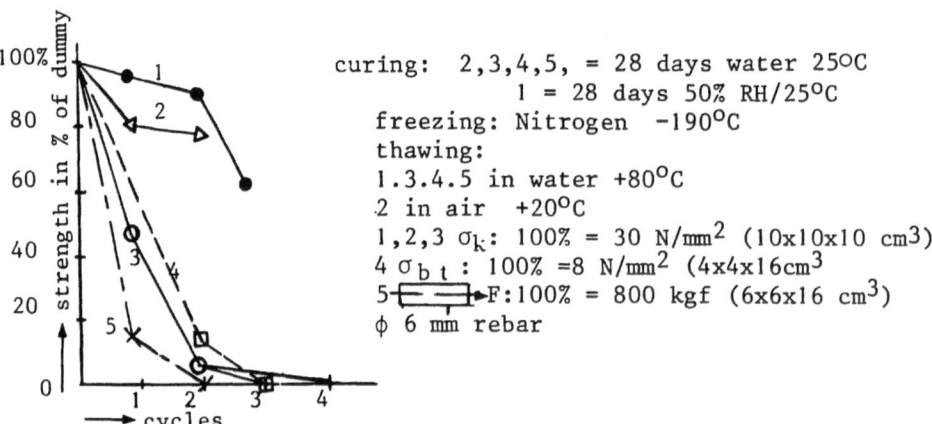

curing: 2,3,4,5, = 28 days water 25°C
                   1 = 28 days 50% RH/25°C
freezing: Nitrogen   −190°C
thawing:
1.3.4.5 in water +80°C
2 in air   +20°C
1,2,3 σ_k: 100% = 30 N/mm² (10x10x10 cm³)
4 σ_{bt} : 100% =8 N/mm² (4x4x16cm³
5⊏══╪►F:100% = 800 kgf (6x6x16 cm³)
ϕ 6 mm rebar

Fig. 1 Effect of freeze-thaw cycling in liquid nitrogen
on strength (concrete, max. particle size 16mm)

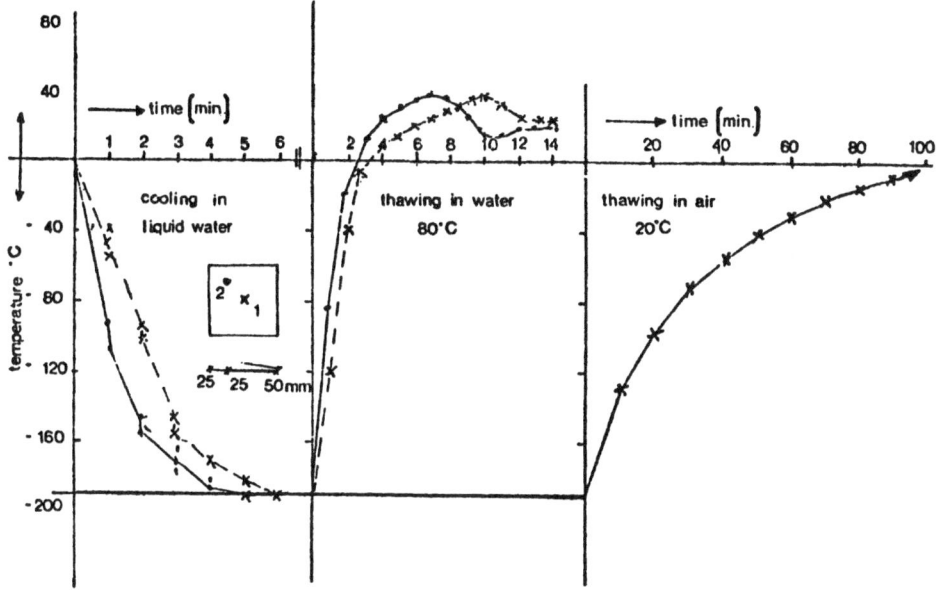

Fig. 2 - Temperature in specimen during freeze-thawing

For test specimen thawed in water (80°C) after 2 or 3 cycles the gravel particles were loosened from the cementmatrix, the same was true after crushing specimen which were cured in air and thawed in water (80°C) or cured in water and thawed in air (20°C) (cube strength about 50-70% of dummy) while water cured specimen which were oven dried and water thawed were heavily cracked with crumbled edges. The pull out force of reinforcement from water cured concrete specimen thawed in water (80°) was reduced to zero after two cycles. Micro-concrete specimen as used for the testing of cement (RILEM-gradation $\frac{W}{C}= 0.50$) pulverized after 2 cycles if freezing took place either in liquid nitrogen (-190°C) or in alcohol (-40°C) and thawing in water of 80°C.

The water-ice transition, extra water intake at thawing in water and probably especially the differences in coefficient of expansion of the water at higher teperatures give extra destroying forces whereas this is much less the case for heat treatment (such as the ASTM method for chemical analysis for cement content) advocated by Mather.

Ishai asked the lecturer to comment on the effect of moisture penetration on the interfacial bonding energy but Tabor did not know of data about this.

Ramachandran reported on densification research, plotting porosity
against strength for densified bodies of various materials, applying different pressures. The materials studied were hydrated portland cement, oxychloride cement, regulated set cement, $Mg(OH)_2$
and $Ca(OH)_2$. An increase in strength appears to be proportional to
the amount of reduction in porosity, each material shows a linear
relationship between porosity and strength. The strengths were compared at equal porosities of all these systems, $Mg(OH)_2$ gave the
highest strengths.

Tabor asked if this does mean that more bonds are formed by compaction and in this connection how the material was prepared and have
aggregates been used?

Ramachandran remarked that no aggregates were used, the powders were
taken in a steel mould and prepared under uniaxial pressure, indeed
he thought of more bonds formed by the compaction Tabor added that
he only could say that hydrogen bonding is stronger than bonding
in Portlandite but the effect on general strength is not known.

Frohnsdorff raised the question of polycrystalline grains. Calcium
hydroxide crystals in hydrated portland cement are quite large
although admixtures change their morphology. More study is needed
to investigate this effect on strength.

4.1   CONTAMINATION PROBLEMS IN THE RECYCLING OF CONCRETE

J. Francis Young

Department of Civil Engineering
University of Illinois
Urbana, Illinois U.S.A.   61801

Concrete rubble has the possibility of being contaminated by a variety of foreign materials. The presence of contaminants should be considered when old concrete is to be incorporated into new, and an assessment made of their potential undesirable interactions, if any, under the anticipated service conditions of the new concrete. It should be said at the outset that there is little published literature giving quantitative information about the level and effects of various contaminants. Nevertheless, one can make a useful qualitative survey based on a general knowledge of the properties of concrete and concrete-making materials.

SOURCES AND TYPES OF CONTAMINATION

A variety of materials may be present in concrete rubble. These can be broken down into two broad categories (see Table 1): those that may be present within the concrete itself; and those that are present in a particular stucture as other building components in addition to concrete, and which may cause contamination during demolition. We will consider each of these categories in turn and attempt to estimate the expected levels of contamination and their potential effects on concrete properties: strength, durability, volume change, and the behavior of the plastic concrete. Finally, we will discuss very briefly possible strategies that could be used to counter possible problems caused by contaminants.

CONTAMINANTS FROM THE CONCRETE

Reinforcing Steel

It is unlikely that significant quantities of steel would

Table 1. Sources and Types of Contaminants in Demolition Concrete

Source	Type
Within the Concrete	1. Reinforcing steel 2. Admixtures used in the old concrete 3. Chemical contamination (e.g., deicing salts, sea salts, oils, etc.)
Within the Structure	1. Metals (aluminum, copper, lead, etc.) 2. Brick 3. Wood 4. Gypsum (plaster and dry wall) 5. Glass 6. Plastics (including coatings and paints)

remain in the concrete. It can be readily removed by magnetic separation and this will be done as completely as possible because of its salvage value, and because of possible damage to machinery used to process the demolition material (crushers, conveyers, etc.). Small amounts of steel should not prove harmful, but might cause staining or surface damage due to rusting of steel close to the surface, particularly if chlorides were present. It is unlikely that enough steel would remain to significantly affect the density of concrete, or its handling characteristics.

Admixtures

A variety of admixtures may have been used in the old concretes (see Table 2). Air-entraining agents will have been invariably used in concrete pavements, but the use of other types of admixtures cannot be predicted with certainty. With the exception of calcium chloride, detection of the admixtures is not an easy matter and requires skill and experience. In most cases interactions will have occurred between the cement paste and the admixture, so that the bulk of the admixture is not likely to be released when the concrete is exposed to water. However, no systematic studies have been published that would substantiate this statement.

The possible harmful effects of previously added admixtures would be the interference with the action of admixtures added to the new concrete, (for example air-entraining agents), or changes in setting behavior. These effects could readily be checked by simple tests undertaken during the development of suitable mix designs. No problems are reported that would indicate detrimental effects of admixtures in old concrete.

Table 2. Principal Types of Admixtures Used in Concrete Construction

Category	Remarks
Air-entraining agents	Pavement concrete, structural concrete in extreme climates
Water-reducing agents	In a variety of structures, principally organic compounds
Set-accelerating agents	Calcium chloride most common, particularly in older concretes
Set-retarding agents	Principally organic compounds

The only long-term effects of admixture contamination would be concerned with corrosion of reinforcing steel or other embedded materials. Demolition concretes containing chlorides should be avoided in cases where prestressing is to be used, concrete cover or rebar will be low, or aluminum or zinc is in contact with the new concrete.

Chemical Contamination

Concrete from pavements or bridges that have been exposed to repeated applications of deicing salts will contain considerable quantities of chlorides, thus the same precautions considered above will hold in these cases also. Concrete regularly exposed to salt spray or certain ground waters may also contain high levels of both chlorides and sulfates. The problems of sulfates are discussed with gypsum contamination.

Oil contamination of concrete can either be deliberate (e.g., a surface treatment with linseed oil to prevent salt scaling) or accidental (from motor vehicles passing over pavements). In either case it is primarily a surface contaminant and will not be more than a trace contaminant in the total concrete. Furthermore, these materials are not water-soluble and so unless heavy contamination is present no problems are likely. Any undesirable effects would be interference with the action of admixtures added to the fresh concrete, or changes in setting behavior. These can be monitored during preliminary testing and steps taken to counteract any noticeable effects.

CONTAMINANTS FROM THE STRUCTURE

Metals

Non-ferrous metals present in demolition concrete would most likely be aluminum from flashings, frames and conduits; lead and copper from plumbing components; brass, bronze, etc. from fixtures. Again, the salvage value of these metals would reduce their likely contamination to low levels and the amounts of any of these metals would not be great enough to affect the properties of the hardened concrete.

The presence of copper and copper-based alloys would present no durability problems since these will not corrode in concrete. However, the corrosion of zinc and aluminum could cause problems through the release of hydrogen in the fresh concrete which could increase the porosity of the concrete in the new concrete. Corrosion of aluminum in hardened concrete could cause cracking if present in large amounts.[1]

Brick

Frondistou-Yannas and Itoh[2] has estimated that about 10% of demolition concrete from buildings will be brick rubble. This material will act as a satisfactory lightweight aggregate with no anticipated durability problems. It will, however, increase the adsorption of the concrete aggregate which can lead to workability problems if not compensated for. The density, modulus and volume changes can be expected to be influenced by the proportion of brick in the demolition concrete and in this regard it will be typical of conventional lightweight aggregates.

Wood

In some structures wood may be a potential contaminant. According to Frondstou-Yannas and Itoh[2] benefication by using density separation procedures should reduce the wood content to negligible levels. Wood is an undesirable contaminant since it swells when moist and is degraded by alkalis, both of which will occur to some degree in concrete. Extensive contamination by wood chips or fibers could interfere with setting or delay early strength gain.

Gypsum

Gypsum ($CaSO_4 \cdot 2H_2O$) will occur in demolition concrete as debris from interior dry wall partitions or plaster surfaces on concrete or brick walls. Because gypsum is relatively weak it is likely to occur predominantly as fine material. Removal of the fines from demolition concrete will not only decrease gypsum levels, but also

reduce water adsorption of the aggregate. Alternatively, the
gypsum content in the fine fraction could be removed by washing,
while larger lumps of gypsum could be removed by density separation
methods.[2]

Gypsum is potentially deleterious because it can react with
the cement paste, either in the recycled aggregate or the new paste,
to form expansive reactions by the formation of ettringite, i.e.
an *in situ* sulfate attack. Gypsum is the only concrete contaminant
that has been systematically studied. Graf added varying levels of
gypsum to concrete rubble and is reported to conclude that about
1% gypsum by weight of aggregate was the maximum tolerable level.[3]
The rate of expansion depends on the particle size of the gypsum
and the moisture contents of the concrete. Gaede on the other hand
suggests lower levels, about 0.5% soluble sulfate by weight of aggregate should be adopted for recycled concrete from both cement and
aggregates. Samarai[5] found that up to 4.5% (by weight of cement)
of total sulfates could be tolerated in concrete (from both cement
and aggregates), and this could be increased to 6.0% if a sulfate
resistant cement were used. These figures correspond to 0.5% and
0.7% respectively of soluble sulfates in the aggregates.

Glass

Plate glass from windows may contaminate demolition concrete,
although no figures are available as to the levels that might be
expected. Since the density of glass is similar to that of
concrete or aggregates separation would be very difficult.

Waste glass could be a potentially dangerous contaminant because
it can take part alkali-silica reactions. It should be remembered
that glass is a particularly reactive material in this regard,
because of its high alkali content.

Plastics

Plastic materials will form only a small fraction of concrete
rubble since most will be removed by density separation. Although
plastics can be diverse in their composition they are generally
inert materials and will not interact chemically with the concrete
components. Some plastics do suffer alkaline degradation and this
could occur in most concrete but such degradation is unlikely to
be harmful to the concrete. Plastic contaminants would not lower
concrete strength and modulus of elasticity significantly when
present in the small amounts one might expect to find in demolition
concrete.

CONTROL OF POTENTIAL PROBLEMS

It can be concluded from the foregoing discussions that, assuming reasonable levels of benefication are used to remove salvagable materials and lightweight materials, very few contamination problems should exist. Although some potentially low levels of contaminants, such as admixtures, could conceivably affect the setting of concrete or the performance of new admixtures (air-entrainment, for example), it would seem unlikely that this would prove a serious problem since the materials must be leached from the concrete aggregate. Any deleterious effects should be noticed during testing of concrete mixes and remedial steps could be taken through use of suitable admixture levels.

Only three contaminants appear to pose potentially serious durability problems: chlorides, sulfates, and glass. The levels of the two former species can be readily monitored by analyses. High levels of chloride would mean paying special attention to reinforcement protection or restricting new concrete to service conditions compatible with existing recommendations (see Table 3). Note that "soluble" chloride is more critical than total chloride since a large fraction of the total chloride ions may be chemically combined in the hydration products of cement. Steps should be taken to keep SO_3 levels below 1.0% in the total aggregate. If the new concrete can be maintained in a dry state higher sulfate levels might be acceptable. The effect of sulfate-resistant portland cements and, pozzolans have been variously reported as beneficial[5] or of no value[7] in combatting sulfate contamination and further investigations are needed. The effects of slag cements should also be evaluated.

Table 3. Recommended Limits for Chloride Ion
in Concrete (Wt % of cement) (Data from Ref. 6)

Category	Upper Limit
Prestressed concrete	0.06%
Conventional reinforced concrete kept moist and exposed to chloride	0.10%
Conventional reinforced concrete kept moist but not exposed to chlorides	0.15%
Conventional reinforced concrete kept dry	No Limit

The amount of glass in aggregate would have to be estimated by microscopic examination. High concentrations of glass would require tests for expansive alkali-silica reactions. These could probably be controlled within safe levels by the use of suitable pozzolans.[8] If concretes will not be used in moist environments, then alkali-silica reactions probably would not prove troublesome.

CONCLUSIONS

Contaminants should not be a problem in demolition concrete originating from "all-concrete" structures pavements, retaining walls, bridges, etc., unless contamination by chlorides or sulfates occur. Concretes from building demolition wastes may contain gypsum or glass contamination which could lead to harmful long-term expansions. If these problems are recognized their presence can be monitored and steps taken to counteract their effects.

REFERENCES

1. H. Woods, "Durability of Concrete Construction," Monograph No. 4, American Concrete Institute, Detroit, Michigan (1968).
2. S. Frondistou-Yannas and T. Itoh, Economic Feasibility of Concrete Recycling, J. Struct. Div., Proc. Amer. Soc. Civil Engrs. 103:885 (1977).
3. O. Graf, quoted by P. J. Nixon, Recycled Concrete as an Aggregate for Concrete - A Review, Mater. Constr. (Paris) 11:371 (1977).
4. K. Gaede, quoted by P. J. Nixon (see ref. 3).
5. M. A. Samarai, The Disintegration of Concrete Containing Sulfate-Contaminated Aggregates, Mag. Concr. Res. 28:130 (1976).
6. ACI Committee 201, Guide to Durable Concrete, J. Amer. Concr. Inst. 74:573 (1977).
7. A. D. Buck, Recycled Concrete as a Source of Aggregate, Proc. Symp. Energy Resource Conserv. Cem. Concr. Ind., CANMET, Ottawa (1976).
8. C. D. Johnson, Waste Glass as a Coarse Aggregate for Concrete, J. Test. Evaln. 2:334 (1974).

4.2 DISCUSSION AFTER THE LECTURE OF J.F. YOUNG

Several participants recorded their experience of disorders
caused by contamination. Ramachandran remarked that although the
lecturer has implied that sulphates in bricks are not a source of
sulphates, in many types of bricks the presence of these salts are
apparent from efflorescence effects.These salts are in a soluble form
and should be taken in account. Moreover it is indicated that metallic
impurities are easily removed from concrete and may be present in
small amounts and that they may not cause ill effects. It should be
noted that even in small amounts if two metals are present, they
create galvanic cells and cause corrosion. Young agreed that there
are important factors to be considered. Ramachandran further remarked
that the presence of MgO should be considered in old concrete in
which cement may not be completely hydrated in making new concrete
(which may use higher cement factor) the total MgO should be kept
within the specified limits. Young answered that since in old concrete MgO has not hydrated, it may also not hydrate when used to make
new concrete. Ramachandran admitted that this will be so but for
the fact that crushing the concrete exposes new sufaces and hence
promotes hydration of MgO. It is thus imperative that either the reclaimed concrete should pass the present specification or new specifications should be drawn. This can be done only after a systematic
work on the various long term durability studies is completed.
Mather underlined the danger of certain types of bricks, especially
with Mg-content and refractory bricks and mentioned a recent serious
problem in the San Francisco Bay Area in California caused by contamination of concrete aggregate with crushed brick. This has been
described in Engineering News-Record of 20 November 1980 (p.13) as
involving concrete provided to more than 100 objects from six ready-
mixed concrete plants that used aggregate from a 35,000 ton pile on
which 4 tons of discarded refractory bricks had accidentally been
dumped. Damage amounting to $250,000 or more was suffered by each
of 20 projects. The brick fragments expand with moisture, causing
"pop-outs" in the concrete ranging up to 15 inch in diameter and
2 inch deep.
Hilsdorf referred to the presence of chlorides in rubble after fire
damage, enquiring of a possible way to transport them into insoluble
salts. The only solution seems to be to refrain from re-using such ce
crete, or to protect reinforcement e.g. by epoxy coatings, use special
cements or water reducers to improve impermeability.
Hansen regretted that the ACI 201 sets no limit to chloride content.

The wording of this specification which refers to "concrete allowed to dry" is considered too vague.

Carpenter gave a case history of recycling pavements with bituminous overlay which affected air entrainment in the new concrete. Sierra remarked as regards   concrete recycled from crushing old pavements, there are other contaminants especially reinforcements and dowel steels, de-icing salts, admixtures etc. but also one could find:

- bituminous products or synthetic resins with various filler contents which have been used as injections for sealing joints.
- subgrade materials which are often slag gravel or soil-cement.
- clays of various fineness located in bared joints, cracks and at joints between the slab and subbase or slab and emergency stopping lane, caused either by erosion of the subbase through the dynamic effect of water trapped under the slab (pumping phenomenon) or deposit after infiltration of extra water (seepage of rainwater, rising of ground water etc.)

5.1   FRACTURE ENERGY OF CONCRETE

H.K. Hilsdorf, S. Ziegeldorf

Institut für Baustofftechnologie

University of Karlsruhe, West Germany

1.   OBJECTIVES

In this presentation problems of fracture of concrete as seen
in relation to the demolition of concrete structures will be dealt
with. In particular, the question arises how much energy will be required to demolish a concrete structure by mechanical means. To limit the extent of the problem we will only consider normal weight,
unreinforced concrete. In the first part a brief overview of the
structure of concrete will be given as far as it is needed in the
context of fracture. Then modes of fracture i.e. propagation of
cracks in concrete will be dealt with. Based on this background information estimates of the fracture energy of concrete as well as
parameters influencing it will be summarized.

Finally the initial question on how much energy is needed to
demolish concrete and approaches to minimize this energy will be
dealt with.

2.   STRUCTURE OF CONCRETE

Since various aspects of the structure of concrete on the
microscopic level have been presented in detail in preceeding papers,
in the following only some relevant aspects of the macroscopic
structure of concrete will be given.

In the simplest case concrete can be treated as a two phase
material consisting of hardened cement paste and of aggregates.
Though this model may give some useful information such as upper and
lower bound values of the modulus of elasticity it fails when applied
to strength and fracture of concrete. Various studies have shown

101

that a three phase model consisting of a matrix, aggregates and the
matrix-aggregate interface will yield more realistic results (Fig 1).
This is so because the structure and the properties of the region
surrounding a coarse aggregate particle differ significantly from
the structure and the properties of the matrix. Finally it was
found that the best results to model concrete can be achieved by
distinguishing between the three phases mortar, coarse aggregates
and interfaces. Though the incorporation of the fine aggregates into the matrix is somewhat arbitrary this model gives a more realistic representation of fracture because interface cracks develop
mainly around coarse aggregates rather than around the fine aggregates in the mortar /1-3/.

Let us now have a brief look at those three phases. The properties of the mortar matrix are controlled by the properties of the
hydrated cement paste. It is composed mainly of calcium silicate
hydrates and calcium hydroxide, the latter being the coarser compound which is more or less uniformly distributed within the hydrated cement paste. From a view point of structure the characteristic
features of hydrated cement paste are its porosity which may be as
high as 5o percent by volume depending on the degree of hydration
and the particular composition, and its specific surface area which
is in the order of 10^2 m^2/g.

The second phase, the coarse aggregates normally exhibit a
strength, a stiffness and a failure strain which are considerably
higher than that of the mortar phase. This can be seen from Fig. 2
in which stress-strain relationships of a typical mortar and of a
quartzitic coarse aggregate loaded in compression are shown.

As already mentioned the structure of the cement paste in the
vicinity of aggregate particles deviates substantially from the
structure of the bulk paste (Fig. 3): Based upon scanning electron
micrographs the interface consists of several layers of distinct
structure and composition (4-6): Immediately adjoining the aggregates there exists a layer of calcium hydroxide crystals with a
thickness of approximately 1×10^{-3} mm. In this layer also some
thin longitudinal ettringite crystals may be incorporated. The subsequent layer has a thickness of $5 \times 10^{-3} - 1o \times 10^{-3}$ mm and initially has a rather high porosity which, however, is reduced in the
course of hydration. It is formed principally of large hexagonal
calcium hydroxide crystals and smaller ones characterized by stacked
platelets. With increasing distance from the aggregate surface the
paste becomes denser so that a transitional layer with reduced po-
rosity and a thickness of $10 \times 10^{-3} - 30 \times 10^{-3}$ mm is formed. The transition phase is followed by the regular hydrated cement paste.

Less is known about the change in structure of the interface
with continued hydration. However, it has been observed that the
strength of the interface ceases to increase at a comparatively
young age.

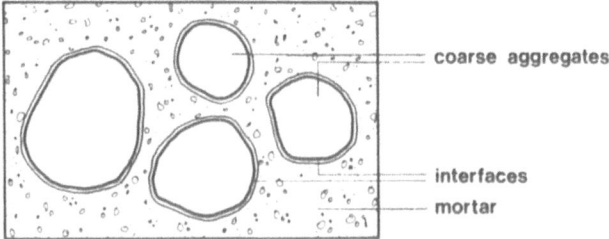

Fig. 1.   Concrete Represented by a Three-Phase-Model

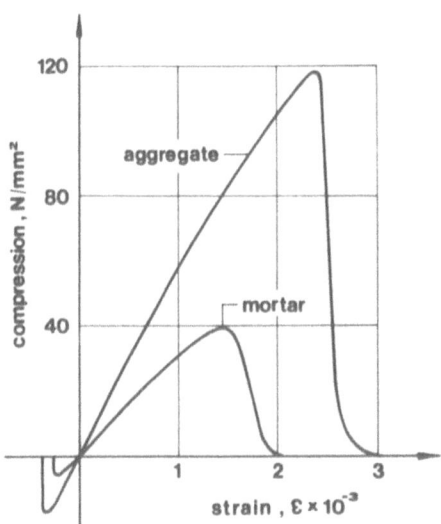

Fig. 2.   Stress-Strain Diagrams of Mortar and of Aggregate

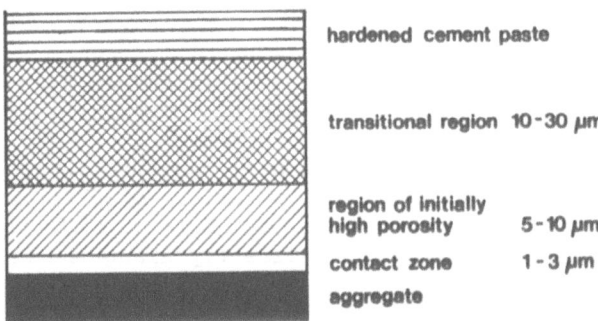

Fig. 3.   Structure of Aggregate-Hardened Cement Paste Interface

3. MODES OF FRACTURE

In numerous studies on concrete specimens it has been found that fracture of concrete is the consequence of stable and unstable crack propagation which occurs in either the interface, the mortar matrix or the aggregates. In the following our knowledge on crack initation and crack propagation in the various phases will be summarized.

3.1 Crack Initiation

When manufacturing concrete, flaws of various sizes are formed within the mortar matrix and particularly at mortar-aggregate-interfaces.

Further on various types of internal stresses are developed already in the young concrete such as stresses due to bleeding, early shrinkage, subsequent drying shrinkage and thermal incompatibility of concrete components when the concrete temperature increases due to the heat of hydration.

These internal stresses often exceed the tensile strength of the mortar or of the interface resulting in the initiation of cracks which start at the flaws mentioned above. Such cracks, however, are observed mainly at aggregates which are larger than a certain limiting size /3/. An explanation for this phenomenon can be given on the basis of the strain energy stored in the vicinity of aggregate particles and in the aggregate particles themselves. Only at aggregates which are larger than a limiting value is the stored strain energy sufficiently high to propagate a crack /2,3/.

As shown in Fig. 4 the initial cracks are formed predominantly along the interfaces though some mortar cracks in radial direction may occur.

3.2 Crack Propagation

When concrete is subjected to external tensile or compressive stress beyond a certain limit the cracks which had been initiated prior to loading will start to propagate.

Initially the interface cracks will continue to grow along the interfaces. In young concrete these interface cracks propagate within the second layer of high porosity as shown in Fig. 3. In older concretes crack propagation occurs directly along the boundary of coarse aggregate particles.

Cracks in a young mortar propagate through the regions with

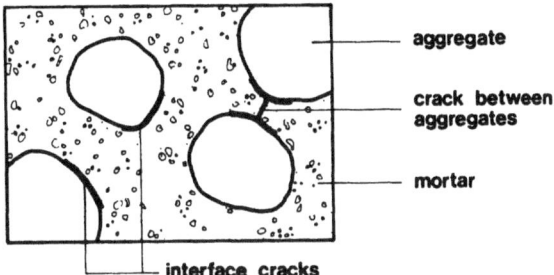

Fig. 4.   Crack Initiation at Interfaces

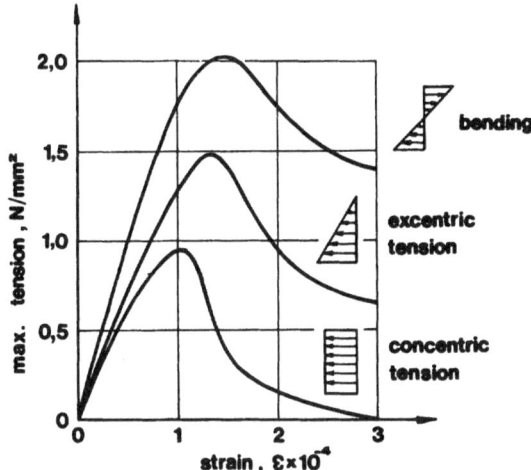

Fig. 5.   Stress-Strain Diagrams of Concrete Loaded in
           Bending and Tension (8)

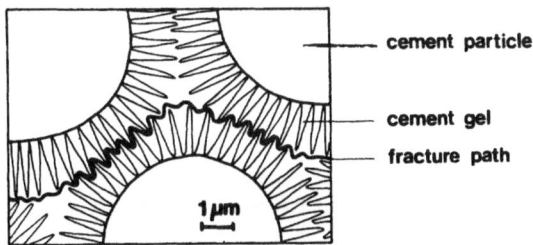

Fig. 6.   Tortuosity of a Fracture Path in Hardened Cement Paste

calcium silicate hydrate particles of high porosity. As hydration
continues and as the total porosity of the paste decreases the
cracks propagate mainly along the boundaries of calcium hydroxide
crystals. This is the reason why in fractured paste surfaces the
calcium hydroxide crystals become visible only after the age of the
paste has exceeded approximately seven days /7/.

The mode of crack propagation in concrete depends on the type
of stress. This will be discussed in the following sections.

3.2.1 Crack propagation of concrete loaded in tension. Crack
propagation in concrete also has a bearing on the stress-strain
relationship as shown in Fig. 5. At low stresses the stress-strain
relationship of concrete is almost linear. No crack propagation
occurs. At a stress level of approximately 8o percent of the uniaxial tensile strength cracks propagate slowly. The extent of slow
crack propagation depends on the stress gradient: as the stress
gradient increases the extent of slow crack propagation as well as
the apparent tensile strength of the concrete increase. When employ-
ing very stiff testing machines the stress-strain diagram of concrete loaded in tension may exhibit a descending portion which is
more pronounced for excentric loading than for concentric loading
/8/. This descending portion is an indication of slow crack propagation at strains beyond the ultimate.

Microscopic observations indicate that even in specimens made
of hydrated cement paste the cracks follow a tortuous path (Fig. 6)
and numerous microcracks rather than a single crack are formed as
the consequence of external tension in the vicinity of an initial
crack as shown in Fig. 7. As the stresses are increased some of
these microcracks join to form a final fracture surface.

In a mortar specimen multiple crack formation occurs. In addition, propagating cracks are arrested by sand particles resulting
in meandering and branching of cracks (Fig. 8). Therefore, the true
crack surface is considerably larger than the apparent fracture
surface.

Multiple fracture is even more pronounced in concrete containing coarse aggregates. As a consequence of this fracture behaviour
the extent of slow crack propagation in concrete loaded in tension
increases with increasing volume concentration and size of aggregate.

Microscopic studies by Moavenzadeh et al. /9/ and Higgins et
al./1o/ lead to the assumption of the following approximate ratios
of true and apparent fracture surface area of
        hardened cement paste : (1 - 2) : 1
        mortar   : (5 - 1o) : 1
        concrete   : (15 - 2o) : 1

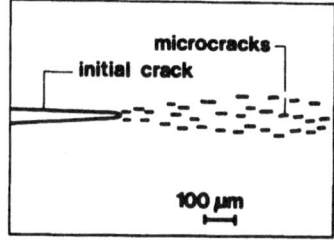

Fig. 7.   Microcrack Formation in Hardened Cement Paste

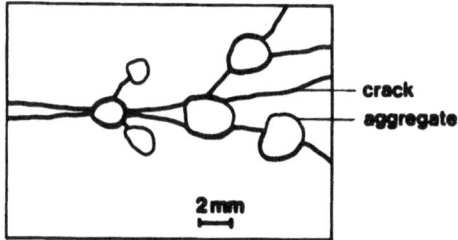

Fig. 8.   Crack Arrest and Crack Branching at Aggregates

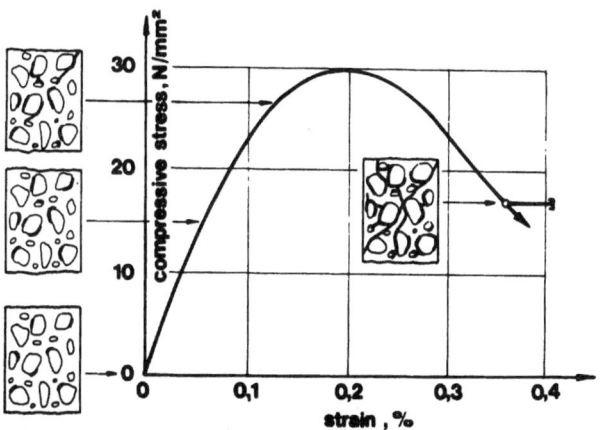

Fig. 9.   Stress-Strain-Relationship and Crack Propagation in
           Concrete Loaded in Compression

Studies of fracture surfaces also indicate that for concretes made
of normal weight aggregates having a strength up to approximately
3o N/mm^2 the cracks normally propagate only in the matrix or in the
interfaces. For light weight aggregate concretes or for concretes
with a higher strength also crack propagation through aggregate particles may occur.

3.2.2 Crack Propagation of Concrete Loaded in Compression. A-
gain, crack propagation of concrete loaded in compression will be
described on the basis of the shape of a stress-strain diagram
(Fig.9).

Though cracks has been formed prior to the application of
an external stress no crack propagation occurs up to an external
stress of 30 - 40 percent of the ultimate. At this stage interface
cracks propagate in a stable manner. At the same time a pronounced
deviation of the stress strain diagram from linearity can be observed. At approximately 8o percent of the ultimate stress numerous
microcracks start to extend into the mortar phase. Their orientation is mainly parallel to the direction of the external stress.
The propagating cracks are hindered by aggregates resulting in increased crack surface area as already described for the case of
concrete loaded in tension. As the stress increases the length of
individual microcracks increases and may eventually reach a critical
length. At this stage smaller microcracks may join to form larger
cracks and sudden unstable crack growth can only be prevented by
decreasing the external load. As long as unstable crack propagation
is prevented the stress-strain diagram shows a pronounced descending portion, the slope of which increases with increasing strain
rate. If the load is kept constant at a certain stage beyond the
ultimate, unstable crack growth will occur resulting in the formation of a continuous, mostly inclined fracture surface.

Some estimates of the crack density, i.e. the length of cracks
per unit cross section of a concrete can be taken from Fig. 1o for
various stress levels.

Similar to concrete loaded in tension, in normal weight aggregate concretes of a strength of less than 3o N/mm^2 cracks propagate around coarse aggregates. For higher strength concretes also
some crack propagation through aggregates may occur.

3.2.3 Effect of type of loading on crack propagation in concrete. From the shapes of stress-strain relationships of concrete
loaded in compression at different strain rates /12/ it may be
concluded that slow crack propagation is the more pronounced the
slower the rate of straining (Fig. 11). The highest crack density
is observed for a concrete loaded to a sustained high overload.
Under such loading conditions failure strains of concrete as large
as 1 percent may be observed.

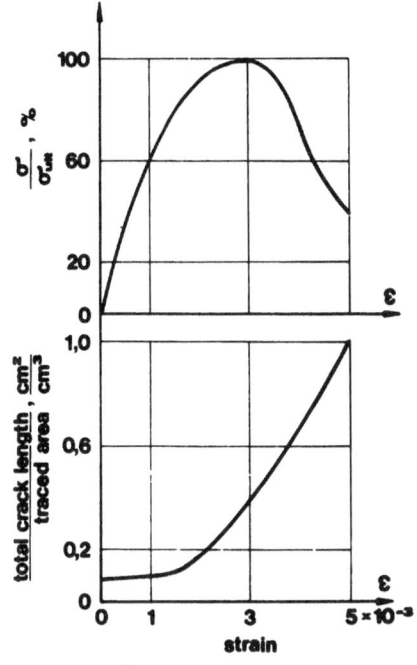

Fig. 10. Crack Density in a Concrete Loaded in Compression /11/

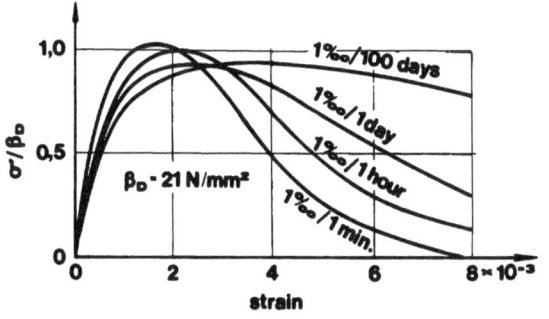

Fig. 11. Effect of Strain Rate on Stress-Strain Diagram of Concrete Loaded in Compression /12/

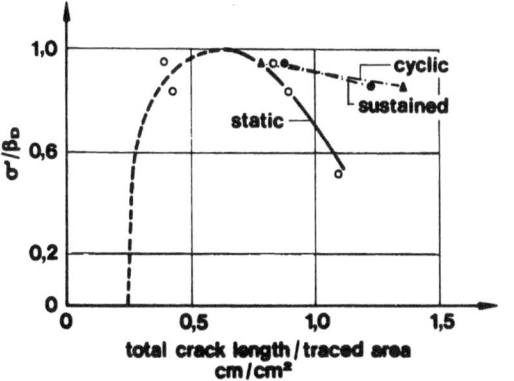

Fig. 12. Effect of Load History on Crack Density of Concrete at Failure /13/

Glücklich has shown that fatigue failure of concrete loaded in
tension also is the consequence of propagation and joining of microcracks similar to the case of concrete loaded in monotonic tension.
However, little information is known on the amount of total crack
density as a function of cyclic loading. Awad et al. /13/ and Diaz
et al. /11/ have shown that crack density at fracture is higher for
repeated or sustained loading than for monotonic loading and that
crack density increases with decreasing maximum stress at failure.
This can be seen clearly from Fig. 12.

For the problem of demolition of concrete, information on the
extent of microcracking and the propagation of cracks for concretes
subjected to dynamic or in particular to impact loads would be of
particular significance. Unfortunately such information is not
available. However, it is well established that both tensile and
compressive strength increase with increasing strain rate. Furthermore it follows from Fig. 12 that the descending portion of the
stress strain diagram of concrete loaded in compression is the
steeper the higher the strain rate, indicating that also the extent
of microcracking is less pronounced for high strain rates than it
is for low strain rates. Nevertheless this tendency is established
only for comparatively slow rates of straining. It is not known
whether this tendency can be   extrapolated towards very high rates
of strain.

4.   FRACTURE ENERGY

4.1 Definitions

Fracture energy is defined as the total work done to fracture
a section divided by the total area of the fractured surface:

$$\gamma = \frac{U}{2A} \tag{1}$$

For a material like concrete where several cracks follow meandering
paths the area of the fracture surface A may be considerably larger
than the cross-sectional area of the member A_0 as shown in section 3.2.

The total energy used to fracture a member consists of various
components:

$$\text{tot } U = U_k + U_f + U_{pl} + U_s \tag{2}$$

In the case of unstable crack growth a certain amount of kinetic energy U_k will be used up and partially transformed into heat.
Friction between individual particles at the crack surfaces will
use up the energy U_f. True plastic deformations may occur within
the solid particles for which the energy term U_{pl} is used. Finally

the term U_s describes the amount of energy which has been used for the formation of new surfaces. The terms U_f, U_{pl} and U_s are difficult to separate and may be expressed by the term U_{eff}, the effective energy. Thus total energy may be described by

$$\text{tot } U = U_k + U_{eff} \qquad (3)$$

where

$$U_{eff} = U_f + U_{pl} + U_s$$

Often the energy U_{eff} is assumed to be a materials characteristic. It can be measured by minimizing U_k, e.g. in a slow bend test on notched or unnotched specimens in which only stable crack propagation occurs. Under these conditions $U_k \rightarrow 0$. Nevertheless there may be dynamic effects occurring on a microscopic scale even if slow crack propagation prevails. Such effects would then be included in the term U_{eff}.

On the basis of these simplifications an effective fracture energy may be defined as follows:

$$\gamma_{eff} = \frac{U_{eff}}{2A_o} \qquad (4)$$

where A_o = cross-sectional area of the fractured member.

An effective energy U_{eff} may be determined experimentally from a load-displacement relationship as determined in a slow bend test (Fig. 13). The effective energy U_{eff} corresponds to the area under the load displacement curve.

The discussion of fracture energy given above refers to a material loaded in tension. Basically the same equations are valid for a material loaded in compression.

However, in this case the cross-sectional area of the member and the area of the fracture surface - if it exists at all - deviate even more than in the case of a material loaded in tension. Therefore, it is suggested to relate the fracture energy U_{eff} to the volume of material under stress V, resulting in a volumetric fracture energy vol γ_c of the material loaded in compression

$$\text{vol}\gamma_c = \frac{U_{eff}}{V} \qquad (5)$$

Similar to concrete loaded in tension, volγ_c for compressive stress corresponds to the area under a stress-strain relationship of a specimen in which no unstabel crack growth had occurred.

4.2 Relation between Fracture Energy and other Fracture Mechanics Parameters

The resistance of a material against brittle fracture may be described by the following well known Griffith relationship:

$$\sigma_c = \sqrt{\frac{2E\gamma}{\pi a}} \tag{6}$$

In this equation E = modulus of elasticity
γ = specific surface energy
α = half length of a crack in an infinite plate

In order to take into account plastic deformations at the crack tip the term 2γ in eq. 6 may be replaced by G_c, the critical energy release rate. E and G_c may be combined as follows (plane stress):

$$G_c \cdot E = K_c^2 \tag{7}$$

The term K_c is referred to as fracture toughness.

Based on these definitions the following relations exist between the effective fracture energy as defined in section 4.1, the critical energy release rate and the fracture toughness (plane stress):

$$\gamma eff = G_c/2$$
$$\gamma eff = \frac{K_c^2}{2E} \tag{8}$$

4.3 Effect of Test Methods on Fracture Energy

Though the effective fracture energy may be a materials' constant it is influenced by a number of test parameters. Such parameters are:

specimen size
crack or notch depth
strain rate
stress state and
internal stresses.

The effective fracture energy increases with increasing specimen size as shown in Fig. 14 for hardened cement paste (the fracture energy values were evaluated from fracture toughness data as presented by Higgins and Bailey (14)). It is likely that the effective fracture energy will approach a "true" and limiting value for

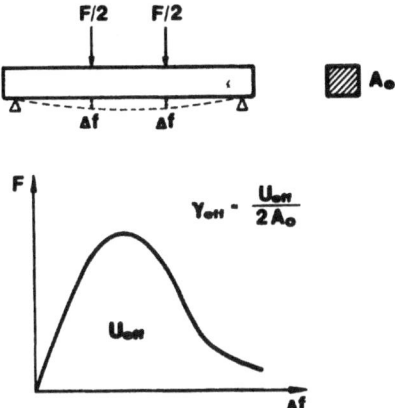

Fig. 13. Experimental Determination of Effective Fracture Energy

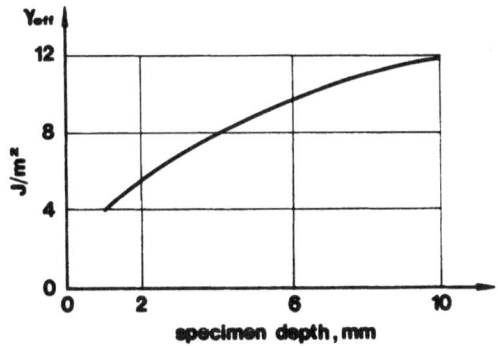

Fig. 14   Effect of Specimen Size on Effective Fracture Energy
of Hardened Cement Paste /lo/

large specimen sizes. Similar behaviour has been proposed for the
size dependence of the fracture toughness of cement paste as well
as of mortar and concrete /15/. A reason for the increase of fracture energy with increasing specimen size is believed to be that
the volume of high stressed material increases with increasing
specimen size. The larger the volume of high stressed material the
more pronounced is multiple crack growth of cracks. This, in turn
will cause an increase of energy required to fracture the concrete.

Figure 15 shows the relationship between effective fracture
energy and relative notch depth. With an increase of notch depth
the effective fracture energy decreases substantially. Studies on
this particular problem are too limited to give a conclusive answer
and to explain the observed behaviour.

No experimental data are available on the dependence of
effective fracture energy on strain rate, stress state and internal
stresses. It is likely that an increase in strain rate will substantially reduce the effective fracture energy as discussed in
section 3.2.3.

No data are available on the effect of stress state on fracture energy. Stress states deviating from uniaxial tension or compression may either increase or decrease the effective fracture
energy.

5.   EFFECTIVE FRACTURE ENERGY OF CONCRETE COMPONENTS

5.1 Hydrated Cement Paste

The effective fracture energy of hydrated cement paste is
controlled primarily by the porosity of the paste and thus by the
water-cement ratio and the degree of hydration or the age at loading, respectively. In Fig. 16 the effective fracture energy of
hydrated cement paste samples as taken from the literature is given
as a function of the water-cement ratio /9/. For ages between
four and fourteen days the fracture energy ranges roughly between
3 and 9 J/m^2. Values of the effective fracture energy of hydrated
cement paste as large as 15 J/m^2 have been reported by Cooper and
Figg /16/.

The fracture energy of Tobermorite has been reported by Brunauer and Kantro to be approx. o,5 J/m^2 /17/. The fracture energy
of calcium hydroxide is approx. 1 J/m^2 /18/. Thus the effective
fracture energy of hydrated cement paste is five to ten times larger than the fracture energy of its components. This is believed
to be partly caused by multiple crack growth and meandering of
cracks as described in section 3.2.1. However, this would account
only for an increase of the fracture energy by a factor of approx.
2. Additional energy is used by friction when seperating hydrated

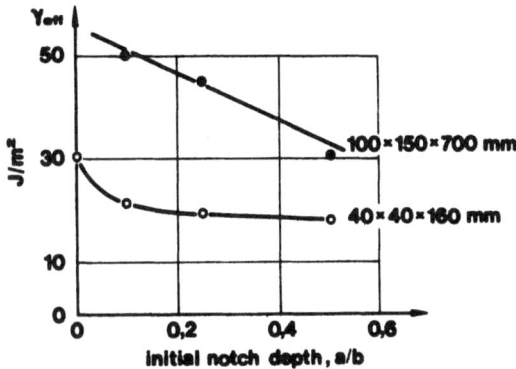

Fig. 15.   Effect of Notch Depth on Effective Fracture Energy of
            Concrete

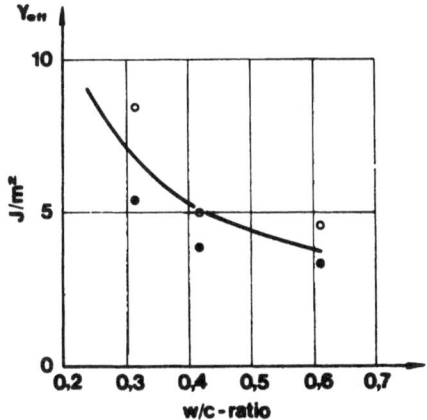

Fig. 16.   Effect of Water-Cement Ratio on Effective Fracture En-
            ergy of Hardened Cement Paste /9/

cement gel particles during the fracture process as defined by eq.2.

5.2 Aggregates

Unfortunately no experiments which would allow a reliable estimate of the effective fracture energy of various aggregates are known to us. From the evaluation of fracture toughness measurements on various types of aggregates we conclude that the effective fracture energy of marble is of the order of 7o J/m^2 whereas the effective fracture energy of quartz is approx. 14o J/m^2 / 19/.

5.3 Interfaces

Experiments on the effective fracture energy of aggregate-cement paste interfaces have been reported by Modéer /15/. He reports values which are approx. twice the effective energy of hydrated cement paste with a water-cement ratio of o,35. This result appears to be rather unlikely. Based upon our own measurements of the fracture toughness of interfaces we suggest the following values for the effective fracture energy of interfaces.

Table 1: Effective fracture energy values of interfaces
(water-cement ratio: o,4)

matrix	aggregate	γ_{eff}, J/m^2
normal	limestone	o,6
cement paste	quartz	o,8
cement paste with shrinkage reducing additive	limestone	2,6
	quartz	1,7

6. EFFECTIVE FRACTURE ENERGY OF CONCRETE

The effective fracture energy of concrete depends primarily on the size and the amount of aggregates as well as on the properties of the particular cement paste i.e. porosity, water-cement ratio and degree of hydration. The effect of maximum aggregate size for mortars and concrete with an aggregate content of 5o percent by volume is given in table 2.

Apparently, the effective fracture energy increases with increasing aggregate size. This is due to the fact that the extent of multiple crack growth and crack branching increases with increasing aggregate size.

The effective fracture energy is also influenced by the amount of aggregates. Table 3 gives estimates of the effective fracture

Table 2: Effect of aggregate size on γ_{eff}

aggregate fraction	γ_{eff}, J/m²
o,25/o,5 mm	27
1/2 mm	34
4/8 mm	39
o/2 mm	33
o/8 mm	4o

Table 3: Effect of aggregate content on γ_{eff}

material	aggregate content and fraction	γ_{eff}, J/m²
cement paste	–	7
mortar	25 percent; o,o8/o,2 mm 25 percent; o,25/o,5 mm	21 24
concrete	5o percent; o/8 mm 5o percent; 4/8 mm 75 percent; o/32 mm	4o 39 57

energy of cement pastes, mortars and of concretes with a maximum aggregate size of 32mm and various amounts of aggregates.

From this it follows that the effective fracture energy of a concrete with a maximum aggregate size of 32mm is roughly eight times larger than the effective fracture energy of the neat cement paste. Modéer /15/ gives the following values for the effective fracture energy of cement paste, mortar and concrete, respectively:

hydrated cement paste	9 J/m²
mortar	61 J/m²
concrete	114 J/m²

No systematic studies on the effect of concrete strength on effective fracture energy of concrete loaded in tension could be found. However, there is some indication that the effect of concrete strength on effective fracture energy is comparatively small once a certain minimum strength has been reached. An increase in strength leading to an increase of fracture energy normally is accompanied by a reduction of ductility which would cause a reduction of fracture energy so that the effective fracture energy is influenced little by concrete strength.

7.   VOLUMETRIC FRACTURE ENERGY OF CONCRETE LOADED IN COMPRESSION

As already pointed out in section 4.1 the definition of fracture energy of a concrete member loaded in compression is more difficult than that for concrete loaded in tension. It appears to be the
least ambiguous to use the volumetric fracture energy as defined
in eq.5.

In Fig. 17 the volumetric fracture energy of concrete is given
as a function of concrete strength while Fig. 18 shows the relationship between volumetric fracture energy and strain rate. These data
have been obtained from the evaluation of stress-strain relationships reported in /8/.

The volumetric fracture energy increases with increasing concrete strength. However the increase of volumetric fracture energy
is much less than the corresponding increase of strength. This is
not surprising since similar to concrete loaded in tension an increase in strength is accompanied by a reduction of ductility.

On the other hand the effect of strain rate is very pronounced:
An increase of strain rate by several orders of magnitude causes a
substantial reduction of the effective volumetric fracture energy.

8.   MEANS TO INFLUENCE FRACTURE ENERGY

From the viewpoint of demolition it is of particular significance to find ways in which the energy required to fracture a concrete member will be minimized. Since there is little substantial
information available on this particular question the following
statements will be more or less speculative.

8.1 Effect of Moisture State

It has been found by various investigators that the tensile
strength of concrete decreases as its moisture content increases.
Setzer /20/ and Wittmann /21/ showed that the reduction of strength
with an increase of moisture content is the consequence of a decrease of surface free energy. Cooper and Figg /16/ reported values
for the effective fracture energy of hydrated cement paste samples
with a water-cement ratio of o,5 which were tested either in a dry
or in a wet state of 14,9 and 12,4 J/m², respectively. This difference is less than would be expected on the basis of differences in
tensile strength.

8.2 Effect of Temperature

In experiments conducted by Saemann and Washa /22/ it was demonstrated that an increase of temperature causes a substantial reduction of tensile strength of concrete. Also data on the energy

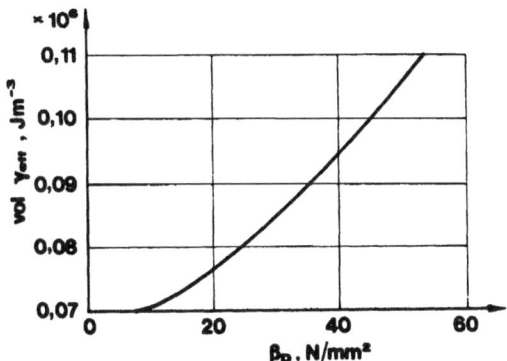

Fig. 17.   Effect of Concrete Strength on Volumetric Fracture Energy of Concrete

Fig. 18.   Effect of Strain Rate on Volumetric Fracture Energy of
Concrete; β_D = 21 N/mm^2 /8/

of rupture are reported which should be an indication of fracture
energy. The following relative energies of rupture are given:

Temperature °C	-18	+1o	+32	+43
relative energy of rupture, percent	1oo	94	69	5o

Thus the fracture energy should decrease substantially with
an increase of temperature.

8.3 Effect of Strain Rate

In section 7 and Fig. 18 it has been shown that the volumetric
fracture energy of a concrete loaded in compression decreases substantially as the strain rate increases. It is likely that similar
behaviour should occur for concretes loaded in tension and that
this reduction will be particularly pronounced if we approach the
range of dynamic loading.

9.   CONCLUSION - HOW MUCH ENERGY IS NEEDED TO DEMOLISH CONCRETE BY MECHANICAL MEANS ?

In the following table the effective fracture energies of concrete components as well as of the various types of concretes are
summarized.

Table 4: Effective fracture energies

hydrated cement paste	$3 - 15$ J/m²
normal weigth aggregates	$5o - 15o$ J/m²
h. cement paste-aggregate interfaces	$o,5 - 3$ J/m²
mortars	$2o - 4o$ J/m²
concretes	$3o - 6o$ J/m²
concrete subjected to compression	$o,o5 \times 10^6 - o,2o \, 10^6$ J/m³

The effective fracture energy of concrete is less than that of
aggregates, however, considerably larger than the effective fracture energy of a hydrated cement paste. It increases substantially
with increasing concrete strength, aggregate size and volume concentration.

The effective fracture energy and thus the energy required to
demolish concrete may be reduced by an increase of moisture content,
of temperature and of strain rate.

However, does the required fracture energy correspond to the energy needed to demolish concrete?

In order to split up a concrete block with a volume of 1 m^3 into cubes of a side length of 5o mm the effective fracture energy required for this process amounts to 46oo J or $1,3 \times 10^{-3}$ kWh under the assumption that the concrete has an effective fracture energy of 4o J/m^2.

If the same concrete block should be demolished by subjecting it to uniform compression the fracture energy necessary would be 15o ooo J or o,o4 kWh if we assume the volumetric fracture energy vol γ_c to be $0,15 \times 10^6$ J/m^3. In reality considerably larger energies are required to demolish a concrete block of that size. This is in part due to the fact that considerably larger fracture areas are formed in a demolition process than assumed in this estimate. Nevertheless it may be concluded from this that the largest fraction of the energy employed to demolish a concrete block is not directly used for fracturing the concrete but is transformed into other forms of energy such as kinetic energy, noise and finally heat.

Demolition of concrete certainly can be faciliated by minimizing the fracture energy, however at least as much emphasis should be placed on means to minimize energy losses during the demolition process.

1o. REFERENCES

/1/ T.T.C. Hsu, Tensile bond strength between aggregate and cement paste or mortar, J. ACI 6o(1963) 371-39o

/2/ S. Ziegeldorf, H.S. Müller and H.K. Hilsdorf, Effect of aggregate particle size on mechanical properties of concrete, 5. Int. Conf. on Fracture, Cannes, 1981

/3/ S. Ziegeldorf, H.S. Müller and H.K. Hilsdorf, Anwendung der Bruchmechanik auf Beton und seine Komponenten, 1oth Session AK Bruchvorgänge, DVM, Freiburg,198o

/4/ D.W. Hadley, The nature of the paste-aggregate interface, PH. D. Dissertation, Purdue Univ. 1972

/5/ B.D. Barnes, S. Diamond and W.L. Dolch, The contact-zone between Portland cement paste and glass "aggregate" surfaces, Cem. and Concr. Res. 8 (1978) 233-244

/6/ G. Rehm, P. Diem and R. Zimbelmann, Technische Möglichkeiten zur Erhöhung der Zugfestigkeit von Beton, Schriftenreihe DAfStb, Heft 283, ed. Ernst und Sohn, Berlin,1977

/7/   R.L. Berger, F.V. Lawrence, Jr. and J.F. Young, Studies on the
      hydration of tricalcium silicate pastes II. Strength develop-
      ment and fracture characteristics, Cem.and Concr. Res. 3(1973)
      497-5o8

/8/   H. Rüsch, Research towards a general flexural theory for struc-
      tural concrete, J. ACI 57(196o) 1-28

/9/   F. Moavenzadeh and R. Kuguel, Fracture of concrete, J. of Mat.,
      JMLSA, 4 (1969) 497-519

/1o/ D.D. Higgins and J.E. Bailey, A microstructural investigation
      of the failure behaviour of cement paste, Hydraulic Cement
      Pastes-Structure-Properties, Univ. Sheffield, 1976

/11/ S.J. Diaz and H.K. Hilsdorf, Fracture mechanisms of concrete
      under static sustained and repeated compressive loads, Univ.
      of Illinois, Urbana, 1971

/12/ C. Rasch, Spannungs-Dehnungslinien des Betons und Spannungs-
      verteilung in der Biegedruckzone bei konstanter Verformungs-
      geschwindigkeit, Schriftenreihe DAfStb, Heft 154, ed. Ernst
      und Sohn, Berlin,1962

/13/ M.E. Awad and H.K. Hilsdorf, Strength and deformation charac-
      teristics of plain concrete subjected to high repeated and
      sustained loads, Univ. of Illinois, Urbana, 1971

/14/ D.D. Higgins and J.E. Bailey, Fracture measurements on cement
      paste, J. Mat. Sc. 11(1976) 1995-2oo3

/15/ M. Modéer, A fracture mechanics approach to failure analysis of
      concrete materials, Univ. of Lund, Sweden, (1979)

/16/   G.A. Cooper and J. Figg, Fracture studies of set cement paste,
      Trans. Brit. Cer. Soc. 71, 1(1972) 1-4

/17/ S. Brunauer, D.L. Kantro and C.H. Weise, Paste hydration of
      beta-dicalcium silicates, tricalcium silicate, and alite,
      Proc. Symp. Struct. Portl. Cem. Paste and Concr., Wash.(1966)
      3o9-327

/18/ G.M. Idorn et al., Morphology of calcium hydroxide in cement
      paste, Proc. Symp. Struct. Port. Cem. Paste and Concr., Wash.
      (1966) 154-174

/19/ B. Hillemeier and H.K. Hilsdorf, Fracture mechanics studies
      on concrete compounds. Cem. and Concr. Res.7(1977) 523-536.

/2o/ M.J. Setzer, A method for description of mechanical behaviour
     of hardened cement paste by evaluating adsorption data, Cem.
     and Concr. Res.6(1976) 37-48

/21/ F.H. Wittmann, Grundlagen eines Modells zur Beschreibung charak-
     teristischer Eigenschaften des Betons, Schriftenreihe, DAfStb,
     Heft 29o, ed. Ernst und Sohn, Berlin, 1977

/22/ J.C. Saemann and G.W. Washa, Variation of mortar and concrete
     properties with temperature, J. ACI, 29(1957) 385-395

5.2 DISCUSSION AFTER THE LECTURE OF H.K. HILSDORF

Attention was drawn by Kreijger to a recent paper by Wittmann
and Mihashi (Stochastic approach to study the influence of rate
of loading on strength of concrete - H. Mihashi, F.H. Wittmann -
Heron vol 25, 1980 no: 3, 54 p.) giving a stochastic approach to
the study of the rate of loading on the strength of concrete.
Ishai asked if the influence of temperature on the reduction of
fracture energy also could be effected by shrinkage cracking during
drying and Hilsdorf answered this may be one of the factors involved.
Shah reported on his own research indicating that impact strength
might reduce free energy and that fracture energy increases at high
strain rate. As regards temperature, water plays an important role -
if it is retained, increasing temperature makes crack formation
easier. If there is no water, other phenomena occur. Three strain
rates must be considered: static, impact and explosive where inertia effects are involved.
Ramachandran reported on research during which the influence of
various solvents and electrolytes were determined on the length
change behaviour of cement paste and vycor glass. Some of them
apparently did not attack the materials. Exposure of these substances to increasing concentrations resulted in an increase in
volume. By steadily decreasing the concentration there was a
hysteresis in many cases. The length change was more than can be explained by the Bingham effect. What could be the reasons for this
phenomenon? Can they be used as an indirect method of assessing
the effects of solvent or absorbents on the fracture energy.
Hilsdorf said he could not explain the hysteresis at first sight.
There is a possibility of dissolution of the material and also the
adsorbtion on surface which affect the charges.
De Pauw asked for the age of the concrete in the showed relationship between the effective fracture energy and the w/c ratio of
concrete. Could the same curve be expected for e.g. 15 year old
concrete? Hilsdorf answered the age was 28 days and for some tests
7 days. The relationship in his opinion will be about the same
for old concrete but further investigation might be necessary.

6.1 ADSORPTION SENSITIVE FRACTURE AND FRAGMENTATION

John J. Mills

Martin Marietta Laboratories
1450 South Rolling Road
Baltimore, Maryland 21227, U.S.A.

INTRODUCTION

Adsorption-sensitive fracture and flow behavior has been known for over 100 years. Indeed, in his excellent review of this topic in the Tewksbury Lecture in 1974,[1] Westwood pointed out that that year was the centennial of Reynold's first discoveries on the embrittlement of iron by hydrogen,[2] the fiftieth anniversary of Joffe's famous publication on ductility enhancement of salt crystals by solvents,[3] and the thirtieth anniversary of the publishing of "Hardness Reducers in Rock Drilling" by Rebinder and his colleagues.[4] This field, in its entirety, is vast, including as it does, stress corrosion cracking of metals or ceramics by liquid or gaseous environments, corrosion fatigue, hydrogen embrittlement, polymer crazing, Rebinder effects, complex ion embrittlement, and chemomechanical effects.

Now, it is a strange fact, but the majority of the western literature on these topics focuses solely on the prevention of failure on fracture by the various mechanisms. And there is a great deal of economic benefit to be made with such an approach. In contrast, Westwood and Mills[1,5,6] have been suggesting that there are comparable economic benefits to be gained by using these phenomena in a controlled way to facilitate fracture so that metal can be machined; ceramics and hard metals ground and shaped; and rocks and ores drilled, fractured, and comminuted more easily, with obvious cost savings. For example, machining costs in the U.S. approximate 5% of the GNP,[7] or ~ $100 x 10^9$. If just 10% of that sum could be saved by such an approach, these savings would pay for the total budget of some of the smaller NATO nations.

This meeting, therefore, is an ideal one at which to present these ideas, and despite the fact that there appears to be very little work done on environment-sensitive fracture of concrete, I look forward to applying some of the principles we have learned over the years to facilitate the destruction of large concrete structures.

As already mentioned, however, the field is vast, and most phenomena do not appear to be relevant. This lecture will concentrate, therefore, on the adsorption-induced flow and fracture of non-metals, with particular emphasis on glasses, ceramics, rocks, and minerals. Even this restriction produces solid-liquid interfacial phenomena which require intimate knowledge of the interface between the scientific disciplines of physics, chemistry, and mechanics. One might say that this field poses two-body problems which require three-body solutions!

First, a description of the various theories of the relevant phenomena will be given. Then, a summary of some of our results in applying these to drilling and fracture of rock will be presented. Finally, I will touch on some of the implications for fracturing concrete, based admittedly on a very limited understanding of concrete fracture.

BACKGROUND AND THEORIES

Environment can influence fracture in a variety of ways. The particular influence depends on the environment itself along with the material and mode of loading and fracture. Basically, the phenomena can be classified into two broad groups (Table I): those in which the environment influences crack propagation directly (Type I), and those in which cracking is influenced through its dependence on the environment-sensitive flow properties of the solid (Type II).

Type I: Direct Environmental Influence on Cracking

The simplest direct influence is that in which existing cracks are blunted by dissolution of the material at the crack tip. The archetype of this influence is the Joffe effect,[3] in which single crystals fracture in a brittle manner in air, but can be bent double without fracture on immersion in a solvent such as water (see Ref. 1 for an illustration). This is perhaps the extreme case of I(a) since the theory is that the surface layer containing the microcracks, which produces the brittle fracture in air, is completely dissolved away in water. However, this hypothesis has never been studied in detail to my knowledge, and other mechanisms have been suggested (see Frank -- Ref. 8).

TABLE I

Types of Environment-Sensitive Fracture Behavior

Direct Influence Type I	Indirect Influence Type II
a) crack blunting by dissolution	a) rate of crack initiation in ductile materials determined by dislocation pileup at barriers
b) cracking facilitated by reaction of crack tip with aggressive medium	b) crack propagation facilitated by increased flow
	c) crack propagation retarded by increased flow

Fox[9] has also suggested that this mechanism operates in silicate glasses in aqueous solutions with pH > 12. He suggests that these alkaline solutions preferentially dissolve the glass around the crack tip. This crack blunting effect would, of course, prevent cracking, since the stress required to propagate a crack is dependent on the crack tip radius. It is only of scientific interest and need not be discussed further.

The second direct phenomenon, I(b) in Table I, in which cracking is directly facilitated by aggressive environments, is perhaps better known to this conference by the name "static fatigue." It was first observed by Grenet in 1899[10] as the decrease in the strength of glass under slow loading conditions compared to that measured under fast conditions. It is also manifested as the delayed fracture of a glass rod or bar under a tensile stress well below its tensile strength; hence, the name "static fatigue." Essentially this phenomenon is caused by the environmentally sensitive slow growth of cracks in glasses, metals, ceramics, plastics, and concrete. Water and its vapor appear to be the most aggressive environment for slow crack growth, as illustrated in Fig. 1 for a silicate glass.[11]

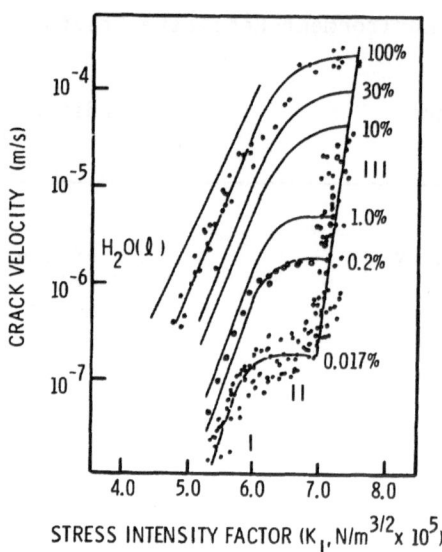

Figure 1.   Dependence of crack velocity on applied force. The
            percent relative humidity for each set of runs is given
            on the righthand side of the diagram. Roman numerals
            identify the different regions of crack propagation.
            Region III shows only half the points used to determine
            the line. The line, therefore, does not seem to fit
            the data as well as it would if all data in region III
            were plotted. (after Ref. 11)

Before the advent of fracture mechanics and its application
to explain static fatigue in terms of slow crack growth, Hillig
and Charles explained static fatigue in terms of the reaction of
water with the silica network.[12] Wiederhorn[11,13] expanded on this
theory and showed that in region I (low crack velocity), the crack
speed is limited by the reaction rate of water with the silica network. In the second region (II) where crack velocity is relatively
independent of K_I, diffusion of the water to the crack tip limits
crack speed. And in region III, the high velocity regime, water
has no influence since it cannot reach the crack tip.[11,13]

Although other liquid environments can also influence crack
growth and cracking phenomena in glasses (see below), even small
amounts of water, present in these media as an impurity, can have
a substantial influence, particularly in region I. There, its

influence depends on its chemical activity, which is proportional
to the partial pressure of water in that medium (or its relative
humidity, rH), not its absolute concentration;[14] these values can
be quite high (e.g., rH ~ 80%).[14] For example, Freiman has demonstrated that the influence of the homologous series of long chain
alcohols in regions I and II is predominantly due to the rH of
these organic liquids.[14] In region III, where water was thought
to have no influence, he showed that the alcohols themselves do
influence crack growth.[14] However, since their influence is
thought to be indirect, the potential mechanisms for this particular phenomenon will be discussed later.

Water appears to be the most aggressive medium for the direct
environmental influence on cracking for a wide variety of materials.
It substantially enhances crack propagation at a given K_I value in
both single crystal[15] and polycrystalline alumina[16] and in piezo-
ceramics,[17,18] and it decreases the strength of these materials
and of rocks[19] and concrete.[20] As long as the pH remains between
5 - 10 (the mildly acidic to mildly alkaline range), the mechanism
appears the same for all materials, i.e. that proposed by Hillig
and Charles[12] and Wiederhorn[11,13] outlined above. As the pH moves
into highly alkaline or acid regions, specific chemical reactivity
undoubtedly comes into play, as discussed by Fox.[9] However, for
practical application in the sense outlined in the Introduction,
such environments are not acceptable from a health hazard viewpoint.
Therefore, these influences are not discussed further.

Type II: Indirect Environmental Influence on Cracking

As illustrated in Table I, there are at least three indirect
influences of environments on fracture behavior and all of these
act on the flow characteristics of the material being fractured.
Hence, it seems appropriate to devote some space to the influence
of adsorbed species on the flow behavior of inorganic non-metals,
phenomena commonly termed chemomechanical effects (CME).

The term "chemomechanical effects" was coined by Westwood after
a) his initial studies of Rebinder effects on non-metals revealed
that environments which do not materially influence the surface
free energy of a solid can substantially influence its dislocation
mobility),[21-23] and b) discussion with Rebinder on these environment-sensitive flow phenomena resulted in Rebinder stating that
these effects could not be Rebinder effects by definition.[24] CME's
are defined as the adsorption-induced alterations in the flow and
flow-induced fracture phenomena in non-metals. Their classic manifestation for MgO immersed in aqueous solutions of differing pH is
shown in Fig. 2.[25] The figure presents the variation of a) hardness; b) the parameter $\Delta L(1000)$ -- the distance dislocations
travel away from the indentation site during a 1000s indentation;

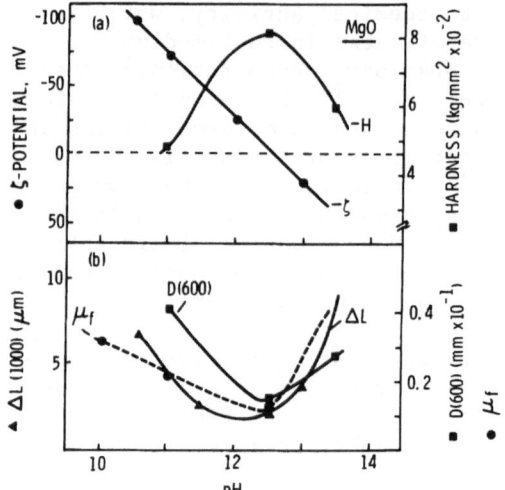

Figure 2. Relationships between pH–dependent ζ–potential and
a) microhardness (H); b) depth of penetration by a
carbide spade bit in 600 s, D(600); c) coefficient of
sliding friction (μ_f); and d) extent of motion of
edge dislocations around an indentation in 1000 s,
(ΔL), for MgO. Note maxima or minima in properties
at ζ = 0. (after Ref. 25)

c) μ_f, the coefficient of friction; and d) D(600), the distance
travelled by a small diameter spade drill bit in 600s. All these
parameters pass through extrema at pH ~ 12, which is also the pH
producing zero surface charge as revealed by ζ–potential measurements* (Fig. 2a) -- a correlation which will be discussed later.
Dislocation motion, which is measured by ΔL(1000), is a minimum,
and hardness, inversely related to dislocation flow is a maximum.
The coefficient of friction is determined using a diamond point
dragged slowly across a freshly-cleaved surface.[27] Since solutions
with pH = 12 minimize dislocation flow, the ease with which the
diamond can plough through the surficial region is also minimized.[27]
According to Westwood and Goldheim,[28] spade bit penetration, D(600)
follows the ease of dislocation motion [i.e. ΔL(1000)], since the
drilling is essentially a fracture phenomenon and the fracture
stress in MgO is essentially controlled by the flow stress[28] -- an
example of Type II(a) and (b) in Table I.

---

* See Ahmed[26] for a review of ζ–potential and its relation to
surface charge.

Of course, H and ΔL(1000) are essentially static measurements, whereas D(600) is a measure of many dynamic fracture events,[28]
and it may be thought that the agreement is somewhat fortuitous.
Not so. Environments can also influence dynamic plastic flow in MgO.
For instance, Fig. 3 presents birefringence patterns of the impact
sites caused by dropping steel balls under identical conditions
onto the (100) surface of MgO that has been freshly-cleaved in the
environments noted.[29] The dark/light contrasting bands reveal
edge dislocation bands, the number of which increases from toluene
to water to dimethyl formamide (DMF), implying that the critical
stress to nucleate near-surface dislocations or the frictional drag
stress is lower for DMF than for water or toluene. Incidentally,
these results correlate well with measurements of ΔL(1000)[22]
and with sliding friction behavior of MgO in the same environments.[27]
Trends in the surface and subsurface cracking behavior (Fig. 3d
presents an illustration of the latter type of crack) are opposite
to what one might expect for flow-assisted fracture. However, the
relationship between cracking and flow behavior in such a dynamic
event is quite complex and this behavior can be explained as follows. The energy of impact under plasticizing environments (DMF)
is dissipated by extensive and fairly diffuse plastic deformation.
As a result, localized stress concentrations are reduced, and the
probability of brittle crack initiation is minimized under DMF as
compared to toluene.[29]

The mechanism for the chemomechanical effect has not yet been
worked out in detail, but the general outline is still as originally
proposed by Westwood, Goldheim, and Lye (WGL).[21] The adsorption
of surface active species induces a charge in the electrostatic
potential of the near-surface region of the solid, thereby causing
a localized redistribution of the charge carriers (electronic and/or
ionic). Since this redistribution results from changes in the occu-
pation of a) electronic bands and b) the energy levels of line and
point defects, the interaction between moving near surface dislocations, and between dislocations and point defects, and between dislocations and the lattice, are changed. These interactions, in
turn, control dislocation mobility in the near surface region
and, consequently, the hardness of the materials.[21]

In considering details of this model, note that the correlation
between ζ-potential and a minimum in dislocation behavior implies
that surface charge is intimately involved in altering near surface
dislocation behavior. Until recently, however, the direct link among
ζ-potential, surface charge, and the flow properties of a surficial
region was lacking. Studies by Ahearn et al.[30-32] addressed this
problem. Using ZnO as a model material in which the relationship
between surface charge and near surface band structure are reasonably well understood, the test crystal was made the working electrode in an electrolytic cell. The surface charge on the crystal

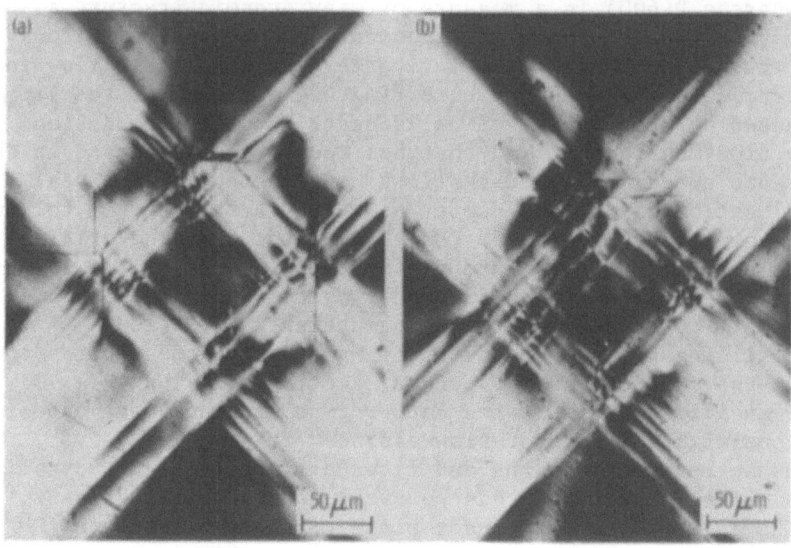

Figure 3. Birefringence micrographs revealing the impact damage on
the (001) surface of MgO under (a) toluene, (b) water,
and (c) DMF. Impact velocity = 2.4 x 10^2 cm/s; ball
mass = 1.0 g. (d) An example of subsurface cracking
under toluene. Impact velocity = 2.4 x 10^2 cm/s, ball
mass = 0.13 g. (After Ref. 29)

was then determined by the applied voltage. Measurements of the
hardness of ($10\bar{1}0$) and (0001) surfaces were made as a function of
applied voltage. The zero surface charge condition was determined
using Dewald's approach.[33] The results of this study a) revealed
that changing the surface charge could alter hardness by 20 to 40%
(Fig. 4), b) confirmed the relationship between surface charge and
surficial dislocation behavior, and c) revealed that the hardness
maximum did not occur at zero surface charge. Further examination
of the data also indicated that these results could <u>not</u> be inter-
preted by the influence of surface charge on the interaction between
dislocations. Rather, the changes in hardness were most likely
attributable to a change in the electrostatic interaction between
the charged dislocations and charged Zn^{2+} interstitial ions.[30,31]

In the original WGL studies,[22,23] the complex interrelation-
ships among environment, dislocation motion, and point defect
concentration in MgO were revealed. This early data has been
reanalyzed by Ahearn and used to demonstrate that, for MgO, both
dislocation-lattice and dislocation-point defect interactions are
affected by surface active environments.[34] This information was
revealed by plotting the dislocation mobility parameter $\Delta L(1000)$

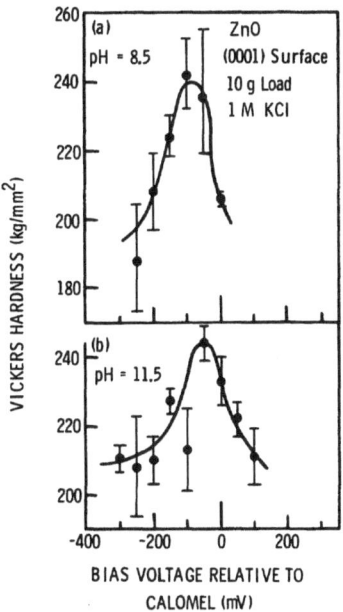

Figure 4.   Hardness of a (0001) ZnO surface vs bias voltage in
            1 M KCl. Indentation time = 10 s, load = 10 g. (after
            Ref. 32)

as a function of the square root of the total impurity concentration in the crystal, $(C)^{1/2}$, for the environments dimethyl formamide (DMF) and dimethyl sulfoxide (DMSO) (Fig. 5). If dislocation-impurity interactions only were environment-sensitive, then the slope of the plot would change, but not the intercept at $(C)^{1/2} = 0$. Conversely, if only dislocation-lattice, or possibly dislocation-dislocation (electrostatic), interactions were environment-sensitive, then the slope would be unaffected, but the intercept would be different. The data of Fig. 5 show that both slope and intercept are environment-sensitive. Further, Haasen[35] has discussed how changes in the electron occupancy of the core states of dislocations in a semiconductor crystal -- caused by changes in the surface charge -- could "destabilize" the core and make kink generation easier. In other words, electrically charged dislocations may be intrinsically more mobile.

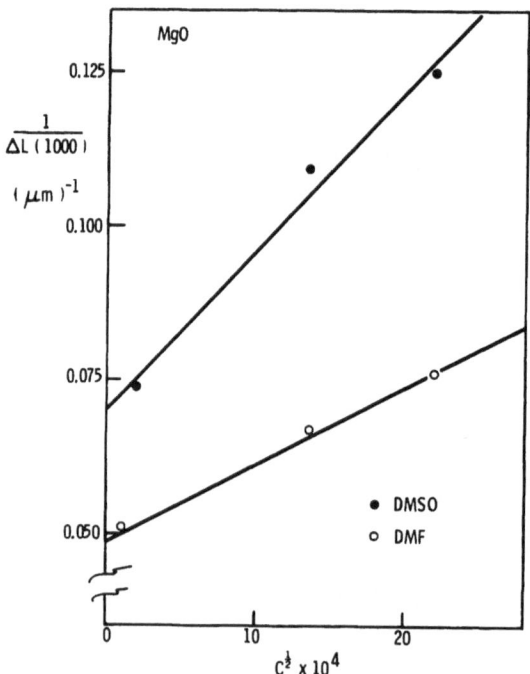

Figure 5.   Effects of impurity concentration, C, and environment on hardness, as indicated by reciprocal of dislocation mobility parameter ΔL (1000). (after Ref. 34, cited in Ref. 24)

Hence, though there seems to be substantial support for the validity of the WGL model for non-metallic materials, it is evident that the relationships between environment and material flow properties is quite complex. Indeed, it is now generally accepted that the mechanisms for CME's in ductile materials such as NaCl, semibrittle materials such as MgO, and brittle materials such as alumina, quartz, and the inorganic glasses, are probably all different.[5] Hence, it is not possible to generalize that a particular environment will influence any material. In this regard, the ζ-correlation turns out to be quite useful since the relationship between zero-ζ-potential and extrema in mechanical properties (e.g., hardness a maximum) appears quite general and has been shown to hold for a wide variety of materials (such as magnesia, alumina, calcite, quartz, and soda-lime glasses)[1] in diverse aqueous and non-aqueous fluids. ZnO seems to be the exception at this time.[30-32]

Having demonstrated that plastic flow behavior in non-metals is dependent on adsorbed species, we turn now to look at how this environment-sensitive flow behavior can control fracture. Table I lists three possible mechanisms of which the second two are most important: increased flow facilitating fracture (IIb) and increased flow retarding fracture (IIc). The first mechanism in which enhanced flow facilitates crack initiation (IIa) is hard to separate from the second (IIb) and will be included in the discussion of that mechanism. In this regard, the correlation between zero ζ-potential and a maximum hardness (called the ζ-correlation) is useful to separate mechanisms IIb and IIc. If the fracture behavior is minimized at $\zeta = 0$, then IIb is predominant. Conversely, if it is a maximum at $\zeta = 0$, then IIc should be regarded as most important.

Cracking Facilitated by Increased Flow

If, as Westwood and Goldheim suggest,[28] the drilling of MgO is regarded as a measure of the fracture behavior of this material, then Fig. 2b presents an example of this mechanism. However, for this meeting, a more relevant example of this mechanism is the influence of the homologous series of alcohols on a) the energy to propagate a crack in;[36] b) the pendulum hardness of;[37,38] and c) the rate of diamond core drilling of soda-lime glasses (Fig. 6).[37] Similar results have also been found in the n-alkanes.[36,37] Since Freiman[14] has demonstrated that the rH of the alcohols has a strong influence on crack propagation of soda-lime glass immersed in such environments, one might suspect that water causes the peaks occurring in the various parameters at heptanol, $N_c = 7$. Indeed, Fox and Freeman[38] -- invoking Freiman's equilibrium rH and crack propagation data[14] and the theory of Evans and Wilshaw[39] relating the volume of material removed under a point indenter to the fracture

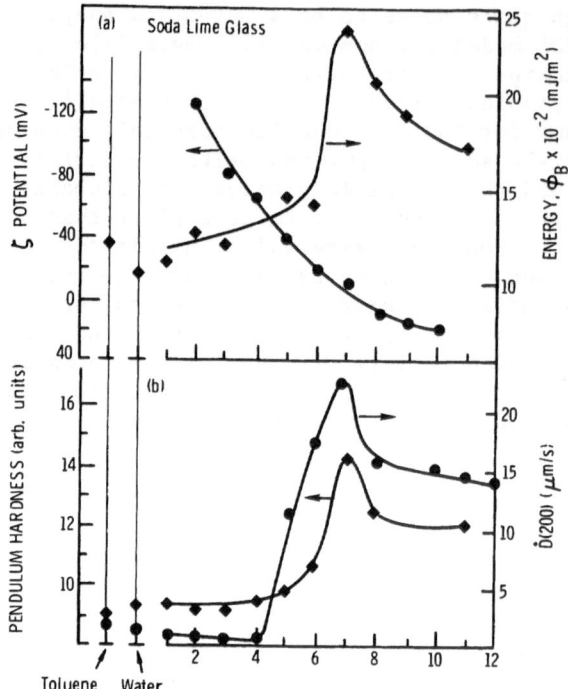

Figure 6. Variation of a) ζ-potential, b) pendulum hardness,
c) energy to slowly propagate a central crack, and
d) drilling rate at 200 s of soda-lime glass immersed
in the homologous series of straight chain alcohols.
(Data taken from Refs. 36 and 37 and replotted.)

toughness of the material -- were able to approximately predict
the pendulum hardness curve in Fig. 6.

However, two other observations complicate Fox and Freeman's
explanation. First, the mixtures of alcohols with chain lengths
on either side of seven (e.g., 5 and 8 or 6 and 10) also produce
maxima in the energy to propagate a crack[36] and in the drilling
rate $\dot{D}(200)$.[37] Second, purposely adding water to the alcohols
decreased the magnitude of the maximum in the drilling rate[40]
(Fig. 7) -- an observation since repeated by Adams.[42] Hence,
although water does have a substantial influence on crack propagation of inorganic glasses, a secondary water independent effect
also appears to exist. This effect was recently noted by Mills,
who argues as follows:[43]

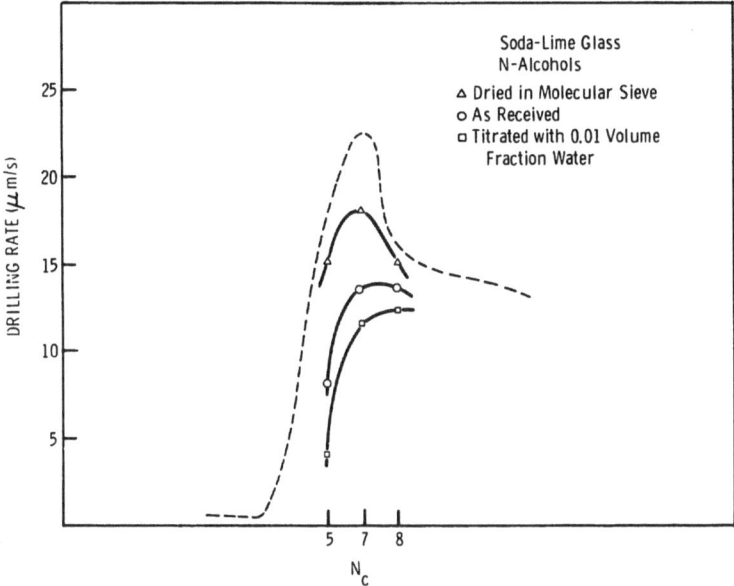

Figure 7. Drilling behavior of s.l. glass (related to pendulum
hardness) as function of water content of n-alcohols
around heptyl alcohol. Note that additions of water to
as-received alcohols reduce drilling efficiency, while
dehydrating the alcohol increases drilling efficiency
...establishing that water either present or created
catalytically is not the cause of the peak at $N_C = 7$.
The dashed line is drawn through earlier data from
Westwood and Latanision.[41] (after Ref. 40)

The alterations in pendulum hardness observed by Westwood et
al. range from ~ 15% for quartz in $Al(NO_3)_3$ (see Fig. 8),[44] to ~ 20%
for soda-lime glass in $Th(NO_3)_4$,[36] to ~ 60% for soda-lime glass in
alcohols.[37] Pendulum hardness is essentially a measure of energy
absorbed by plastic flow and cracking under the rocking diamond
indenter.[38] Also, the kinetic energy of a propagating crack is
roughly proportional to the crack velocity squared,[45] assuming a
constant, initial crack length and applied stress. Furthermore, if
we assume that a) all the absorbed energy goes into propagating
cracks in the material b) any plastic flow present serves only to
control the crack velocity, and c) that the environment influences
this crack velocity, then the crack velocities in the three examples
above are changed by 7%, 10%, and 30%, respectively. However,
measurements of crack velocity versus stress intensity are notor-
iously inaccurate even when great care is taken, e.g., Wiederhorn's
studies on the influence of $Th(NO_3)_4$ solutions on crack velocity

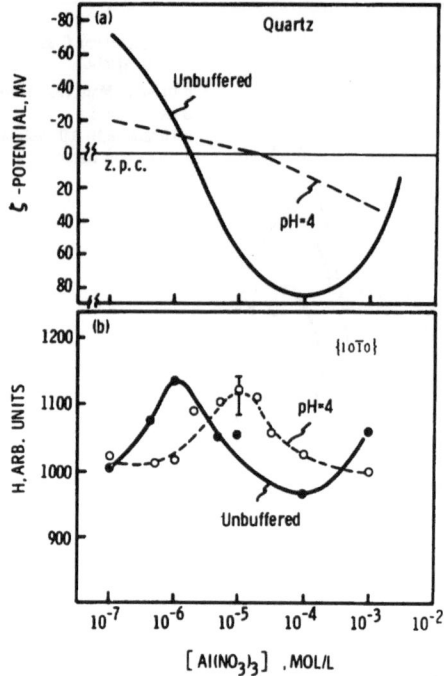

Figure 8.   Variation of (a) ζ-potential and (b) pendulum hardness
of (10$\bar{1}$0) surface of single crystal quartz in buffered
and unbuffered Al(NO$_3$)$_3$ solutions. (after Ref. 44)

of glass.[46] Using special techniques, Wiederhorn was able to reduce
the scatter in his data to only ~ 30%, well above the size of the
effect observed by Westwood et al. for glass in the same environments. Hence, he was unable to detect the influence observed by
Westwood et al. because of the scatter of his data. And, although
the magnitude of the effect on pendulum hardness of glasses immersed
in alcohols might be larger,[36] (i.e. 60%), the scatter in Freiman's
data -- not atypical for log V-K curves in homogeneous materials
such as glass -- is practically a complete order of magnitude
(Fig. 2).[14] As a result, Freiman was only able to detect the
influence of alcohols on crack velocity in the high velocity region,
III, and any variation in crack velocity with N_c is lost in the
experimental scatter in other K-V regions.[14] Interestingly, however, in regions I and II where no effect of alcohols is supposed
to be apparent, the results for heptanol seem to lie towards the
top of the scatter in Freiman's data,[14] in agreement with Westwood
et al.'s observations.[36]

Thus, this suggestion that a second independent effect exists[43] appears to resolve a fairly long-standing controversy between West- wood and his coworkers and Wiederhorn and Freiman. However, while the mechanism for the effect of water is fairly well understood, what is unclear is how the alcohols or aqueous solutions of $Th(NO_3)_4$[36] or $Al(NO_3)_3$[44] produce a maximum in pendulum hardness or energy to propa- gate a crack when $\zeta = 0$.

A possible explanation for this second order effect is that the original explanation by Westwood is valid.[36] Namely that the active species adsorb onto and modify its surface charge. When the surface charge is altered, i.e., when $|\zeta| = 0$, the electric fields that are created cause an excess of cations for $\zeta < 0$ or anions for $\zeta > 0$ to diffuse towards the surface. Since an excess of non-network ions weakens glass,[47] a hardness maximum will occur at the point of zero charge.

Given that species other than water on surfaces (i.e., H^+ and OH^-) can influence the flow behavior of glass, Westwood suggests that the enhanced flow obtained in heptanol influences the ease with which cracks are initiated during drilling into this notch-brittle solid.[36,37] An alternative explanation is that the environment does not directly facilitate cutting of the material but influences bit performance through its wear rate as is found for diamond drilling in rocks.[48] Although some evidence supporting both these explana- tions exist, just how evironment-controlled flow behavior enhances diamond drilling in glasses is controversial.

In MgO, the situation is somewhat more clear cut and there is considerable evidence for increased flow facilitating crack- ing.[6,27,29] MgO and soda-lime glasses and perhaps calcite appear to be the only materials studied in which enhanced flow behavior is suggested to facilitate fracture.[1,27,29,36,37] Most other materials, e.g., alumina and rocks, appear to exhibit the opposite behavior, namely, increased flow retarding fracture by crack blunting and energy dissipation as discussed below.

Cracking Retarded by Increased Flow

One of the more spectacular examples of flow retardation of fracture is complex-ion-embrittlement,[49,50] which now is recognized as another manifestation of the chemomechanical effect.[6,51] When polycrystalline AgCl is exposed at room temperature to aqueous environments containing highly-charged, complex ions, such as 6N NaCl presaturated with AgCl, its fracture mode changes from ductile and transcrystalline to brittle and intergranular. Both positively and negatively charged complexes can cause this embrit- tlement. Also, Westwood et al. have demonstrated in a series of publications[49,50] that the degree of embrittlement a) increases

with charge and concentration of the critical complex species
present in the environment and b) is a function of the distribution of charge on the complex. Other work has revealed that
whereas unnotched single crystals are essentially immune to cracking, notched monocrystals behave in a manner similar to the polycrystalline material.

The mechanism proposed by Westwood and his coworkers,[49,50]
which is substantiated by an impressive body of evidence, is that
this embrittlement phenomenon is caused by the repeated formation
and rupture of point-defect hardened charge double layers in the
crystals. Essentially, the charging of the surface by the complex
ions causes diffusion of Ag_i^+ interstitial ions -- for negatively
charged complexes -- and Ag_v vacancies -- for positively charged
complexes -- to the surface region to form a space charge double
layer. Because an excess of both of these point defects retards
dislocation mobility, the flow stress is raised to the point where
brittle fracture can occur and the crack propagates through the
hardened region until it is blunted out in the softer material
beyond the defect hardened layer. Fresh complex ions then chemisorb
onto the new crack surface, extend the space charge layer beyond
the crack tip and the whole cycle starts again.[49,50]

As an example of type II(c) behavior that is more relevant to
this conference, Mills, Westwood, and Huntington have measured the
work done in removing a unit volume of that rock by a wedge indenter
forced into a rock surface -- the specific chipping energy or
E_{sce}.[52] This parameter is a minimum for granite rock immersed in
solutions producing $\zeta = 0$ (Table II). In another example, Lewis
and Dunn[53] have shown that the time for failure of hollow quartzite
cylinders (internally pressurized to produce a hoop stress 94% of
the static strength) was 45-50% sooner in solutions producing $\zeta = 0$
than in water. Both examples illustrate that minimizing flow
behavior (i.e., at $\zeta = 0$) can facilitate the ease with which a crack
can propagate. In particular, the minimum in E_{sce} was produced by
an increase in volume of rock measured for the same amount of
energy.[52] Since the indenter is pushed slowly into the rock at
the same speed and to the same depth each time, $\zeta = 0$ environments
enhance slow crack growth under the indenter.

Whereas the studies mentioned above emphasize the influence
of environment-retarded flow facilitating slow fracture under
essentially static conditions, several other investigations present
evidence for such phenomena under the dynamic conditions of drilling
and grinding quartz, alumina, and rocks.[54-65] In particular, the
Al_2O_3 drilling and grinding studies are quite revealing[54-56] and
demonstrate the importance of plastic flow in some material removal
processes but not in others. Thus, Swain et al. demonstrated that
in abrasive grinding using diamond bits, it is necessary to enhance

TABLE II

Specific Energy Data from Wedge Indentation Studies[52]

Rock	Westerly granite				Seattle granite	
Wedge	Sharp		Blunt		Sharp	
DTAB Conc. (mol/l)	E_3 (MJ/m^3)	Number of Tests	E_3 (MJ/m^3)	Number of Tests	E_3 (MJ/m^3)	Number of Tests
0(H$_2$O)	184 ± 8*	26	268 ± 18	11	182 ± 14*	10
10^{-5}	128 ± 12	9	221 ± 15	14	150 ± 18	4
10^{-4}	112 ± 5	11	144 ± 9	8	142 ± 10	6
2.5x10^{-4}	136 ± 8	7	196 ± 19	18	---	--
10^{-3}	133 ± 11	7	230 ± 26	15	175 ± 11	6

* Values shown are means ± standard errors.

flow (i.e., hardness a minimum; $|\zeta| \gg 0$) to optimize material removal.[55] In contrast, diamond core drilling requires a hard brittle solid for maximum material removal efficiency.[55]

The importance of matching the cutting action of the tool to the deformation characteristics of the solid to be fractured can be appreciated by recalling that a twist bit designed for use with soft metals will be ineffective in drilling granite, and an impact-bit designed for use with granite will not be very successful in penetrating soft aluminum. Recognizing these limitations, some fracture-inducing devices or tools are designed to cut most efficiently through solids that exhibit a significant amount of plasticity. Their cutting action usually depends on a ploughing and shearing of the material. Such tools include twist and spade bits. In contrast, other tools are designed for use with more brittle solids, and their effectiveness depends on their ability to initiate and propagate brittle cracks in the material. Such tools include pneumatic hammers, percussive drills, and diamond-loaded bits.

With this knowledge, one can sense intuitively how a given environment could produce opposing effects on the cutting efficiency of a given solid. Consider the situation illustrated in Fig. 9. Suppose that the environment selected softens the surface of the solid by 20% or so, not an unreasonable amount.[36,44] As a consequence, the cutting action of the ploughing type of tool shown on the left will be facilitated, while that of the impact tool on the

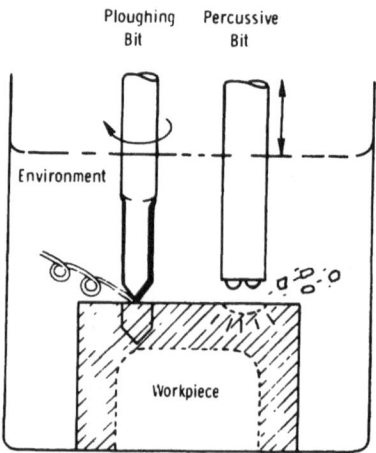

Figure 9.   Illustrating how a given environment can cause opposite effects on drilling efficiency depending on cutting action of bit. (after Ref. 5)

right will be hindered, because relatively more of the former's (impact) energy for fracturing will be dissipated by plastic flow processes, and less will be available to initiate and propagate brittle cracks.

Given this simplified characterization of all fracture-producing processes as either predominantly plastic flow (PF) -- or brittle fracture (BF) -- dependent, and given an awareness of the relationship between ζ-potential and hardness, one may conclude that the fracturing efficiency of PF-dependent tools will be facilitated by chemical environments which produce large surface potentials on the solid (i.e., $|\zeta| \gg 0$). Conversely, the efficiency of BF-dependent tools will be enhanced by chemical environments which produce a zero surface potential on the surface of the workpiece.[5]

Finally, some of these drilling and grinding studies also suggested that in attempting to apply any flow controlled fracture mechanisms to practical situations, the importance of the physical properties of the active fluid must be recognized,[55] a theme which is discussed in more detail in the next section.

PRACTICAL APPLICATIONS

In the previous section, drilling and grinding behavior was employed as a measure of the fracture behavior of brittle and ductile materials. Accordingly, drilling is one of the most important applications of both environment-sensitive fracture and flow-controlled fracture. However, the mechanisms of drilling are complex, depending on bit type, drilling method, and material being drilled. If one adds the complexities of environment-sensitive fracture to these intricacies, it is no wonder that many of the studies in even one aspect of this field (e.g., diamond drilling hard rocks) produce conflicting, if not contradictory, results (e.g., 57 and 58, 59, 60 and 61).

Given the understanding of the interaction of the tool, the material, and the environment, outlined in the previous section, the first difficulty one encounters when trying to apply the adsorption-dependent fracture phenomena is that the circulating fluids (air, oil, water, etc.) used in either percussive, rotary, or diamond drilling of rock, already perform several critical functions (Table III). These functions are dependent mainly on the physical properties of these fluids. Hence, when anything that alters one of these fluid physical properties is added to a drilling medium, then the efficiency of one or more functions of the liquid also will change and overall drilling performance may be affected -- detrimentally.

TABLE III

Influence of Physical Properties of Circulating Fluid on Factors Involved in Diamond-Bit Drilling on Hard Rocks

Function[†]	Relevant Physical Property		
	ρ*	η*	κ*
1. Drill string lubrication	yes	yes	no
2. Bit lubrication	yes	yes	no
3. Bit cooling	yes	yes	yes
4. Debris removal up the annulus	yes	yes	no
5. Debris removal from under bit face	yes	yes	no
6. Corrosion inhibition	no	no	no
Influence On:			
1. Rock properties	none	none	none
2. Bit life	yes	yes	yes

---

* ρ = density of fluid
  η = viscosity of fluid
  κ = thermal conductivity of fluid

[†] In drilling for oil in soft rocks, other functions, such as hole wall stabilization and control of subsurface pressures, are also involved.

Here, however, we are concerned with the influence of the chemical properties of a coolant on drilling performance. To distinguish these phenomena, note that, while a high concentration of additive usually is required to change the physical properties of a drilling fluid (say 10-20% by volume), the relevant chemical properties can be markedly altered by the presence of surfactant concentrations as low as 10^{-5} mol/liter.[1] Although some of the functions listed in Table III can be altered chemically (e.g., fatty acids can influence functions 1 and 2), in this section we are concerned only with altering the flow, fracture, and frictional characteristics of rocks by chemical means.

As an example, replacing air with water will certainly beneficially change the drilling performance of all types of drill bits, but more because the heat transfer and debris flushing functions of water are superior to air and any chemical influence water may have on fracture is masked by this function improvement. Since water is generally the base for many drilling fluids, the trick is to find additives which, by chemical means, can comparatively improve the bit performance.

We have done such a study on diamond core drilling of hard rocks, for which the drilling fluid usually is simply water, and hence an ideal standard for such studies.[62] Although the following discussion is specific to this type of drilling, the factors found to be important (Table IV) are probably also important for other types of drill bits (e.g., percussive) in other, different, hard and brittle materials (e.g., concrete).

TABLE IV

Factors Influencing Application of Adsorption-Controlled Fracture

Charge on the surface to be drilled
Polarity of surface active species
Concentration of surface active species
Type of test
Type of drilling tool
Adsorption dynamics
Tool failure mode

Surface Charge, Polarity, and Concentration Additive

The charge on the surface of the material being drilled using some standard liquid (e.g., H_2O) is important because the ζ-correlation indicates that this charge must be changed in such a way that the hardness, necessary to optimize the particular drilling performance is altered in the most appropriate way.[1] For instance, most silicate-containing rocks, and probably cement, are negatively charged in water. Consequently, cationic surfactants are required to produce $\zeta = 0$, the condition required for maximum diamond drilling performance in hard rocks.

On the other hand, many surfactants are long-chain organic molecules known to enhance boundary layer lubrication and so are intrinsically, likely to improve bit performance. It might also be thought that lubrication would be the principal origin of their influence. However, it now appears that a maximum in bit performance will occur whenever the ζ-potential of the rock is zero, regardless of the type of chemical used to achieve that condition. For example, Appl found that concentrations of Al^{3+} -- an inorganic cation -- similar to those that produce $\zeta = 0$, maximized normal and tangential forces on a single point of diamond cutting granite and minimized diamond wear,[63] incidentally reproducing the results obtained by Rebinder et al. in their earlier work.[4] In another case, Gielisse et al. found that the forces on a single diamond cutting Al_2O_3 were maximum in the long chain alcohols which produced $\zeta = 0$ in this material.[56] Of course, neither $AlCl_3$ nor the alcohols are practical additives, the first being corrosive and the second noxious, but the results obtained with $AlCl_3$ at low concentrations (10^{-5} mol/liter) on rocks, and with the alcohols on Al_2O_3 emphasize the generic nature of the "ζ-correlation."

Such chemical effects should be distinguished from the effects caused by changing the physical properties of the fluid, as mentioned above. For example, those reported by Selim et al.[57] from experiments using anionic surfactants, glycerine, and ethylene glycol, and by Unger et al.[58] from studies using anionic, non-ionic, and cationic surfactants. Since these effects occur at much higher additive concentrations, they are more likely to be due to changes in the physical of water and hence are less likely to be cost-effective.

Type of Test

The test strategy adopted to reveal any chemically enhanced drilling behavior must also take into account the nature of a) bit cutting action, b) the material being drilled, and c) the action of the chemical on the fracture behavior as discussed in the previous sections.

For example, in the diamond drilling of hard rocks, Mills and Westwood have demonstrated that chemomechanically active effects probably occur primarily by reducing the rate of diamond wear.[48]

Although various workers, including Rebinder et al.,[4] Selim et al.,[57] and Westwood and Goldheim[27] had long ago noted that surface active environments can improve bit life, at present this factor is recognized as the most important one in surfactant-aided hard rock diamond drilling. This view can be supported by the data of Fig. 10.[48] In Fig. 10a, the penetration rate obtained by using a freshly-dressed (i.e. sharpened) diamond-studded bit for drilling granite under water is compared to the rate of a similar bit drilling under an active fluid, namely, 10^{-3}M aq. DTAB (dodecyl trimethyl ammonium bromide). (This concentration is approximately that which produces a zero ζ-potential on the rock phases.) Note that, for any time after the first few seconds, the rate of penetration obtained using the active fluid is greater than that for water. The critical experiment is that illustrated in Fig. 10b, however. Using one freshly dressed bit throughout, and without redressing at any time, the drilling environment was alternated between water and the aq. 10^{-3}M DTAB solution. Note now that the data do not reveal any abrupt change in drilling rate with change of environment such as would be expected if the cause of the differences in Fig. 10a were environmentally-induced changes in rock fracture behavior. On the other hand, this result can be understood if the primary role of the active fluid is to reduce the rate of wear of the diamonds in the bit. Then, the explanation is the following: when exposed to solutions which produce $\zeta \simeq 0$, the hardness of granite is at a maximum. Thus, the coefficient of ploughing friction between the rock surface and the diamonds in the bit is a minimum. Consequently, the heat generated by their impingement also is a minimum, and so the diamonds run cooler. Since temperature is a prime factor in determining diamond wear, they now wear less, and stay sharp longer. Therefore, both the average rate of bit penetration and bit life are enhanced.[48] A somewhat similar argument has also been expressed by Cooper and Berlie.[61]

These comments, of course, only apply to diamond drilling of hard rocks, which is particularly relevant to drilling in concrete. With percussive drilling, however, our body of knowledge of how chemically-active fluids influence bit performance is not as well developed. Hence, while data showing large increases in the percussive drilling rate of hard granite rocks 200 s after starting, $\dot{D}(200)$, have been published,[59,60] we do not know whether this performance results from enhanced brittle fracture facilitating cutting or from reduced blunting of the bit by a mechanism similar to that suggested for diamond bits.

Of course, it should be pointed out that this mechanism is valid only when diamond wear is the most important factor in determining drilling efficiency ... as it certainly is in drilling hard rocks with rotary diamond-studded bits. However, when drilling softer solids, such as calcite, abrupt changes in drilling rate

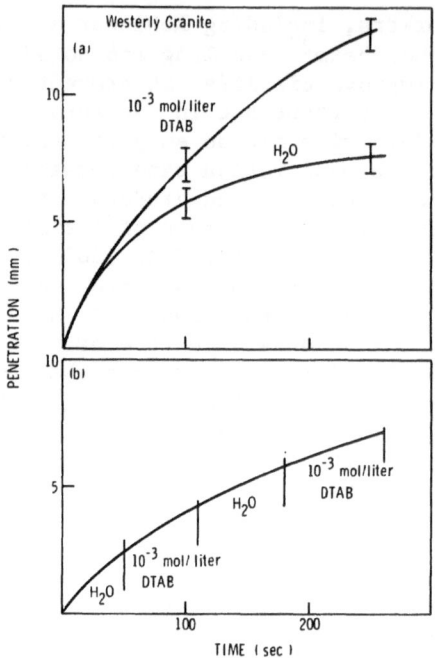

Figure 10.   (a) Penetration vs time curves for freshly redressed,
             0.6-cm o.d. diamond-impregnated bits drilling into
             Westerly granite under water or 10^{-3} M DTAB, 2000 rpm,
             40 N thrust; (b) similar data for one freshly dressed
             bit. Note, however, that on changing from water to
             the 10^{-3} M DTAB environment, no significant change
             in slope occurs. (after Ref. 48)

upon changing the environment <u>can</u> be observed, as shown in Fig. 11.[5]
In such cases, it is presumed that environment-sensitive, flow-
dependent, fracture behavior is involved.

     Thus, as was often the case in earlier work,[58,61,66,67] experi-
ments which measure diamond drilling rates over only a small fraction of the total bit life (as an extreme example, over the first
10 s in Fig. 10), will <u>not</u> reliably reveal differences in the
effectiveness of various environments. Tests must be conducted
over a significant fraction of the life of the diamonds. Of course,
such extended experiments could be very expensive if full-size
bits were used in field-scale tests. Fortunately, plots of bit
penetration rate vs. penetration distance indicate early on what
influence the drilling environment has.[68,69] For example, consider
the data presented in Fig. 12. Note that, after a settling-in

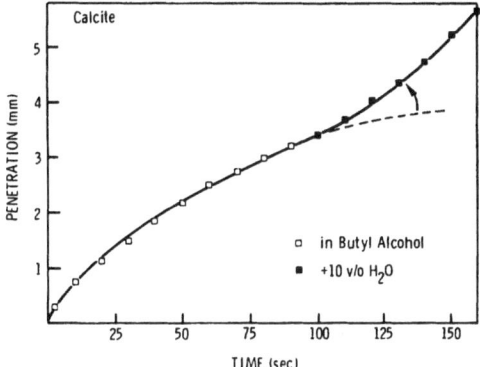

Figure 11.   Penetration vs time curve for calcite in butyl alcohol
to which 10 v/o water is added. Carbide spade bit.
Note abrupt change in penetration rate. (after Ref. 5)

distance of about 0.3 in., the rate of decay of penetration rate
with distance is markedly less in the DTAB solution, a cationic
surfactant, than in water ... but the initial drilling rates are
not significantly different.

    In general,[68,69] the rate of penetration (D) of a bit is
related to the distance drilled (D) by the expression:

$$D = V_0 - WD \tag{1}$$

where V_0 is the initial drilling rate and W, a wear rate parameter.
As Figs. 12 to 14 illustrate, V_0 is not significantly environment-
sensitive, but W is.[48] Thus, it is possible to define a parameter
R, where

$$R = W(water)/W(surfactant\ solution). \tag{2}$$

Since V_0 is essentially environment-independent, R is also defined
as

$$R = D_0(surfactant\ solution)/D_0(water) \tag{3}$$

where $D_0 = V_0/W$, and is termed the "life" of the bit. Thus, R can
be taken either as the factor by which the wear rate of a bit is
decreased, or as that by which life of a bit is increased. This

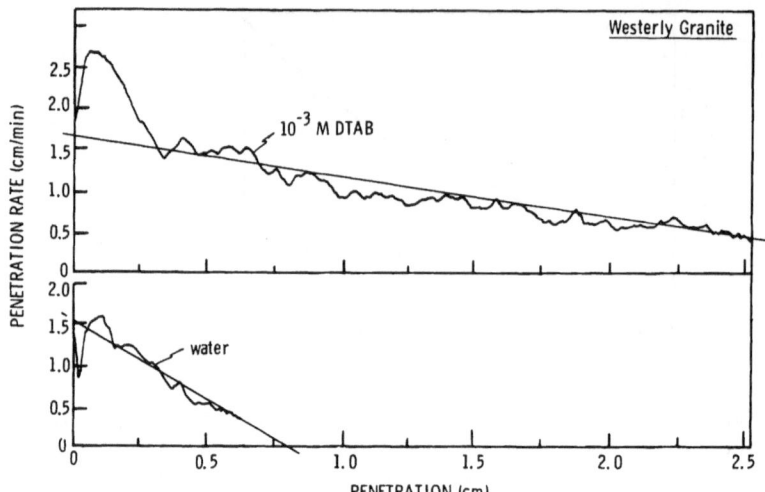

Figure 12. Penetration rate vs penetration distance data for
            Westerly granite drilled with a 0.95-cm o.d. diamond-
            impregnated core bit under (a) 10^{-3} M DTAB and (b)
            water. The thrust was 60 N, and the bit rotation
            speed 5000 rpm. (after Ref. 48)

quantization of the environmental influence is especially useful
because, empirically, R appears to be largely independent of both
the flow rate of the circulating fluid and the size of the bit.[48]

    That the importance of bit wear has not always been recognized
is realized by a comparison of the results of several authors (e.g.,
57 and 58; 59, 60, and 61; 64 and 65). Unfortunately, a lack of
space precludes explaining these differences. However, one comparison will be discussed as an example. In the work of Selim et
al., plots similar to those in Figs. 12 and 14 were employed,[57] and
the chemicals used produced significant increases in bit life,
largely due to changes in the physical properties of the drilling
fluid rather than to any chemomechanical effect. Unger et al.,
on the other hand, measured the average penetration rate over a
significant, but not large, fraction of the life of the diamonds.[58]
That is, if Fig. 10 is representative of their results, they compared the slopes of the secants intersecting the curves at 50S.
Consequently, the effects they observed were smaller than those
reported by Selim and coworkers.[57]

TYPE OF DIAMOND DRILLING BIT

    The data in Fig. 13 indicate that the interval between re-
dressings for full-sized, impregnated bits can be substantially

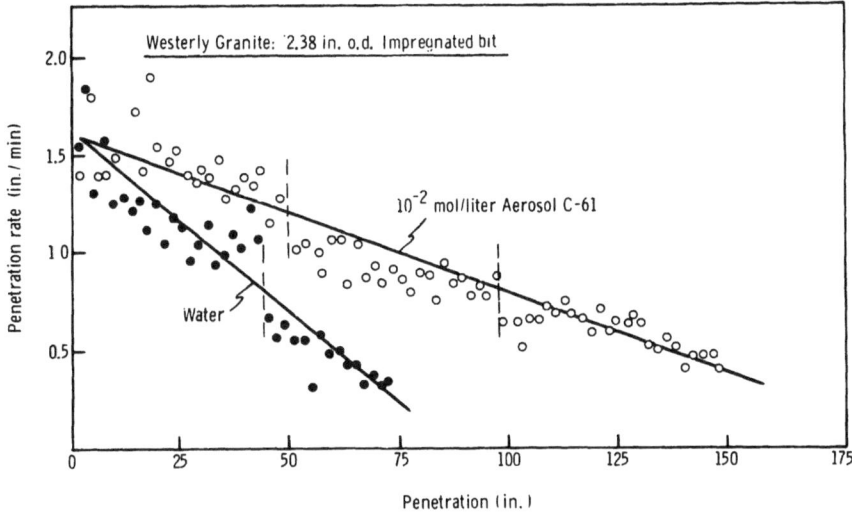

Figure 13. Influence of a 10^{-2} mol/liter Aerosol C-61 solution on
wear rate of a BX impregnated bit rotating at its design
speed of 300 rpm; 3000 lb thrust. (after Ref. 70)

increased by chemomechanically active fluids.[70] The curves in
this figure were produced using the same 2.375-in. o.d. bit,
freshly redressed before each test. If we assume that redressing
would be called for whenever the penetration rate fell below, say,
0.5 in./min., then we can see that the interval permitted when
drilling under 10^{-2} mol/liter Aerosol C-61 is more than twice that
for water.

However, nowadays impregnated bits are frequently fabricated
with a matrix that wears at about the same rate as the diamonds so
that new, sharp diamonds are continually exposed. These are called
self-sharpening bits, and for these, even if diamond wear rate is
decreased chemically, bit life is not substantially improved because
the wear rate of the metal matrix is not affected. To take advantage of the chemomechanical effect with these bits, they must be
fabricated with a more abrasion-resistant matrix to match the
chemomechanically enhanced diamond life.

Surface-set bits, on the other hand, use relatively few, large
diamonds embedded in the surface of a metal matrix. When these diamonds are worn out, the bit is returned to the manufacturer and
the diamonds are either reset in a new matrix or discarded. Chemomechanically active fluids can have a profound influence on the
life of such bits. For example, the results presented in Fig. 14
were obtained with two comparable surface-set bits of 2.375-in.
o.d., one drilling under water, the other under a 0.25 w/o solution
of Marvansoft FBH, a cationic surfactant.[70] Note that in these

full-scale tests, the initial drilling rates of the two bits were
similar (~ 1.5 in./min) and typical of the rates employed in the
field for this rock. However, the FBH solution enhanced the life
of the bit 2-3 fold over that when using water (i.e., from ~ 4 to
~ 10 ft). Moreover, after drilling 3-4 ft, the rate of penetration
was about three times greater under the FBH solution than under
water.

To eliminate the possibility that the differences illustrated
in Fig. 14 result from using different bits, the FBH solution was
replaced by water at point A without altering any of the other
drilling parameters. The penetration rate promptly decreased at a
more rapid rate than under FBH solutions, and the bit drilled only
a few inches more instead of the several feet that might have been
achieved if drilling continued in the FBH solution.

ADSORPTION DYNAMICS

It is possible to select a surfactant concentration producing
$\zeta = 0$, and to carry out the appropriate test with the right kind of
bit, and yet still observe no significant influence on drilling

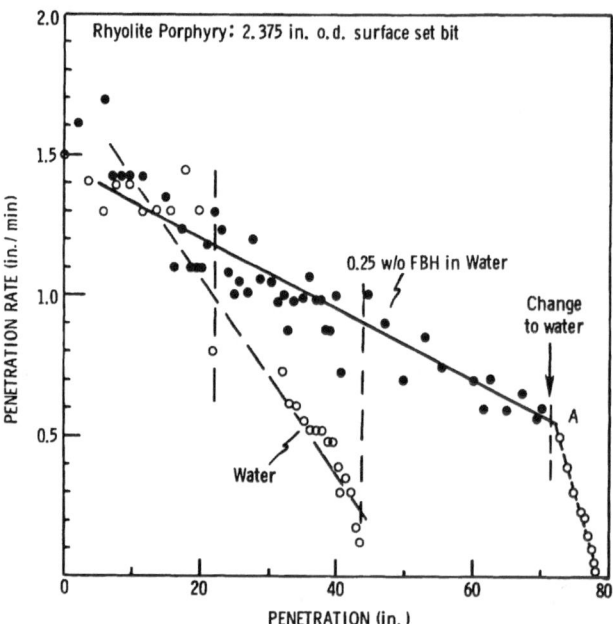

Figure 14. Penetration rate vs penetration distance for BX surface-
set bits drilling into a rhyolite porphyry rock. Test
conditions were: 500 rpm, 15 gpm of either water or
0.25 w/o Marvansoft FBH solution, 3000 lb thrust.
(after Ref. 70)

behavior...either because the surface-active molecules do not
have time to arrive at a freshly fractured surface before the next
cutting edge arrives or because, having arrived, they do not chemisorb quickly enough.[70]

The importance of this factor was first appreciated when it was
observed that, although certain very long-chain organic surfactants
can, under equilibrium conditions, produce $\zeta = 0$ and hardness maxima
at concentrations of $10^{-5} - 10^{-6}$ mol/liter, the concentrations
necessary to maximize drilling efficiency might be as large at
10^{-3} mol/liter.[70] This shift was attributed to the inability of the
$10^{-5} - 10^{-6}$ mol/liter solution to produce $\zeta = 0$ under the dynamic
and distinctly non-equilibrium conditions of drilling.

The influence of adsorption dynamics was later studied[70] by
drilling Westerly granite with bits of three sizes at different
rotational speeds and under various concentrations of Aerosol C-61
bracketing the concentration that produces $\zeta \approx 0$ under equilibrium
conditions, i.e., $10^{-3.5}$ mol/liter. It was found that when the
bit life parameter R (see Eq. 2) was plotted against the peripheral
speed of the diamonds, a "master" curve could be obtained for each
surfactant concentration by shifting the curves for each bit along
the abscissa, Fig. 15.[70]

The shape of these curves can be explained by comparing the
rate of adsorption of the active cations at a particular concentration with the time during which fresh surface, created by the
drill bit, is exposed to these active species. In brief, at slow
speeds the active ions can adsorb to their equilibrium surface
concentration and the ζ-potential is as measured. As the drill
speed increases, fewer molecules adsorb, and for those concentrations
where the ζ(equilibrium) > 0, the ζ-potential is decreased towards
zero. Under specific speed conditions for each bit, $\zeta = 0$ and the
drill life is maximized at that speed.[70]

Consequently, these experiments reveal (a) the importance of
bit rotational speed in optimizing the use of chemomechanical
effects in diamond-bit drilling, and (b) the existence of characteristic or "master" curves for given rock-environment combinations.
The significance of this latter finding is its implication that
lab-scale screening tests with small diameter bits may reasonably
and reliably simulate full-scale drilling tests. Thus, surfactant
solutions that can produce $\zeta = 0$ and also meet acceptable criteria
for toxicity, corrosivity, and biodegradability could be rapidly
and inexpensively screened for effectiveness in the field.

Such an evaluation procedure has been used to select a drilling fluid, which is inexpensive, biodegradable, corrosion inhibiting, and which increases the life of surface-set bits in a rhyolite

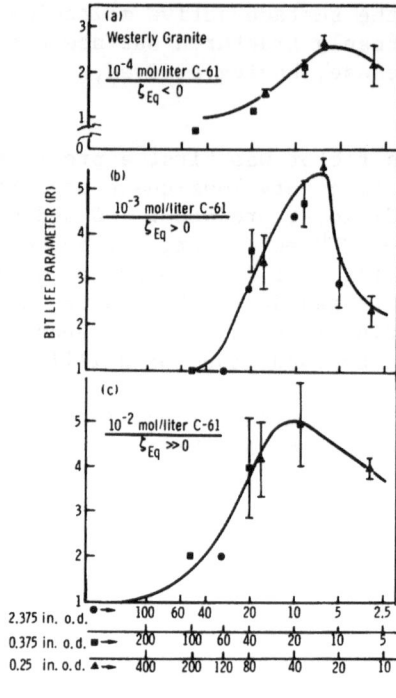

BIT PERIPHERAL SPEED, v (in/min)

Figure 15.   Variation in bit life parameter, R, for diamond bits
             drilling into Westerly granite induced by three dif-
             ferent concentrations of surfactant Aerosol C-16 as
             function of bit rotation speed. (after Ref. 70)

porphyry up to a factor of four.   Figure 14 presents some of the
results obtained with this fluid, 0.25 w/o Marvansoft FBH in water.

DIAMOND FAILURE MODE

        Another factor may confuse the proper interpretation of drilling
data: The mode of diamond failure.   During drilling, diamonds can
fail either by wear, a process involving abrasion, oxidation, and
graphitization, or by fracture and disintegration.   If the diamonds
fail by fracture, chemomechanically active fluids will have only
limited influence on bit life.   For example, compare the data
presented in Fig. 14 with those in Fig. 16.   In the former set of
experiments, the bits drilled about 50 in. into the porphyry under
conditions of gradually increasing thrust before the test thrust
of 3,000 lb was applied.   This procedure rounded off any sharp
diamond points, allowing them to better bear the 3,000-lb load
without fracture.   In contrast, the bits used to obtain the data
presented in Fig. 16 were first drilled only about 6-10 in. under

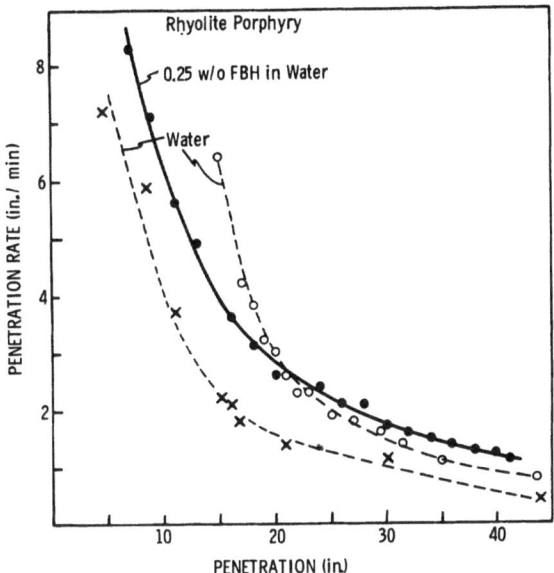

Figure 16. Variation of penetration rate with penetration of a BX surface set bit into a rhyolite phorphyry rock illustrating the rapid, non-linear decrease due to diamond fracture. Test conditions were: 500 rpm; thrust 4,000 lb; water on a 0.025 w/o solution of Marvansoft FBH; flow rate 15 gpm. (after Ref. 62)

a thrust of 1,000 lb and then the bit load was increased to 4,000 lb. After only a short settling-in, this high load caused shattering of the diamonds and a rapid, environment-insensitive, nonlinear decrease in drilling rate.[62]

It follows from the foregoing discussion that drilling strategy -- which includes all the factors discussed above -- should be carefully planned if the maximum beneficial influence of chemomechanically active fluids on diamond-bit life is to be reliably achieved.

IMPLICATIONS FOR THE FRACTURE OF CONCRETE

In view of the siliceous nature of cement, sand, and many aggregates, one might expect that many of the phenomena discussed above might also be found in concrete or cement pastes. However, a computer literature search for information on environment-sensitive fracture or cracking of concrete over the last eight years revealed zero citations on this subject. Discussions with a more reliable data base, namely, an expert in the field of concrete fracture,

Dr. Mindess, revealed that, indeed, there has been very little
work done in this general area.[71] Only static fatigue appears to
have been studied.

Gluklich appears to have been one of the first to document
static fatigue in concrete.[72] Krokosky reviewed the evidence
and theories for it in 1973.[73] More recently, Barrick and Krokosky[20] have presented evidence in support of the application of
the Charles-Hillig-Wiederhorn model to concrete, but with some
modifications. In essence, they propose that the silicate bonds
in the cement are attacked by hydroxyl ions, just as in the model
for glasses. However, the increase in time to failure with increasing temperature -- the opposite to that found in glasses -- is due
to the fact that the hydroxyl ions are released from the $Ca(OH)_2$
present.[20,73] Since the solubility of this material decreases with
temperature, an increase in test temperature results in a reduction
of OH ions available for attack on the silicate network and hence,
an observable slowing of crack velocity.[20]

Other than these static fatigue studies, there appear to be no
reports of other adsorption-sensitive fracture phenomena in cement
or concrete. Yet, several should be applicable given that the components are formed mostly from silicates. For example, one would
expect that the ζ-correlation would apply and environment-sensitive, flow-controlled cracking might promote strength degradation
beyond that caused by water alone. That is, the minimum in E_{sce} and
time to failure at $\zeta = 0$ should appear if the slow crack growth
mechanism in concrete for adsorbed species other than H^+ or OH^- is
even roughly similar to that observed in glass.

On both a scientific and economic basis, perhaps the most pro-
fitable application of the previously discussed phenomena might be
in the drilling and sawing of concrete. Presumably, when destroying
large concrete structures, the drilling of blast holes and the sawing
of slots to weaken the structure will be necessary. The effects observed by Westwood and his coworkers [59,60] on the percussive drilling
of rocks should be transferable to concrete. However, before this
can be done, the exact mechanism by which such bits more easily
penetrate rock and presumably concrete must be determined. If the
wrong mechanism is assumed, much time and effort can be lost in
attempting to transfer these effects to the field.

Finally, the influence of surface active agents on the life of
diamond loaded drill bits and saw blades should transfer directly,
especially if the factors discussed above for diamond drilling are
considered.

ACKNOWLEDGEMENTS

The author wishes to express his appreciation of the helpful comments freely given by Drs. J. Ahearn and J. Green of Martin Marietta Laboratories, and the assistance of Dr. S. Mindess of the University of British Columbia, in the preparation of this manuscript. The patience and unfailing good humor of Mrs. Lois Craig in typing the manuscript is also gratefully acknowledged.

REFERENCES

1. A.R.C. Westwood, J. Mater. Sci. 9:1871 (1974).
2. O. Reynolds, Manchester Lit. Phil. Soc. 13:93 (1874).
3. A.M. Joffe, M.W. Kirpitschewa, and M.A. Lewinsky, Z. Physik. 22:286 (1924).
4. P.A. Rebinder, L.A. Schreiner, and K.F. Zhigach, "Hardness Reducers in Rock Drilling," Acad. of Sciences, Moscow, 1944. Trans. C.S.I.R.O., Melbourne (1948).
5. A.R.C. Westwood, and J.J. Mills, in: "Surface Effects in Crystal Plasticity," (R.M. Latanision and J.F. Fourie, ed.) (1977).
6. A.R.C. Westwood, in: "Mechanisms of Environment-Sensitive Cracking of Materials," (P.R. Swann, F.D. Ford, and A.R.C. Westwood, ed.) Metals Soc., London, 283 (1978).
7. M.E. Merchant, private communication, Cincinnati Milacron Inc. (1975).
8. F.C. Frank, in: "Surface Effects in Crystal Plasticity," (R.M. Latanision and J.F. Fourie, ed.), Noordhof, Leyden, 933 (1977).
9. P.G. Fox, in: "Mechanisms of Environment-Sensitive Cracking of Materials," (P.R. Swann, F.D. Ford, and A.R.C. Westwood, ed.) Metals Soc., London, 298 (1978).
10. L. Grenet, Bull. Soc. Enc. Indust. Nat. Paris, Series 5, 4:838 (1899).
11. S.M. Wiederhorn, J. Amer. Ceram. Soc. 50(7):407 (1967).
12. W.B. Hillig and R.J. Charles, in: "High-Strength Materials," (V.R. Zaday, ed.) John Wiley and Sons, Inc., New York, 685 (1965).
13. S.M. Wiederhorn, J. Amer. Ceram. Soc. 55(2):81 (1972).
14. S.W. Freiman, J. Amer. Ceram. Soc. 57(8):350 (1974).
15. S.M. Wiederhorn, Int. J. Frac. Mech. 4(2):171 (1968).
16. A.G. Evans, J. Mater. Sci. 7:1137 (1972).
17. J.G. Bruce and B.G. Koepke, First Technical Report on Project CG103, Contract Number N00014-76-C-0025/P00002 to ONR (July 1977).
18. S.W. Freiman, K.R. McKinney, and H.L. Smith, in: "Fracture Mechanics of Ceramics," (D.P.H. Hasselman and F.F. Lange, ed.), Plenum Publishing Co., New York, 659 (1974).
19. See E.M. Van Eeckhout, Int. J. Rock Mech. Min. Sci. 13:61 (1976).

20. See e.g. J.E. Barrick II and E.M. Krokosky, J. Test. Eval. 4:61(1976).

21. A.R.C. Westwood, D.L. Goldheim, and R.G. Lye, Phil. Mag. 16(141): 505 (1967).

22. A.R.C. Westwood, D.L. Goldheim, and R.G. Lye, Ibid. 17(149):951 (1968).

23. A.R.C. Westwood, and D.L. Goldheim, J. Appl. Phys. 39(7):3401 (1968).

24. Cited by A.R.C. Westwood in J. Colloids and Surfaces 2:1 (1981).

25. A.R.C. Westwood, R.K. Viswanadham, and J.A.S. Green, Thin Solid Films 39:69 (1976).

26. S.M. Ahmed, in: "Oxides and Oxide Film's," (J.W. Diggle, ed.) Marcel Dekker, 486 (1972).

27. N.H. Macmillan, R.D. Huntington, and A.R.C. Westwood, J. Mater. Sci. 9:697 (1974).

28. A.R.C. Westwood and D.L. Goldheim, J. Amer. Ceram. Soc. 53(3):142 (1970).

29. J.S. Ahearn, J.J. Mills, and A.R.C. Westwood, J. Appl. Phys. 50(5):3699 (1979).

30. J.S. Ahearn, J.J. Mills, and A.R.C. Westwood, J. Appl. Phys. 49:96 (1978).

31. J.S. Ahearn, J.J. Mills, and A.R.C. Westwood, J. Appl. Phys. 49:614 (1978).

32. J.S. Ahearn, J.J. Mills, and A.R.C. Westwood, J. de Physique, Colloque C6, supplement to No. 6:C6 (1979).

33. J.F. Dewald, Bell Syst. Tech. J. 39:615 (1960).

34. J.S. Ahearn, unpublished work, reported in Ref. 6.

35. P. Haasen, Phys. Status Solidi 28:145 (1975).

36. A.R.C. Westwood and R.D. Huntington, in: "Mechanical Behavior of Materials," Soc. Mat. Sci. Japan IV:383 (1972).

37. A.R.C. Westwood, G.H. Parr, and R.M. Latanision, in: "Amorphous Materials," John Wiley, London, 5332, 1971.

38. P.G. Fox and I.B. Freeman, J. Mater. Sci. 14:151 (1979).

39. A.G. Evans, and J.R. Wilshaw, Acta. Met. 24:939 (1976).

40. W.H. Mularie, and A.R.C. Westwood, Martin Marietta Laboratories, unpublished work, (1973): cited in Ref. 24.

41. A.R.C. Westwood and R.M. Latanision, in: "Science of Ceramic Machining and Finishing," National Bureau of Standards SP 348: 258 (1973).

42. R. Adams, University of Warwick, G.B., private communication (April 1978).

43. J.J. Mills, to be published.

44. A.R.C. Westwood, J.H. Macmillan, and R.S. Kalyoncu, SME Trans. 256:106 (1974).

45. D. Broek, "Elementary Engineering Fracture Mechanics," Noord-hoff Int. Pub., Leyden, 140 (1924).

46. S.M. Wiederhorn and H. Johnson, J. Amer. Ceram. Soc. 58(7-8) (1975).

47. S.M. Cox, Phys. and Chem. Glasses 10:226 (1969).

48. J.J. Mills and A.R.C. Westwood, J. Mater. Sci. 13:2712 (1978).

49. A.R.C. Westwood, D.L. Goldheim, and E.N. Pugh, Mat. Sci. Res. 3:553 (1966).

50. A.R.C. Westwood, D.L. Goldheim, and E.N. Pugh, Phil. Mag. 15: 105 (1967).

51. N.H. Macmillan, in: Surface Effects in Crystal Plasticity," (R.M. Latanision and J.T. Fourie, ed.), Noordhoff Int. Pub., Leyden, 629 (1977).

52. J.J. Mills, R.D. Huntington, and A.R.C. Westwood, Int. J. Rock Mech. Min. Sci. 13:289 (1976).

53. L.W. Lewis and D.E. Dunn, (Abstract) Proc. 1976 Annual Meeting of Geol. Soc. Am., 978 (1976).

54. A.R.C. Westwood, W.H. Macmillan, and R.S. Kalyoncu, J. Amer. Ceram. Soc. 56(5):258 (1973).

55. M.W. Swain, R.M. Latanision, and A.R.C. Westwood, J. Amer. Ceram. Soc. 58(10):372 (1975).

56. P.J. Gielisse, T.J. Kim, L.F. Goyette, and R.V. Nagarkar, Final technical report to the Naval Systems Air Command, Contract Number N00019-72-C-0202, University of Rhode Island (1974).

57. A.A. Selim, C.W. Schultz, and K.C. Strebig, Soc. Pet. Eng. J. 9:425 (1969).

58. H.F. Unger, B.S. Snowden, and W.H. Englemann, Trans. SME/AIME 258:185 (1975).

59. R.E. Jackson, N.H. MacMillan, and A.R.C. Westwood, in: Proc. Cong. IRSM, 3rd Vol. IIB, Advances in Rock Mechanics, Nat. Acad. Sci., Washington, D.C. (1974).

60. N.H. MacMillan, R.E. Jackson, W.M. Mularie, and A.R.C. Westwood, Trans. SME/AIME 258:278 (1975).

61. G.A. Cooper and J. Berlie, J. Mater. Sci. 11:1771 (1976).

62. J.J. Mills and A.R.C. Westwood, in: "Fundamentals of Tribology," (N.P. Suh and N. Saka, ed.), The MIT Press, Boston (1980).

63. F.C. Appl, B.N. Rao, and B.H. Walker, Paper number 78-PET-39 ASME, presented at Energy Technology Conference, Houston (November 5-9, 1978).

64. A.C.T. Joris and G. McLaren, Mining and Mineral Eng., 3:190
     (1967).
65. W.H. Engelmann, H.F. Unger, and B.S. Snowden, Trans. SME/AIME
     258:185 (1975).
66. A.E. Long and W.G. Agnew, U.S. Bureau of Mines, Report No.
     RI3793 (1945).
67. W.H. Engelmann, D.R. Tweeton, G.A. Savanick, and D.I. Johnson,
     U.S. Bureau of Mines, Report No. 8186 (1970).
68. C.E. Tsoutrelis, Trans. SEM/AIME 244:365 (1969).
69. J.J. Mills and A.R.C. Westwood, Ind. Diam. Rev., 264 (August
     1977).
70. J.J. Mills and A.R.C. Westwood, J. Mater. Sci., in press.
71. S. Mindess, private communication, University of British
     Columbia (October 1978).
72. J. Gluklich, Rheologica Acta, Band 1(4-6):366 (1961).
73. E.M. Krokosky, Mater. et Constr. 6(30):447 (1973).

6.2 DISCUSSION AFTER THE LECTURE OF J.J. MILLS

   Shah asked if research had dealt with the effect of adsorbtion
on drilling rate with the crack velocity/stress intensity factor
k_{ic}. The k_{ic}-curves had been measured in glasses by Mills showing
that even very brittle materials can have plastic flow.
Tabor considered that the lecturer's approach was explicative, but
not predicative.The only predicative factor for Mills is the
zêta correlation. Whatever the material and the environment, hardness tends to be maximum at zero zêta potential. Tabor suggested
similarly with electro-capillary effects where changes in surface
energy are function of charge. The Dutch participants described their
experiment to change zêta potentional by measuring the surface charge
of hydrated cement paste by electro osmosis; after choice of adsorbtion ions in the same solution the k_{ic} values were taken.
Ramachandran reviewed his studies with cement paste and porous glass
in various solutions. There is no chemical reaction involved, but
lenght changes greater than can be explained by Bingham effects.
What is the connection with adsorbtion and crack propogation?
The more potassium chloride added to water, the greater the length
change. There is a subsequent decrease but not back to the initial
state. Mills wondered if ions are being injected into the surface
layer, setting up compression stress. Tabor saw the possibility of
electrically charged double lagers which provide repulsive forces
between the particles as in colloïdal systems.
Answering Pomeroy on the possible effect of permeability and porosity, Mills underlined that adsorbtion dynamics is a critical factor because of the new surfaces created. The bit life is found to
improve with greater rotational speed because adsorbtion time is
shorter, so zêta potentional surface charge decreases and moved to
zero

7.1 ECONOMICS OF CONCRETE RECYCLING IN THE UNITED STATES

S. Frondistou-Yannas

President
Management and Technology Associates, Inc.
Newton, MA

ABSTRACT

The economic parameters associated with use of recycled concrete debris as aggregate in the U.S. economy have been studied.
It was found that recycled concrete debris can compete economically
with conventional aggregate in areas where the latter is locally
unavailable as well as in areas where large quantities of concrete
debris are generated every year so that economies of scale in the
processing of these debris can be realized.

Concrete produced with recycled concrete debris as aggregate
has lower strength than concrete of similar composition produced
with natural aggregate. In order to retain the same strength, a
higher cement content should be used with the recycled material.
To compensate for the increased cost of cement, recycled aggregate
should sell for less than natural aggregate. At a price of $2.40
U.S. dollars per tonne for recycled aggregate versus a price of
$4.70 U.S. dollars per tonne for natural aggregate, the two raw
materials would yield a concrete which would cost the same and perform identically as construction material.

INTRODUCTION

The successful recycling of concrete from demolition wastes
as a substitute aggregate can make a contribution to the solution
of two problems of increasing magnitude that several countries,
including the United States of America[1], Canada[2], and Great Britain[3],
are facing.

There exists, first, an aggregate availability problem.[1,4,5,6,7]

163

The large volume of construction activity during the past 30 years has depleted a significant fraction of the known reserves of good quality aggregate. Additionally, urban expansion has led to closing of some aggregate plants while stricter environmental laws have led to closing of still others. Consequently, it becomes necessary to transport aggregates from increasingly longer distances. This creates a serious economic problem since aggregates are bulky and heavy, and the cost of their transportation is correspondingly high.

Secondly, there is a waste disposal problem. Concrete is the most popular construction material and the most abundant one in demolition debris: In the United States it accounts for 67% by weight of all demolition debris.[8] The disposal of such massive quantities of concrete waste poses a difficult problem. Indeed, while torn-up pavements and demolished buildings have found use as landfill for many years, the need for fill today, especially in the highly developed metropolitan centers, suffices to take care of only a small fraction of the rubble generated to make way for new construction. It has become increasingly difficult and expensive to dispose of construction debris within the bounds of the increasingly critical environmental requirements.[9]

Under the pressure of the problems outlived above, highway contractors have already started recycling concrete pavement debris and using it typically as an aggregate base or subbase for pavement[10,11,12] and in a few instances as an aggregate for new concrete pavement and new bituminous pavement[13,14]. Additionally, recent research in universities and government laboratories (for example see Refs. 15 to 21) deals with the properties of Portland cement and/or bituminous concrete that is produced with recycled concrete as aggregate while also dealing with the economic feasibility of concrete recycling.

Published research in the area of economics of concrete recycling has demonstrated the existence of economies of scale.[16] Concrete used in streets and highways -- the only concrete recycled currently in the U.S. -- accounts only for about 15 to 20 percent of total concrete consumption.[16] In order to operate recycling plants at high capacities, thereby realizing economies of scale, the larger quantities of concrete debris generated from the demolition of buildings are also required.[22] This introduces the problem of concrete contamination as the latter concrete debris is mixed with quantities of gypsum, wood, plastics, brick and steel.

Furthermore, since the market for highway bases and subbases, the currently existing market for recycled concrete aggregate, is relatively small, new markets have to be found for the larger quantities of the recycled aggregate that have to be produced in order to realize economies of scale. These new markets have to include the relatively large market of aggregates for concrete.

For the wider scale recycling of concrete debris discussed above, the following two questions must be answered: From the point of view of technological attractiveness, is recycled concrete a satisfactory substitute for natural aggregate in the production of new concrete? and under what conditions is recycled concrete economically attractive?

The technological feasibility of concrete recycling has been assessed by several researchers. It is reported in the literature[16] that concrete produced with uncontaminated concrete debris as aggregate has a somewhat lower strength than concrete of similar composition produced with natural aggregate. On the other hand, the mix design can always be manipulated to yield a product of similar strength. Additional properties, summarized in Table 1, also confirm the technological adequacy of uncontaminated recycled concrete as aggregate for new concrete.

Of the common contaminants of concrete debris gypsum is the most harmful to new concrete. When concrete debris which has been contaminated with gypsum is recycled as aggregate for new concrete the strength and stiffness of the latter suffers a reduction, the magnitude of which is positively related to the amount of gypsum in the mix. In the extreme case, where gypsum comprises 100% of coarse aggregate, the strength can drop to 15% of the value of natural aggregate concrete.[16] It has been concluded from the above that concrete recycling plants should be able to remove concrete contaminants.

The present report is limited to the economics of such plants in the United States.

A prerequisite for the economic justification of concrete debris recycling is the presence of sufficiently large quantities of concrete debris so that a recycling plant of optimal size can be operated at high utilization factors. The above quantities should be estimated at the local (as opposed to the national) level, because the market for aggregate is a local market. Accordingly, the next section of this report is devoted to an assessment of the quantities of concrete debris produced locally in the U.S. Following this comes a section on recycling technology which is in turn succeeded by a section on recycling economics.

ASSESSMENT OF THE QUANTITIES OF CONCRETE DEBRIS PRODUCED LOCALLY

Wilson and his co-workers[8,22] have made a detailed assessment of the generation of concrete debris in the various American cities. One of his findings is that, on the average, 0.27 tonnes of concrete debris per capita are generated each year in the United States. It follows that in urban areas with population greater than half a million people, the amount of concrete debris generated annually is

TABLE 1 – Comparison of Properties of Uncontaminated Recycled
Aggregate Concrete and Natural Aggregate Concrete
of Similar Composition (Control)

Type of Property	Uncontaminated Recycled Aggregate Concrete
Aggregate-mortar bond strength aggregate primarily gravel from the old concrete	comparable to that of control[15]
aggregate primarily mortar from the old concrete	55% that of control[15]
Compressive strength	64 to 100% that of control[15,16,17,18]
Static Modulus of Elasticity in compression	60 to 100% that of control[15,16]
Flexural Strength	80 to 100% that of control[18]
Freeze-thaw resistance	comparable to that of control[17,18,19]*
Linear Coefficient of Thermal Expansion	comparable to that of control[17,19]
Length Changes of Concrete specimens stored for 28 days at 90% R.H. and 23°C	comparable to that of control[17,19]
Slump	comparable to that of control[15,16,17]

*In the special case where the original concrete had low resistance
to frost action because of the frost susceptibility of its aggregates, the new concrete using the former as aggregate had improved
frost resistance.[17,19]

of the order of a few hundred thousand tonnes. By contrast, a single
highway demolition project produces only a few tens of thousand
tonnes of debris.

A special and very important case is the case of a natural
disaster, for instance an earthquake. In the latter case, several
million tonnes of concrete debris may be produced. For example,
the writer has estimated[16] that the San Fernando (U.S.A.) earthquake
of 1971, produced close to 5 million tonnes of concrete debris.

The recycling plants studied in this report have the capacity
to recycle a few hundred thousand tonnes of concrete debris per year.
Accordingly, they are assumed to operate in a fixed location near a
large city.

RECYCLING TECHNOLOGY

The design of four concrete recycling plants that include
sorting systems for uncontaminated debris appears in Refs. 16 and
21. The designed plants have capacities of 110-275 TPH; 275-410 TPH;
410-545 TPH; and 545-680 TPH. They can thus process a few hundred
thousand tonnes of concrete debris per year.

The design of the above plants has benefited from experience.
Indeed, there are at least 10 concrete recycling plants that are
operating profitably in the U.S. today. In the overwhelming majority of cases, these plants recycle uncontaminated concrete debris
that comes from the demolition of pavements[12,23,24]. In the area
of sorting facilities needed to remove contaminants such as gypsum,
plastics and wood, the design of the above plants benefited from
the experience of the natural aggregate industry in which similar
problems had to be solved. Moreover, of considerable help were
producers of new equipment which was designed specifically for the
recycling of constructed facilities.

It has been estimated[16,21] that it is economically attractive to
combine a concrete recycling plant and a sanitary land fill (SLF). In
this case, trucks that bring in debris on their way back will carry
aggregate. Additional transportation savings accrue from the fact that
concrete contaminants do not have to be transported to a distant dump.

Portable units have been specified for all designed plants so
that the plant can be relocated to a different site next to a new
SLF when the old SLF is filled.

An overview of the various steps in a concrete recycling plant
which accepts both contaminated and uncontaminated concrete debris,
together with a materials balance, is presented in Figure 1, while
a schematic design of a recycling plant with a 110-275 TPH capacity
appears in Figure 2. The first step in the processing sequence
involves preliminary cleaning and site reduction. This is followed

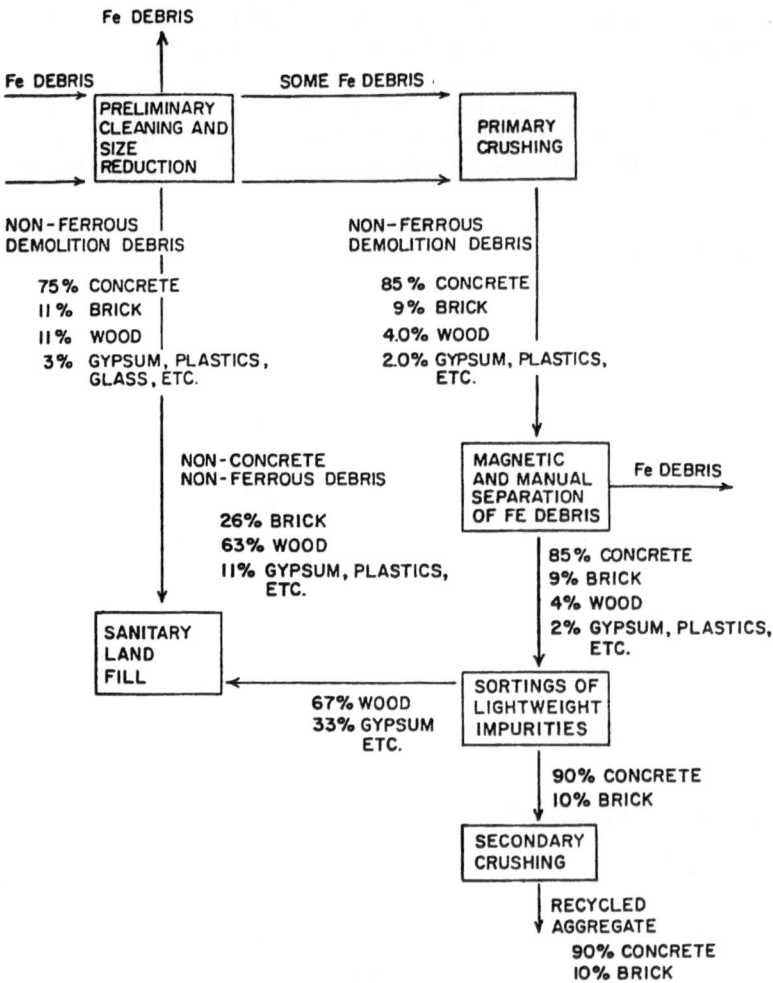

Figure 1 – Material Balance of Plant for Recycled
Concrete Aggregate.   After Ref. 16

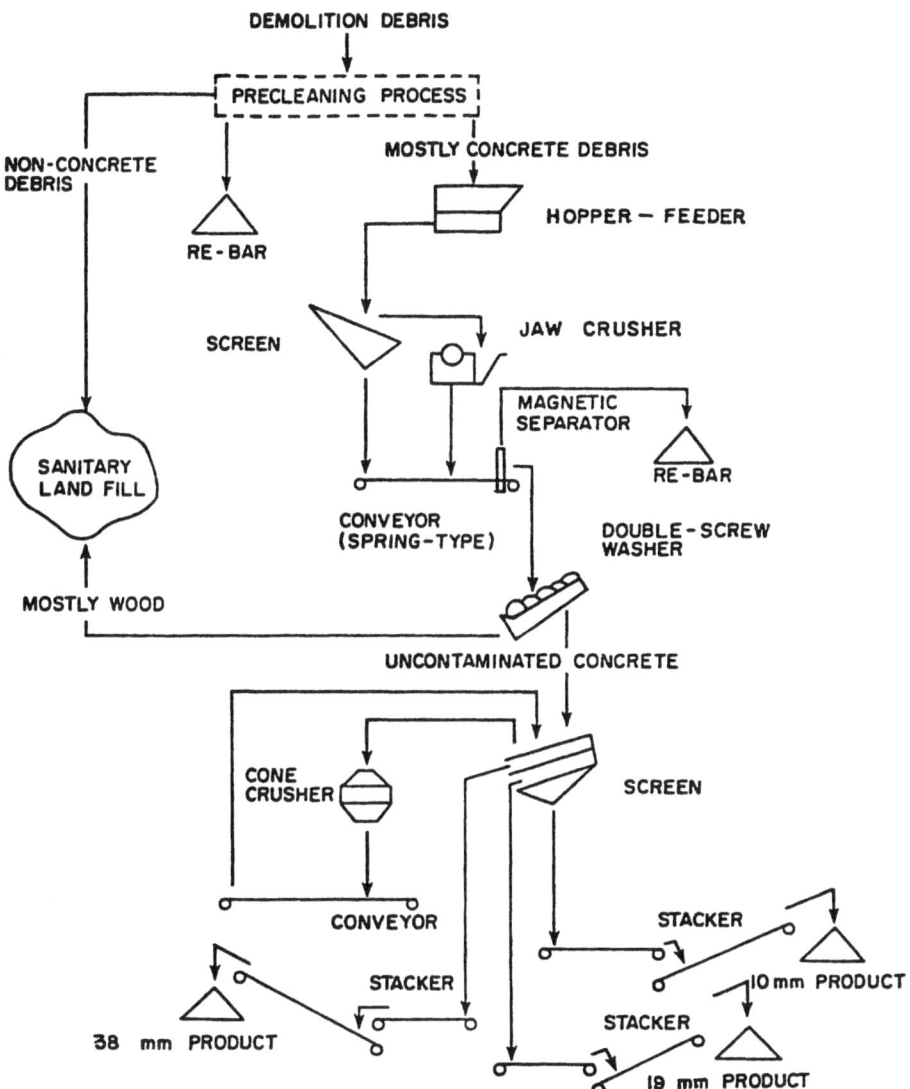

FIGURE, 2 – Schematic Design of a Recycling Plant
           with a 110–275 TPH Capacity.   After Ref. 16

by primary crushing, magnetic and manual separation of ferrous debris, sorting of lightweight impurities and, finally, secondary crushing.

Preliminary Cleaning and Size Reduction

Debris brought to the recycling system mostly consists of concrete pieces with embedded steel re-bars or wire meshes. Additionally, there are considerable quantities of wood and brick, together with small quantities of gypsum, plastics and glass (Figure 1). At the preliminary cleaning stage, one or more bulldozers are used to pick up large pieces of non-concrete debris.

Concrete pieces too large to be fed into the recycling system have to be reduced in size. For this purpose, the designed plants use one or more hydraulic hammers mounted on backhoes (with buckets removed). Steel rods longer than two feet are unacceptable with most of existing systems[16] and are therefore cut into shorter lengths by re-bar cutters.

Primary Crushing

After preliminary cleaning and size reduction operations, the debris is fed into a hopper-feeder and through the latter into a screen which separates it into two categories: larger than 10 cm debris which has to go through primary crushing, and smaller than 10 cm debris which by-passes primary crushing.

Feeding equipment used in all 4 plant designs includes front-end loaders. Additionally, in the two large plants, a dragline crane with bucket has been included to assist in the feeding operation. Following the above equipment comes a vibrating feeder and hopper which regulates the flow of debris into a screen. In the three larger plants designed in Ref. 16 the latter piece of equipment is a perforated plate, which sorts out steel rods unattached to concrete before they are fed into a jaw type primary crusher.

Concrete debris entering the jaw crusher still carries attached steel bars. For this reason, heavy duty jaw crushers that also contain some type of tramp-iron-release device have been selected[16].

In the jaw crusher steel rods are physically separated from concrete and are discharged lengthwise through the discharge opening of the crusher to the under-crusher belt conveyor. If the head-room between the discharge opening and the under-crusher belt conveyor is not large enough, long steel rods may just stay half-way through the discharge opening and block the opening. A previous operation at Taylor, Michigan[27] elevated the jaw crusher 1.8-2.4m above the under-crusher belt conveyor and also installed a turning type chute below the discharge opening so that discharged steel rods hit the conveyor belt at a less damaging angle. With standard

portable systems one cannot obtain 1.8-2.4m of headroom below the
discharge opening. For this reason, the recycling plants designed
in Ref. 16 make use of an under-crusher belt conveyor that has a
spring adjustment and can therefore be moved downward when a long
steel rod forces its way through and blocks the discharge opening.

Magnetic and Manual Separation of Ferrous Debris

After the steel rods have been effectively separated from the
concrete pieces in the jaw crusher, they are sorted out manually,
when sufficiently long, or else they are sorted by an overhead magnetic separator and magnetic head-pulley installed at the end of a
long and wide belt conveyor which serves as a "picking table". This
is the same belt conveyor with the spring adjustments immediately
following the jaw crusher.

In order to avoid complete shut-down of the system in case of
malfunctions caused by the steel rods, a surge pile, in the design
on the three larger plants, has been used to serve as a relay so
that downstream operations (e.g., secondary crushing, washing and
screening) can operate independently of the upstream operations
(e.g., feeding, primary crushing and magnetic sorting). The surge
pile system consists of a trench in the ground with belt conveyor
and feeder installed.

Sorting of Lightweight Impurities

The latter mostly include gypsum, in the forms found in construction, wood chips and plastics.

To sort out the above materials, one can adopt one of the many
processes used by the aggregate processing industry. In Ref. 16
a screw type washer dewaterer has been selected which simultaneously
separates and sorts lightweight impurities and dewaters the washed
aggregate so that the latter can be sent directly to a secondary
crusher.

Secondary Crushing

Following the screw washers-dewaterers is a screen which directs
the larger than 3.8 cm aggregate to further size reduction in a cone
crusher. The latter is of the short-head type operating in closed
circuit.

Potential of New Technologies for Concrete Recycling Plants

At plant capacities of more than 700 TPH more sophisticated
technology becomes attractive. For instance, it has been suggested[16,28] that a water jet cutting system is potentially attractive
for preliminary size reduction of concrete debris. At this point

the economic attractiveness of water jet systems for concrete
cutting cannot be assessed because no such system has yet been
developed on a commercial level.

ECONOMIC FEASIBILITY OF CONCRETE RECYCLING

In this section the economic attractiveness of the four concrete
recycling plants described in the previous section is estimated.

Initial Investment in the Recycling Plants and SLF's

The total required initial investment in a recycling plant is
the sum of purchase cost and set-up cost of the equipment. The
total purchase cost of equipment for the four designed plants
appears in Table 3, line 1.1, while a detailed analysis of the
derivation of these costs for one of the plants, that with a 110-275
TPH capacity, appears in Table 2. The set-up costs for equipment
is assumed to be 15% of their purchasing cost and includes engineering and erection expenses (Table 3, line 1.2).

The relationship between the required initial investment in
the recycling plant and plant capacity is shown graphically in
Figure 3. It can be seen that in the case of initial investment
there are no economies of scale.

The required total initial investment in a SLF is the sum of
land acquisition, excavation, lining, engineering and facilities
costs (Table 3, Part II).

The assumption has been made that the SLF is located at a
distance of 20 to 25 kilometers from the center of the city[16] where
land sells for $3200/per 1000 m^2 and that the space at the SLF
suffices to accomodate the non-concrete debris that will be dumped
there throughout the life of the adjacent plant. Using the data
of Figure 1 as a base, the tonnage of non-concrete debris that will
be dumped at the SLF has been derived. To convert this tonnage
into volume a fill density of 0.71 tonnes/m^3 has been assumed.
Once the volume of dumped debris was determined, the required area
was estimated by assuming that the SLF is a rectangle with a square
floor area equaling the required area and a constant depth of 30.5m.

It was assumed that the required land is a valley (for example,
an old quarry) and that only 6.0 out of the 30.5m of required depth
will be produced by excavation at a cost of 1.00 U.S. dollars per
m^3.

To prevent leachate problems the total inside surface area of
the landfill rectangle will be covered with 5.1 cm of asphalt and
15.2 cm of sand, at a lining cost of 10.00 U.S. dollars per m^3.

TABLE 2 - Equipment for a Recycling Plant
of 110 to 275 TPH Capacity

Equipment	Quantity	Estimated Purchase Price 1980 (in US dollars)
1. 107cm x 4.9m hopper-feeder	1	200,000
2. 1.5m x 4.3m single-deck screen (10.2cm openings)	1	13,000
3. 76.2cm x 106.7cm jaw crusher (discharge 10.2cm; max feed ∼ 61cm x 61cm, capacity 180 TPH)	1	90,000
4. 106.7cm x 127cm belt conveyor (spring type)	1	20,000
5. magnetic separator	1	16,000
6. 111.8cm x 6.10m double-screw washer (max feed ∼ 10.2cm, capacity 360 TPH)	1	44,000
7. 1.53m x 4.88m triple-deck screen (openings: 3.8cm, 1.9cm, 1.0cm)	1	20,000
8. 5100 short-head cone crusher (discharge 2.5cm; max feed ∼ 14cm, capacity 180 TPH)	1	175,000
9. 61cm x 18.3m recirculating belt conveyor	1	17,000
10. 61cm x 15.2m radial stacker	3	14,000@
11. 500 KW power unit	1	17,000
12. 15.2cm diameter pump and piping	1	4,000
13. 3.45m³ front-end loader	1	120,000
14. 77 KW bulldozer	1	70,000
15. backhoe with hydraulic hammer	1	74,000
	TOTAL	922,000

NOTE: Equipment 1 to 12 requires setting up.

TABLE 3 - Economics of the Recycling Plants at SLF Sites

	Plant Capacity (Tonnes per Hour)			
	110-275	275-410	410-545	545-680
Part I - Initial Investment in the Recycling Plants				
1.1 Purchase Cost of Equipment U.S. dollars	922,000	1,415,000	1,910,000	2,276,000
1.2 Set-up Cost (15% of Purchase Cost of Equipment that Requires Set-up[1]) U.S. dollars	99,000	155,000	195,000	237,000
1.3 Total Initial Investment (1.1+1.2) U.S. dollars	1,021,000	1,570,000	2,105,000	2,513,000
Part II - Initial Investment in the SLF				
2.1 SLF area, in 1000m^2	29	52	73	94
2.2 Land Acquisition Cost (at $3200/1000m^2) U.S. dollars	93,000	166,000	234,000	301,000
2.3 Excavation Cost for 6.0m, excavation at $1.00/m^3, U.S. dollars	173,000	308,000	432,000	556,000
2.4 Lining Cost at $10.00/m^3, U.S. dollars	503,000	805,000	1,066,000	1,320.000
2.5 Engineering Cost U.S. dollars	6,000	6,000	6,000	6,000
2.6 Facilities Cost U.S. dollars	7,000	7,000	7,000	7,000
2.7 Total Initial Investment (2.2+2.3+2.4+2.5+2.6) U.S. dollars	782,000	1,292,000	1,745,000	2,190,000
Part III - Production Cost for the Recycling Plant				
3.1 Depreciation Cost of Purchased Equipment (line 1.1 over 15,000 hours) U.S. dollars/hr	61	94	127	152
3.2 Write-off of Set-up Costs (line 1.2 over 15,000 hrs) U.S. dollars/hr	7	10	13	16
3.3 Maintenance and Repair Cost (90% of line 3.1) U.S. dollars/hr	55	85	114	137
3.4 Labor Cost (after Table 4) U.S. dollars/hr	93	137	153	184
3.5 Fuel and Lubrication Cost U.S. dollars/hr	60	104	126	136
3.6 Overhead (½% of line 1.3) U.S. dollars/year	5,105	7,850	10,525	12,565
PLANT PRODUCTION (tonnes/year)	204,000	318,000	408,000	476,000
HOURS OF OPERATION PER YEAR (at 100% capacity to produce the stated output)	750	778	750	700
The figures that follow in the rest of the Table are pertinent to the above assumed annual production.				
3.7 Overhead (line 3.6 over no. of hours of operation) U.S. dollars/hr	7	10	14	18
3.8 Interest (9%) and Insurance (1%) U.S. dollars/hr	107	137	215	274
3.9 Total Production Cost (3.1+3.2+3.3+3.4+3.5+3.7+3.8) U.S. dollars/hr	390	577	762	917
3.10 Total Production Cost (line 3.9 times no. of operating hours per year) U.S. dollars/yr	292,440	448,688	571,500	641,900
3.11 Depreciation and Write-off Costs (line 3.1+3.2 times no. of operating hours per year) U.S. dollars/yr	51,000	80,912	105,000	117,600
3.12 Total Production Cost (line 3.10 over plant production) U.S. dollars per tonne	1.43	1.41	1.40	1.35

TABLE 3 - (cont.)

	Plant Capacity (Tonnes per Hour)			
	110-275	275-410	410-545	545-680
Part IV - Operating Costs of SLF				
4.1 Production Cost (after Fig. 4) U.S. dollars/yr	149,000	191,000	233,000	246,000
4.2 Rental Cost of Scrapers (1 scraper for every 365 metric tonnes of fill per day of operation @ $3,780/month) U.S. dollars/year	19,000	38,000	57,000	51,000
4.3 Depreciation Cost (straight line depreciation of items in lines 2.3,2.4 and 2.5 over life of plant) U.S. dollars/year	34,000	56,000	76,000	94,000
4.4 Property tax (7.5% on value of land and facilities) U.S. dollars/year	7,500	13,000	18,000	23,000
4.5 SLF Operating Costs (sum lines 4.1 to 4.4) U.S. dollars/year	209,500	298,000	384,000	414,000
Part V - Annual Revenues and Expenses				
5.1 Revenue from the Sale of Recycled Aggregate (Annual production sold at $2.40/tonnes) U.S. dollars/year	490,000	763,000	979,000	1,142,000
5.2 Revenue from Sale of Re-bars ($0.36 per tonne of concrete debris processed), U.S. dollars/yr	73,000	114,000	146,000	171,000
5.3 Revenue from Dumping Charges ($7.16/tonne of non-concrete debris) U.S. dollars/year	292,500	455,000	585,000	682,000
5.4 Operating Revenues (sum of lines 5.1-5.3) U.S. dollars/year	855,500	1,332,000	1,710,000	1,998,000
5.5 Operating Expenses (sum of lines 3.10 and 4.5) U.S. dollars/year	501,940	746,688	955,500	1,055,900
5.6 Operating Income (line 5.4-5.5) U.S. dollars/yr	353,560	585,312	754,500	942,100
5.7 Income Tax (50% of Operating Income) U.S. dollars/year	176,579	292,656	377,250	471,050
5.8 Net Income - U.S. dollars/year	176,579	292,656	377,250	471,050
Part VI - Cash Flow and Net Present Value Analysis				
6.1 Cash Inflow from Operations Without Cash Shield (lines 5.8+3.11+4.3) U.S. dollars/year	261,579	429,568	558,250	682,650
6.2 Tax Shield (lines 3.11+4.3 over 2) U.S. dollars/year	42,500	68,456	95,500	105,800
6.3 Cash Inflow from Operations with Tax Shield (lines 6.1+6.2) U.S. dollars/year	304,079	498,024	653,750	788,450
6.4 Cash Inflow from Sale of Land at end of Life of Plant (assumes 6% per year land appreciation and 30% capital gains tax) U.S. dollars	239,204	405,812	598,176	809,589
6.5 Total Cash Outflow for Capital Investment at Beginning of Operations (sum of lines 1.3&2.7) U.S. dollars	1,803,000	2,862,000	3,850,000	4,703,000
6.6 Net Present Value at 16% Discounted Rate U.S. dollars	-0-	-0-	-0-	-0-
6.7 Required Output for a 16% Return on Investment tonnes/year	204,000	318,000	408,000	476,000

[1]With the exception of cranes, bulldozers, loaders and backhoes, all plant equipment requires setting-up (Ref. 20).

SOURCE: Refs. 16, 22 and Engineering News-Record Journal, various issues.

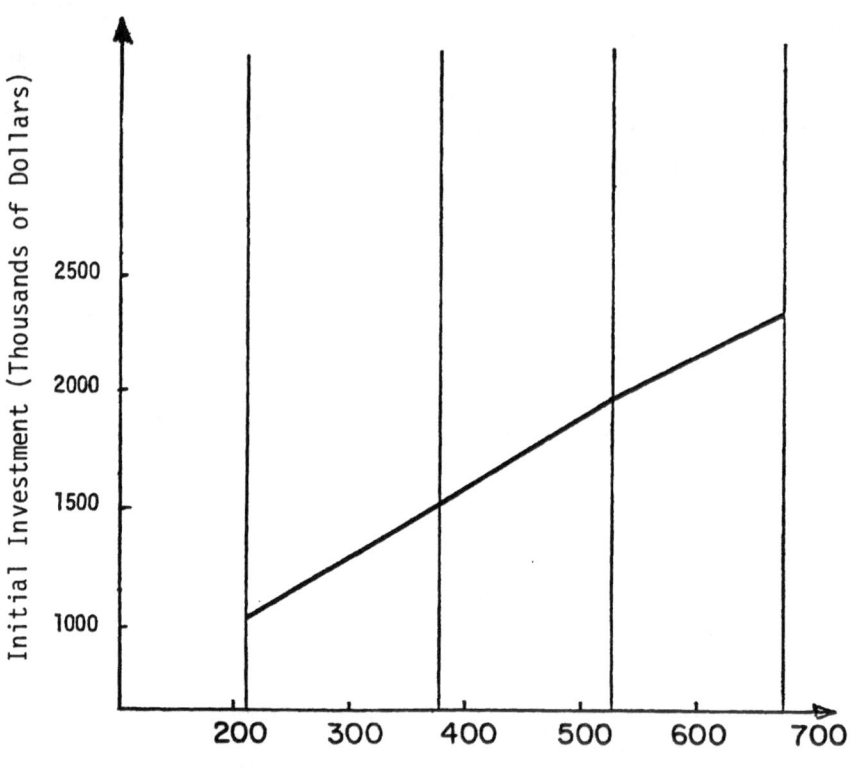

FIGURE 3 - Relationship Between Required Initial
          Investment in a Concrete Recycling Plant
          and Plant Capacity.

Additionally, there are engineering costs (e.g., initial study and surveying) at 6,000 U.S. dollars as well as costs for the various facilities (roads, fences, etc.) at 7,000 U.S. dollars.

Production Cost Estimate for the Recycling Plants at the SLF Sites

The production cost of recycling aggregate is the sum of the production costs of the recycling plant and of the SLF. The production cost of the recycling plant is the sum of the following items: depreciation cost of equipment, write-off of set-up costs, maintenance and repair, labor, fuel and lubrication, overhead, and interest and insurance costs. The above costs for the four designed plants appear in Table 3, Part III. Straight line depreciation and write-off charges are based on a life expectancy of 15,000 hours for equipment, while maintenance and repairs have been assumed to be 90% of depreciation charges. Overhead cost, in dollars per year, has been estimated at 1/2 of one percent of total initial investment. To derive the hourly overhead cost, the above number has been divided by the actual number of hours the plant is in operation each year. Similarly, interest (at 9% of the required capital) and insurance (1% of this capital) charges are yearly charges and the actual hourly expense depends on the number of hours of plant operation during that year. On the other hand, labor cost is an hourly cost on the assumption[16,21,27] that a fairly flexible labor market exists so that operating engineers and workers can be hired for part of the year only, whenever existing market conditions do not warrant full time plant operation. Labor costs in Table 3 are based on the labor wages and requirements listed in Table 4, while fuel and lubrication costs in Table 3 have been derived from equipment requirements in Ref. 16.

To derive specific production costs in Table 3 the assumption was that the four plants under study produce 204,000, 318,000, 408,000, and 476,000 tonnes respectively, of recycled aggregate per year. It has been selected to present in detail the economic analysis of these specific annual productions because these are the outputs that will yield a 16% return on investment, as commonly required in this business sector.

There are economies of scale in the area of labor, fuel and lubrication costs which reduce the total production cost per ton as plant capacity increases (Table 3, line 3.12).

There are also economies of scale at the SLF, in dollars per ton of debris that is stored. These can be seen in Figure 4. Excluded from the costs in the above figure are the rental cost of scrapers, depreciation costs and property taxes.

The rental cost of a 8.5m^3 capacity scraper has been assumed to be 3780 U.S. dollars per month, and one such scraper is needed

TABLE 4 - Manpower Requirements and Cost for Recycling Plants

Description of Worker	Hourly Wage in US Dollars per Man-Hour	Manpower Requirement, in Numbers of Men Plant Capacity-Tonnes per Hour			
		110-275	275-410	410-545	545-680
Laborer	13.00	2	4	4	4
Loader Operator	18.00	1	1	1	1
Crane Operator	16.00	-	-	1	1
Bulldozer Operator	16.00	1	1	1	2
Backhoe Operator	18.00	1	2	2	2
Crusher Operator	15.00	1	1	1	2

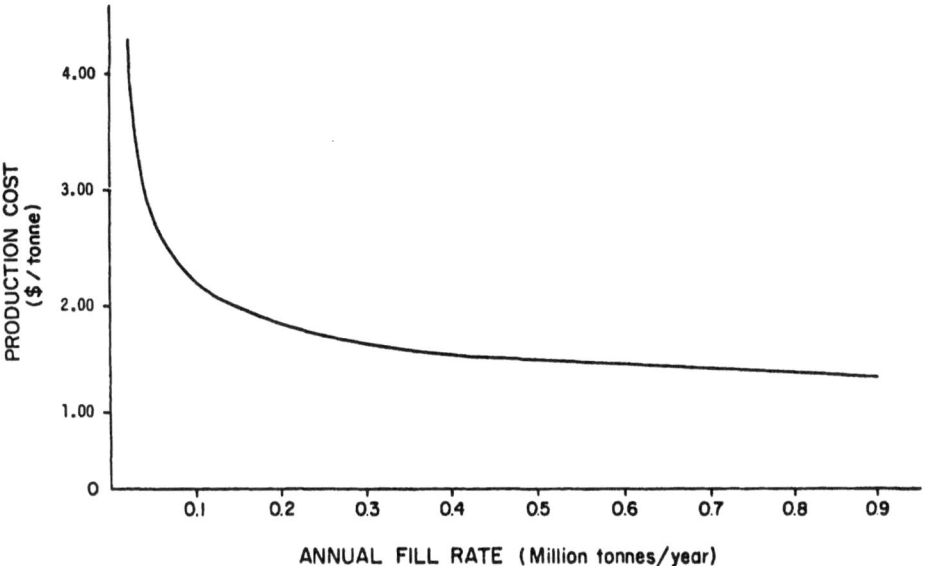

FIGURE 4 - Production Cost of the SLF System. After Ref. 16

for every 365 tonnes of fill per day of operation[16].

All items of initial investment (Part II of Table 3) with the
exception of land are depreciable and a straight line depreciation
has been assumed. The plant has been assumed to have a 20 year
life expectancy. Additional expenses included a 7.5% annual tax
on the value of land and facilities[16].

Annual Revenues and Expenses of the Recycling Systems

In the four recycling plants revenue is generated from three
sources:
   a) Sale of recycled aggregate. The assumption has been made[16]
      that 100% of the processed aggregate will be sold at a
      price of 2.40 U.S. dollars per tonne.
   b) Sale of rebars at 0.36 U.S. dollars per tonne of concrete
      debris processed[16,21,27].
   c) Dumping charges generated from the following charging
      system: 7.16 U.S. dollars per tonne for the dumping of
      non-concrete debris. Dumping of concrete debris is free.

Annual expenses at the recycling plant-SLF system are the
following:
   a) Production cost of the recycling plant (line 3.10, Table 3).
   b) Operating costs of SLF (line 4.5, Table 3).

The difference between the sum totals of the above revenues
and expenses is the operating income. A 50% income charge is imposed on the latter so that net income is the remaining 50% of the
operating income (line 5.8, Table 3).

Attractiveness of Investing in the Recycling Systems

To investigate whether the profit derived is satisfactory when
investment requirements and other investment opportunities are taken
into account, a net present value analysis was performed at several
assumed annual production levels for the four plants under discussion.
In such an analysis, a cash flow of the studied recycling systems is
required. The cash flow analysis for the four plants under consideration at annual productions that would yield a 16 percent return on
investment, appears in Part VI of Table 3.

Cash outflow in any year includes capital investment (if any),
the production cost of the recycling plant, excluding depreciation
and write-off of set-up costs, and the total production cost of the
SLF, including property tax and rental cost of equipment but excluding depreciation costs.

All cash outflow for capital investment occurs in the beginning
of operations and equals the sum total of investment in the re-

cycling plant and SLF.

Depreciation and write-off charges have been consistently excluded from cash outflows, since they do not represent actual cash
payments.

Cash inflow in any year includes sale of capital (if any),
operating revenues and a tax shield which is 50% of total depreciation. At the end of the operations the SLF area, after having been
completely filled with refuse material and properly treated, is
assumed to have been sold at a price which reflects a 6¢ annual
land appreciation. The difference between the purchase price of
land, at $3200/1000m², and the sale price of land is subject to a
30% capital gain tax. Therefore, the net receipt from the sale of
land at the end of operations is 70% of the profit plus the original
purchase price of land (Table 3, line 6.4).

All cash flows were discounted by 16% to obtain their present
values and the algebraic sum of all present values gave the net
present value of the investment. After several trials it was
estimated that the net present value is zero, or that the investment
yields a 16% return, at an annual production of 204,000 tonnes for
a plant of 110-275 TPH capacity, 318,000 tonnes for a plant of 275-
410 TPH capacity, 408,000 tonnes for a plant of 410-545 TPH capacity
and 476,000 tonnes for a plant of 545-680 TPH capacity (Table 3,
line 6.7). To put it in other terms, if a 110-275 TPH plant is to
produce at least a 16% return on investment, this plant should process and sell no less than 204,000 tonnes of recycled aggregate a
year. The required output increases as plant capacity increases.

The above findings imply that urban areas of at least 1 million
people are needed to support the operations of a concrete recycling
plant.

Comparison of Recycled Aggregate Concrete and Natural Aggregate Concrete

In previous sections the implicit assumption has been made
that the recycled aggregate producer will be able to sell 100% of
his product. For this assumption to be correct, recycled aggregate
must compare favorably with its competitor, natural aggregate.

A fair comparison between the two types of aggregate would
involve comparison of two concrete members of equal performance,
one made with recycled and the other made with natural aggregate.
To compensate for the reduced strength the member using recycled
aggregate would have to have a 10% higher cement content.

To compensate for the higher cement cost, recycled aggregate
should sell for less than natural aggregate. In Ref. 16 it has

been estimated that for a price of 4.70 U.S. dollars per tonne for
the natural aggregate, recycled aggregate should sell for 2.40 U.S.
dollars per tonne in order to yield concrete of equal cost and performance. In other words, an unprejudiced person would be indifferent
between natural aggregate that sells for 4.70 U.S. dollars per tonne
and recycled aggregate that sells for 2.40 U.S. dollars per tonne.

However, there are good reasons why a person can be prejudiced
against recycled concrete. For one, experience with it is limited;
secondly, there are no design aids for recycled aggregate similar to
the design aids (for instance, in the form of tables of properties)
that exist for natural aggregate.

For these reasons, recycled aggregate would sell best in cases
where there is a cost advantage in its favor, and this can happen in
cases where economies of scale are realized so that recycled aggregate can sell for less than 2.40 U.S. dollars per tonne, or in cases
where there is a transportation advantage in favor of recycled aggregate. Indeed results of a recent survey reveal that most recycling
plants now in existance in the U.S. operate in areas where natural
aggregate is not locally available.[16] Moreover, 10 out of 12 existing
concrete plants cited reduced transportation costs as one of their
main advantages.

In the area of energy, a recent study[26] estimates that the processing energy for both natural and recycled concrete aggregate is
approximately the same. Energy savings through recycling can only be
realized in the form of reduced consumption of transportation fuel
when natural aggregate is locally unavailable.

CONCLUSIONS

The economic attractiveness, as well as energy savings potential
of recycled concrete as a substitute for natural aggregate in concrete has been reviewed in this report. It has been concluded that:

1. The existing technology of the natural aggregate industry (aggregate beneficiation processes) can adequately eliminate detrimental concrete contaminants such as gypsum, wood, plastics, glass,
   etc.
2. Recycling plants of large capacity (up to 700 TPH) can be totally
   based on standard, widely used equipment. At present, in the
   United States, there are more than 10 concrete recycling plants
   of small to medium capacity in operation. These are mostly portable plants quickly assembled at the site of debris accumulation.
3. In economic terms, recycling of concrete debris by a commercial
   plant that will operate continuously in the area is attractive,
   provided that at least 204,000 tonnes of concrete debris is annually produced in the area. The above amount of concrete debris
   will be produced in an area with a population of about 1,000,000

people. As the tonnage of generated debris increases, so does
the attractiveness of investing in a recycling plant.

4. At the price of 2.40 U.S. dollars per tonne for recycled aggregate -- vis a vis a price of 4.70 U.S. dollars per tonne for
natural aggregate -- an unbiased customer will be indifferent
between recycled and natural aggregate, since in either case the
final product, concrete, will cost the same and will perform
identically as construction material. Even in the presence of
consumers who may be biased against it, however, prospects for
recycled aggregate sales are good because the economics in several cases are in favor of the recycled product. For instance,
natural aggregate is locally unavailable in several areas to the
point where recent research is focusing on exotic solutions, such
as digging aggregate from the ocean floor.[27] In all these cases,
there is a significant transportation advantage in favor of recycled aggregate. Furthermore, the potential returns realized
at high levels of recycling operations are large enough that
recycled aggregate producers can afford to lower their prices to
undersell their competitors (natural aggregate producers) while
still realizing significant savings.

5. Energy savings through concrete recycling are realized in the
form of reduced consumption of transportation fuel when natural
aggregate is locally unavailable.

REFERENCES

1. "Policy Statement on Aggregate Availability", Internal Memoran-
dum, National Crushed Stone Association, Silver Spring,
MD, January 1975.
2. J.J. Emery, and S.K. Change, "Trends in the Utilization of
Wastes for Highway Construction", Proceedings, Fourth
Mineral Waste Utilization Symposium, Illinois Institute
of Technology, Chicago, IL, (1974).
3. W. Gutt, "Aggregates from Waste Materials", Chemistry and
Industry, (London), 49:439 (1972).
4. C.R. Marek, B.M. Gallaway, and R.E. Long, "Look at Processed
Rubble - It's a Valuable Source of Aggregates", Roads
and Streets, 114:82 (1971).
5. C.R. Marek, "Supplemental Aggregates for Construction",
Journal of Materials, 7:50 (1972).
6. T.White, "Recycling of Concrete & Asphalt Rubble", Marketing Research Rpt. Nordberg Machinery Group, Rexnord, Inc., Wis.
7. Proctor and Redfern Limited, Toronto, Ontario, "Towards the
Year 2000 - A Study of Mineral Aggregates in Central
Ontario", Ontario Ministry of Natural Resources, Toronto,
Ontario, 1974.

8. D.G. Wilson, P. Foley, R. Wiesman and S. Frondistou-Yannas,
    "Demolition Debris: Quantities, Composition and Possi-
    bilities for Recycling", Proceedings, 5th Mineral Waste
    Utilization Symposium, (Chicago, April 1976) Aleshin, E.,
    Editor, U.S. Bureau of Mines, Chicago, IL. (1976).
9. "Recycling Rubble for Highway Purposes", Public Works,
    103:87 (1972).
10. "Crushing Concrete Rubble into Subbase Aggregate", Roads and
    Streets, 114:44 (1971).
11. "Recycled Rubble Saves Contractors Money", Roads and Streets,
    116:80 (1973).
12. J.G. Dresser, "Rehabilitation of Runway 13-31. A Concrete
    Recycling Project", paper presented at the 57th Annual
    Meeting of the Transportation Research Board, Session 57,
    Washington, D.C., January 17, 1978.
13. "P.C. Pavement Recycled", Mid-West Contractor, Oct 6, 1976.
14. G.K. Ray, "Recycled Concrete-Savings Potential", The American
    Transportation Builder, 55:12 (1978).
15. S. Frondistou-Yannas, "Wast Concrete as Aggregate for New
    Concrete", Journal of the American Concrete Institute,
    74:373 (1977).
16. S. Frondistou-Yannas and H.T.S. Ng, "Use of Concrete Demolition
    Waste as Aggregates in Areas that Have Suffered Destruction.
    A Feasibility Study", Report R77-37, Department of Civil
    Engineering, Massachusetts Institute of Technology, Cambridge
    MA, Nov. 1977 (NTIS No. PB275888/AS).
17. A.D. Buck, "Recycled Concrete", Highway Research Record, No.
    430 (1973).
18. V.M. Malhotra, "Use of Recycled Concrete as a New Aggregate",
    Report 76-18, Canada Centre for Mineral and Energy Tech-
    nology, Ottawa, Canada, (1976).
19. A.D. Buck, "Recycled Concrete as a Source of Aggregate",
    Journal of the American Concrete Institute, 74:212 (1977).
20. M. Takeshi, K. Torao, N. Muneo, and K. Masafumi, "Study on
    Reuse of Waste Concrete for Aggregate of Concrete",
    Japan-U.S. Science Seminar on Energy and Resource Conser-
    vation in Concrete Technology, Sept. 10-13, 1979, San
    Fransisco, U.S.A.
21. S. Frondistou-Yannas, and T. Itoh, "Economic Feasibility of
    Concrete Recycling", Journal of the Structural Division,
    ASCE, 103:885 (1977).
22. D.G. Wilson, T.A. Davidson, and H.T.S. Ng, "Demolition Wastes:
    Data Collection and Separation Studies", Massachusetts
    Institute of Technology, Department of Mechanical Engineer-
    ing, Cambridge, MA (1979).
23. "Recycling Roads and Buildings with Portable Plants", Pit and
    Quarry, 65:91 (1973).

24. G. Calvert and R. Butson, "Iowa Experience with Concrete
     Pavement Recycling", paper presented at the 57th Annual
     Meeting of the Transportation Research Board, Session 57,
     Washington, D.C., January 17, 1978.
25. T. Itoh, "An Assessment of the Economic Attractiveness of Waste
     Concrete as Aggregate Material", Thesis presented to Massa-
     chusetts Institute of Technology in Cambridge, MA, in 1976,
     in partial fulfillment of the requirements for the degree
     of Master of Science.
26. G.K. Ray and H.J. Halm, "Energy Savings through Concrete Re-
     cycling", paper presented at the 57th Annual Meeting of
     the Transportation Research Board, Session 57, Washington,
     D.C., (1978).
27. "Report Cites Potential in Ocean Floor Sand and Gravel", Tech
     Talk, Massachusetts Institute of Technology, Cambridge, MA,
     February 2, 1977.

7.2 DISCUSSION AFTER THE LECTURE OF S. FRONDISTOU-YANNAS

All participants were interested in the forecasting of the evolution of amounts of concrete debris. Frondistou-Yannas stated that
her figures were based on work by Prof. D. Wilson of M.I.T. who
found population the best basis for calculating this.
Hansen and Kreijger referred to the material balance which is drawn
from data on demolition permits and estimated expectancy of buildings
also plants design Frondistou-Yannas stressed the need for a pilot
plant which would give practical figures. Hansen then raised two
technical questions concerning a heavy duty jaw crusher which should
contribute to grinding gypsum to fines, thus eliminating it from
course aggregate; also the double screw classifier which should help
against the gypsum content hazard. Mather recalled that natural aggregates also contain contaminants which engineers have learnt to
counteract by using low-alkali cement or adding pozzolan .
Shah enquired about density to learn that recycled concrete behaves
like lightweight aggregate concrete (high porosity, water adsorbtion,
preferential fracture) so this might be an economical asset.
The discussion moved to the possiblity of comparing fragmentation
problems with those solved by the mineral dressing industry, whereas
Frondistou-Yannas research was restructed to the aggregate industry.
Tabor pointed out that if the cost,of energy needed is not an important factor in the total overall cost the adhesion problems of recycled concrete are not crucial. Frondistou-Yannas traced consumer
bias to reluctance and fear of recycled concrete for large structures.
Significant economic development is necessary to make recovered aggregate saleable. Mather was convinced that development depends on
decision-making. If society and specifier require contractors to use
these materials, the market will improve.Kreijger illustrated the
important role of governments.

Little is know of the strenght characteristics of recycled
concrete under dynamic loading Frondistou-Yannas considered that for
static conditions careful mix design will ensure satisfactory
strenght. De Pauw finally remarked that the economic balance of a
recycling plant could be seriously disturbed by the primary cutting
that needs to be done to the concrete elements in order to bring
them back to dimensions acceptable for feeding in the jaw-crusher
while Frondistou-Yannas answered that cutting was calculated in the
total economic balance of the plant. It was suggested by De Pauw
that one could have bigger plants so that elements coming from
demolition sites could go into the fragmentation equipment without
primary cutting.

8. GENERAL DISCUSSION AFTER THE LECTURES

The session began with a statement by Hansen on EEC interest in recycling. Since the construction industry produces more waste than any other, the Waste Management Committee of EEC has recognized the necessity of including construction waste in overall planning of waste disposal. A working party on "Demolition, re-use and dismantling of concrete" under the chairmanship of Prof. Hansen has been set up to prepare proposals for joint community action and has reported to the EEC in February 1980. In the report, the forecasts of concrete waste arising in (8) member states of the EEC are stated to be 55.10^6 ton in 1980, 68.10^6 ton in 1990, 162.10^6 ton in 2000 and as much as 302.10^6 ton in 2020 (prediction of a private firm, Environmental Resources Ltd.) while some 500.10^6 ton of sand and gravel are used as aggregate for road building and concrete production within the Community every year. With regard to the problem of the recycling of concrete, cooperation is recommended within the EC for the purpose of developing safer, more practical and less energy-intensive techniques for demolition of reinforced and prestressed concrete which at the same time meet modern standards for protection of the environment. It is suggested that EEC on community basis should support projects where actual demolition sites are used for full-scale testing purposes. Research is recommended to develop improved and less-energy-intensive methods of fragmenting, cleaning and sorting demolition rubble and concrete to meet current or revised standards for various end-uses. In order to stretch the dwindling resources of natural sand and gravel within the Community it is recommended that one should support research with a view towards replacing natural sand and gravel with reprocessed demolition rubble and old concrete for as many applications and to the greatest extend possible. Regarding further recovery of concrete waste it is suggested that EC should support research and development of improved

and less energy-intensive methods of seperating reinforcing steel
from concrete while simultaneously fragmenting the concrete into
fractions suitable for various end-uses (meeting the strictest requirements to protection of the environment). Further the EC should
support studies of properties of demolition rubble and old concrete
which have been processed by various methods, in order to establish
for what end-purposes the recovered materials are effective substi-
tutes. For example in this respect it will be necessary to explore
the impact on workability, mechanical properties and durability
of recycled concrete. Also support of research is suggested to
development and improved methods of separating from demolition
rubble and old concrete such low-density materials, soluble salts
nd other constituents which may prove to be harmful from the point
of view of the end-use of the processed material. The EC should also
support development of methods of improving the homogeneity and general quality of reprocessed demolition rubble and old concrete for
higher grade construction purposes (meeting modern standards for
protection of the environment) which finally it is suggested that
EC should support research aimed at reducing the detrimental effects
on the general environment associated with increased resource recovery from demolished concrete in centralized plants near urban
areas, as well as from mobile plants for road construction purposes.
Support for design of one or more pilot plants and studies of problems in such plants is suggested.
Although there have been presented research proposals, for example
the 3-country project (from Belgium, Netherlands and W-Germany)
Hansen stressed the necessity of disposing of many possible jointed
research projects from as many EC-countries as possible and called
for proposals to the various research items indicated.

    After an exchange of opinion concerning methods to predict the
development of the amount of debris, it was generally accepted that
this is very difficult to access in a world under constant change in
living patterns, energy needs etc. But certainly urban renewal, rehabilitation and demountable construction are all contributing pura-
meters. Mindess enquired into role of basic science in recycling,
since progress seems concentrated on the technological level. Pomeroy
considered that it is a matter of refinements and turning, but the
opportunity for this in the first stages of recycling are small.
Kreijger emphasized the need to examine the mechanisms of existing
and new techniques for demolition and fragmentation. For example
do microwaves only heat water and, in high pressure water jetting
which is used already in practice nowadays, is pressure its only
destroying force? And regarding water, could cavitation, which is
said to cause implosion, be used as a fragmentation technique ?
Discussing this, Mather called attention to the English translation
of a Russian paper on cavitation:
"Inozemtsev, Iu. P., 1980, Cavitation Erosion Resistance of Concrete
    (k. voprosu kavitatsionne-erozionnoi stoikosti betona).
    Gidrotekhnicheskoe stroitel'stvo, No. 2, 1980, Energiia,
    Moscow, pages 30-31 (translation by WP&RS, USDI, Denver, Colorado,
    Book No. 12,303, Paper No. 7)."

That appears to be relevant since it notes (in para 2) that
"the influence of aggregate particle shape on its interface strength
with the cement paste is common knowledge. The disruption of cohesion
along the coarse-aggregate graindissolved concrete ingredient interface has been verified experimentally under cavitation conditions
(2), and this permits the suggestion that cavitation tests be used
as a method to determine the adhesion and cohesion of binders."
(a copy of the translation was sent to the chairman later on).
Apart from that all basic science has not get dealt with normal
aggregates in Mather's opinion and specifications by ASTM make
no distinction between aggregate requirements for different structural purposes, whereas ACI stipulates "properties appropriate to
use in a particular project "but application is not enforced. Both
Frohnsdorff and Pomeroy underlined the economic advantage of concrete,
following comparative input-output analyses in relation to energy
embodied in all building materials per unit weight, especially in
view of its load bearing capacity. In this respect Kreijger referred
to his publications on energy content per unit weight and unit volume of material per unit of property (strength, modulus of elasticity), see e.g. reference 3 in 1.2 of this part of the Proceedings.
Much remains to be done to use this potential. Science should work
to help predict the performance of concrete.

   After this more general discussion, the Chairman proposed to
review the various lectures in the sequence as given. Regarding
Pomeroy's lecture on the structure of concrete, Lambotte referred
to the various rheological models which are used to explain mechanical behaviour and especially Burger's model is a useful model.
In this and most other models viscosity also enters in the deformation behaviour and to his opinion this property may be important
for its influence on fragmentation as well as for the properties
of the demolished concrete aggregates and consequently the recycled
concrete. Pomeroy fully agreed that considering viscosity is important. Lambotte then raised the problem of the influence of recycled
aggregates on standard deviation of strength of the new concrete
since a designer has to consider characteristic strength which implies not only average strength but standard deviation as well.
Pomeroy regretted that no specific data on variability are known
to him at the moment. In this opinion there are new parameters introduced by using demolished aggregates and the coefficient of variation
of recycled concrete can be expected to be higher. The consequence
for design should be that larger safety coefficients have to be applied. Kreijger and Mather stated that particle shape is a function
of the reduction ratio. Fines from the primary crusher have unsatisfactory shape and should not be used, but instead the coarse aggregate should be recrushed.
Frondistou-Yannas pointed out that these factors have than also to
be considered by stadging the economics of recycling plants.

The topic of strength gave rise to further informative discussion between several participants, especially Hansen who has a few
results, not really statistical, that seems to indicate that standard
deviation of low strength recycled concrete is not much higher than
that of normal concrete of the same strength. Recycled concrete of
average strength should still have an acceptable standard deviation
but high strength recycled concrete should be dealt with carefully
in practical use because of its high variability.

Pomeroy pointed out that the input of a recycling plant will be different because of types of concrete coming from different demolition
sites and so the output, i.e the aggregates will be of a quality with
perhaps rather high variability.
Results of Hansen's tests left him more concerned about variability
in standard deviation then the problem of achieving a certain strength
level. There is already a simular problem which natural aggregates in
Mather's opinion. Available aggregate must be matched to the concrete properties required and he saw no aspect of the problem which
cannot be attacked by existing technology.

Discussion than moved to the influence of the characteristics
of the original concrete on the recycled product and Kreijger mentioned as first possible parameter the strength of the old concrete.
De Pauw had traced this and the effect of the type of cement in 15-
year old concrete and also the cement content (see contribution to
wotkshop 2 - recycling of concrete). But Frondistou-Yannas believed
that most of the original mortar is lost as fines from the final recycled product which is mainly aggregate. This depends on the method
of fragmentation. De Pauw emphasized, especially with explosive demolition, the effect of demolition and fragmentation techniques on size
and shape of aggregates should not be underestimated. Shah agreed
and called for eliminating intermediate processes by judicious demolition methods which would produce appropriate sized fragments.
At present the criterion used is based on convenience in transport
of debris but this attitude is detrimental to recycling.

Hansen remarked he has produced recycled aggregate concrete of
higher strength then the original material. He explained the test
procedure, using rapid hardening cement, curing to equivalent of
80-90 days, crushing to produce new concrete. Frondistou-Yannas
questioned the possibilities of obtaining higher strength because
the recycled aggregate is porous and unless used in surface dry condition it adsorbs water, the w/c ratio falls but if the w/c ratio
is identical, improved strength seems doubtful. Hansen indicated
that higher adsorbtion of recycled aggregate was taken into account
and the concrete produced had the same effective w/c ratio. Higher
strength could be explained by probable elimination of weak aggregate in the jaw crusher. So failure of recycled concrete does not
originate at the "old" aggregate-mortar interface because the "new"
cement mortar has the same or higher strength compared with its

interfacial strengths with the old concrete rubble.
Ishai thought of a more basic explanation:
when breaking old concrete it will fragment either along lines of
weak interfacial bonding or in areas of high residual stresses
Hence such critical areas are eliminated in the fragmented rubble,
the average strength of which is consequently higher compared with
the source material. This is similar to the statistical effect of
decreasing the probability of the occurence of critical flaws by
breaking a glass fiber into shorter pieces: In Mill's opinion the
Weibull theory of fracture concerning the statistical distribution
of flaws could be relevant since with crushing the distribution is
modified. Also the fact of using new sand might imply that the weak
link was eliminated.

A more complete understanding of energy dissipation in fragmen-
taion was called for by Hilsdorf who already in his lecture indicated
that very little effective fracture energy is needed. German experts
give only 1% efficiency ratio of fragmentation, the remainder goes
into elastic strain energy transformed into heat at very local
points as was e.g. found by Schönert (referred to by Kreijger in
1.1 p. 16)

Tabor turned the discussion towards surface treatment of aggregates and its influence on bond strenght. Ziegeldorf reported on
research which has not yet lead to a final solution confirming that
bond strength can be related to concrete strength. Cracks develop
in matrix or mortar even in sealed specimens. Propogation is slower
with low w/c ratio and faster with high alumina content of cement.
Without cracks the bond strength between mortar and aggregate is
about the same order of magnitude as the mortar strength. Shah considered that cracks in the matrix are indicated by bond cracks.
Polymer can be used to improve bond strength stated Pomeroy but the
treatment is expensive. Tabor asked what the main reason is for the
improved strength achieved by the introduction of polymer inter-
layers-increased adhesion or improved ductility which prevents cracks
spreading. Pomeroy confirmed improved flextural strength and Ishai
was convinced of the improved ductility through including a source
of plastic flow at the interface in an otherwise brittle system.
It is known that even a small percentage of ductile filler may lead
to a significant improvement in the fracture toughness of a brittle
material system such as concrete.Mather ended this discussion by
saying that the same advantage is offered by using natural aggregate
of lower modules of elasticity.

In connection with demolition of prestressed concrete elements,
Mather noted that a series of post-tensioned prestressed concrete
test beams had been demolished at his laboratory to permit examination of the metallic elements for corrosion. The results of his
work are described in a report by E. F. O'Neil which is cited in
the paper by Mr. Schupack, entitled "Behaviour of 20 Post-Tensioned

Test Beams Subject to up to 2200 Cycles of Freezing and Thawing in the Tidal Zone at Treat Island, Maine." (pages 133-156 in "Performance of Concrete in Marine Environment," American Concrete Institute SP-65, 1980.)

PART 3 - WORKSHOPS

1. Workshop 1 - Fragmentation of plain and reinforced concrete
2. Workshop 4 - Fragmentation of prestressed concrete
3. Workshop 5 - Contamination effects on fragmentation, fibre- and polymer concrete
4. Workshop 8 - Fragmentation of all types of concrete (research needs)
5. Workshop 2 - Recycling of concrete (aggregates for use in concrete)
6. Workshop 3 - Re-use of concrete (other than as aggregates for concrete)
7. Workshop 7 - Contamination effects on recycling and reuse
8. Workshop 6 - Future demolition-friendly materials

1. WORKSHOP 1

Fragmentation of plain and reinforced concrete

Chairman: J.J. Mills,   secretary M. Geudelin

Attendance: R. Brepson, P. Cormon, G. Frohnsdorff, M. Geudelin,
            H.K. Hilsdorf, P. Lindsell, J.J. Mills, S. Mindess,
            C. Molin, U. Neck,A.T.F.Neerhoff, D.M. Roy, S.P.Shah,
            D. Tabor, J.W. Weber, S. Ziegeldorf.

1.1.   Minutes of workshop 1 - M. Geudelin

1.2.   Contributions:

1.2.1.   Demolition techniques for concrete structures -
            P. Lindsell

1.2.2.   Observed energy-dissipative features of crack pro-
            pogation in mortar - S. Mindess, S. Diamond

1.1 MINUTES OF WORKSHOP 1

M. Geudelin

Direction de la Recherche, UTI, Paris

The session opened with a presentation by Mindess of the fracture path from micrographs to study energy dissipative mechanisms on mortar test pieces similar to ASTM notched fracture specimen (water-cured 3 days, air-cured 46 days,polished) under the scanning electron microscope. Under load, new cracks followed existing shrinkage cracks, preferably around aggregates: satellite or secondary cracks contributed to form a discontinuous crack pattern on the surface similar to those in brittle ceramics. The fracture surfaces were far from smooth. It was thought that the tortuous crack pattern would be similar in dry concrete and since the micrograph loading was almost only tensile, under crushing certainly cracks would form in every direction.

Questioned if the fracture path followed the aggregate-cement interface, Mindess considered that it lay slightly in the cement-paste. Ending his presentation he emphasized the need to bridge the gap between fracture mechanisms and the practical problem of crushing concrete, also study of a strain criterion as well as energy criterion.

The discussion then turned to the macroscale problem especially as regards high crack velocity as necessary for concrete demolition, pointing to the research topic of measuring fracture energy as compared to microscale in the laboratory. As a designer Lindsell supported this approach, underlining the interest of experimenting with a range of geometrics to find energies necessary for different cracks patterns set up by drilling, bursters etc. When compressive force is used (ball and chain) the energy is dissipated or lost in elastic rebound before it is transformed into tensile force. The problem is made more complex by plastic deformation, friction,

oisco-elastic damping and multiple cracking creating surfaces but
only locally. When applying tensile forces (Cardox, notches holes)
other mechanisms are involved such as the kinetic energy in flying
debris.

Experiments have already shown the discrepancy between the surface energy measured in small samples compared with the energy
necessary to fracture large concrete members. The question of where
the energy dissipates entailing low efficiency in crushing seemed
difficult to answer without the complete background and Hilsdorf
amongst others recommended that experts in the field of tunnelling,
rock dressing etc. be consulted. Tabor seconded this, convinced that
fundamental studies in rock mechanics could provide academic knowledge and challenging problems to engineering departments (use of
cavitation, hydrostatic pressure etc.) Meanwhile there remained the
problem of practical methods to optimize breaking in terms of existing knowledge and the economic appraisal of the part played by the
energy of fragmentation in recycling concrete. Frohnsdorff also
stressed the need to examine the quality of the aggregate obtained.

Having agreed that the energy consumed for demolition on site
seemed of relatively minor concern, compared with other aspects
(noise, dust, safety etc.), participants stressed the problems of
fragmenting large debris, to be studied through the factors which
influence propagation of cracks in concrete (Shah), the optimization
of fracture to obtain the most valuable final product (Frohnsdorff)
and the energy balance involved in each method of demolition
(Ziegeldorf)

Shah placed the first priority on identifying and establishing
a flow diagram (table 1) for each method of breaking including demolition and fragmentation, thus defining the areas where knowledge
was available or needed. The modes of deformation and fracture should
be assessed, using stress analysis of the application of forces.
The relevant material properties could then be defined (for example
dynamic compressive strength useful for ball and chain demolition,
presence of corrosion, thermal conductivity if heat is to be used,
etc.). These preliminary steps would make it possible to undertake
an energy balance analysis and as a final goal improve the quality
of the processed product, whilst gaining full knowledge of material
properties essential for effective reuse

Table 1 - Flow diagram   regarding methods of breaking

1. Modes of deformation and fracture

2. Stress analysis of application of forces

3. Material <u>properties</u> to be known: static, dynamic

        3.1 environmental, corrosion, chemical
        3.2 fracture, surface energy
        3.3 plastic deformation
        3.4 thermal properties, conductivity, ultrasonic

4. Energy balance of method of breaking

5. Improvement of processed product

6. Final materials properties, optimization of end-product

1.2.1 DEMOLITION TECHNIQUES FOR CONCRETE STRUCTURES

Peter Lindsell

Department of Civil Engineering
University of Surrey
Guildford, Surrey, GU2 5XH

SYNOPSIS

Conventional methods for demolition are discussed and their
performance compared in terms of their efficiency, safety and
general levels of noise, vibration and dust produced. More recent
techniques for the cutting and fragmentation of concrete structures
and pavements are reviewed, and the relative advantages and
limitations for their use in the demolition industry are examined.
It is concluded that more research into the use of explosives and
bursting techniques for partial demolition of both reinforced and
prestressed concrete members could lead to considerable savings in
time and reductions in demolition costs.

INTRODUCTION

Traditional methods of demolition rely upon some form of impact
to break up a structure and have been used extensively on brick and
masonary buildings. A major disadvantage of any impact tool is the
high level of noise produced, while the quantities of dust and flying
debris are a direct threat to the health and safety of the
operatives. In Britain, the methods of working are controlled to
some extent by the Factories Act[1] and other legislation[2,3] in order
to protect both the demolition workers and the pollution of the
environment. There are also restrictions imposed by Codes of
Practice[4,5] to control the levels of noise, vibration and pollution
produced on demolition sites.

Conventional techniques for demolition are sometimes unsuitable
for reinforced and prestressed concrete structures. The presence of

the reinforcing steel makes the demolition operation more difficult
and often more hazardous for the demolition operatives. Consequently,
methods of demolition have been introduced in recent years in order
to speed up the demolition process and they are aimed at reducing
costs, providing better safety and minimum disturbance to the public.
The inherent hidden dangers in the demolition process are reflected
by the very high insurance rates. Third party cover in the
demolition industry is not less than £100,000, but most of the
large contractors carry insurance cover of £1 million for any one
accident.

TRADITIONAL METHODS

     The efficiency of a demolition technique may be examined in
terms of the energy input, the manpower required and the time taken
to complete the operation. Concrete is relatively weak in tension,
so that any method that induces tensile stresses directly into the
material will, in theory, be more efficient. Unfortunately for the
demolition contractor, the majority of concrete structures contain
reinforcing steel which has been specifically designed to compensate for the low tensile strength of the concrete. Therefore,
although some conventional methods can induce tensile cracks quite
effectively, a separate operation is often required in order to cut
or burn through the reinforcement. Severing the reinforcement is a
time consuming process, which is frequently performed by hand while
the structural element has already been reduced to an unstable state
by the fracturing procedure.

Hand Demolition

     Demolition of high-rise buildings is frequently carried out
using hand-held tools, such as pneumatic concrete breakers, in order
to reduce the structure to third or fourth storey level. Heavy
equipment may then be introduced to demolish the remainder of the
structure if the safety of the public is not at risk. Partial
demolition is also carried out by hand in order to maintain good
control and to minimize damage to adjacent parts of a structure.

     The overall noise levels produced by a compressor and a
pneumatic breaker can be reduced by fitting a silencer to the breaker
and using a sound deadened tool. However the noise, vibration and
dust produced by the process are particularly unpleasant for the
operators who may work for 8 to 10 hours a day with this equipment.
Concrete used in reinforced and prestressed structures has a high
compressive strength, so that progress in the demolition of such
structures is relatively slow and expensive.

     Safety is a prime consideration for the operatives, but there

is a frequent temptation in practice to avoid the use of temporary
working platforms and scaffolding. There is also the practical
problem of removing large quantities of broken concrete and
reinforcement from upper storeys to ground level. A purpose made
chute or skips may be used for this purpose, but it is also possible
to cut structural members into convenient sections to be removed by
lifting gear or by a crane.

Wire Rope Pulling

     This technique is commonly used on low-rise buildings or
structures that have been reduced to a low level by other methods.
A steel wire rope is used with a circumference of at least 38 mm
and sections of a building are simply pulled down using a well-
anchored winch or a tracked site excavator.

     The forces developed in a wire rope can be very high and the
tension in the rope should be increased in a gradual manner.
Demolition workers are kept well clear during the pulling operation
to avoid injuries from flying debris and rope breakage. A horizontal
distance at least twice the height of the section being removed is
normally allowed as a safety margin during collapse.

     Where a structure is being pulled down in sections, there is
always a chance that the previous pulling operations have reduced
the stability of the remainder. This possibility can be easily
overlooked and it is necessary to understand the structural interaction between adjacent sections before this approach is adopted.

Demolition Ball

     This method is commonly used on isolated structures less than
30 m in height. A large steel ball is attached to a crane and is
dropped vertically onto roof and floor slabs. Alternatively, the ball
may be swung against the side of a wall or framed building using two
methods to impart momentum to the ball. A slewing movement of the
crane's jib can be used or the ball can be swung in line with the jib
by first pulling it backwards with a tag-line.

     To operate this method in a safe manner, the crane has to be
sturdy and the dragline excavator type is often employed. In
addition, the jib head should be at least 3 m above the section being
demolished and the angle of the jib to the horizontal is limited to
a maximum of 60°.

     Where structures are several storeys high, there is a strong
possibility that falling debris within the perimeter of the building
can exert large lateral pressures on the lower walls and cause

premature collapse. In reinforced concrete structures, some of the
reinforcement will inevitably remain intact so that removal of the
debris often requires cutting by an oxy-acetylene torch. It is at
this stage that the structure is in a potentially hazardous state
and presents the most danger to the site workers.

Reinforced and prestressed concrete structures are by nature
relatively resilient to impact forces. Consequently, much of the
initial energy produced by the demolition ball is lost in elastic
rebound. Although this technique appears to be inefficient it is
economically viable in many situations. The method is more commonly
rejected because of restrictions on ground vibration, noise and
possible damage to adjacent property.

Explosives

The large scale demolition of a major structure can be effectively
carried out by explosives, but it must be successful. Occasionally,
a half-demolished structure is obtained in an extremely dangerous
condition which makes completion of the operation very difficult.

The potential energy of tall structures is several orders of
magnitude greater than the explosive charge necessary to cause
collapse. The explosive expert attempts to exploit this energy to
maximum advantage using the minimum amount of explosive charge. His
main objective is to obtain the best fragmentation of the structure,
but at the same time it is often vital to restrict ground vibrations
to a minimum.

Many types of explosives are available ranging from high energy
explosives to slow burning powders. In addition, shaped charges are
produced for specific purposes such as ring charges to cut directly
through reinforced concrete members. Explosive charges are fired by
electric detonators, either simultaneously or in sequence using delay
detonators. Correctly charged and well tamped holes are essential
in order to minimize the effects of air-blast, ground vibration and
noise.

The successful use of explosives on any structure depends largely
on experience and trial charges. Frequently a structure is weakened
initially by experimental charges that are needed to find the best
positions and size of charge required. Where the position of a
structure is relatively isolated, the cost and speed of demolition by
explosives is far more attractive than other methods. However, it
should be borne in mind that, in the case of large structures, the
mountain of concrete rubble and tangle of reinforcement obtained in
a few seconds may take up to 6 months to clear away from the site.

MODERN METHODS

There are a number of techniques that have been specifically developed to provide rapid cutting and fragmentation of structural concrete. These methods are relatively efficient in terms of manpower, time, cost and energy consumed in the demolition process. However, each method has a limited field of application and there is a wide range in the initial capital cost of the equipment.

Thermic Boring

A thermic lance is used primarily to burn holes in both reinforced[6] and prestressed concrete members[7]. It can be used to cut reinforced concrete members that are in an unsafe condition, as the operators can stand away from the member and there is no vibration produced by the process. Various sizes of hole may be produced according to the diameter of the lance, and burnt holes have been used to allow insertion of explosive charges, cardox shells and hydraulic bursters into concrete structures.

The lance consists of a steel tube which is packed with small diameter steel rods. Oxygen is passed from a cylinder through a pressure regulator and flexible hose into the lance, and the open end is ignited by heating the tip with an oxy-acetylene torch. Combustion of the lance at the tip occurs at a temperature of 3500°C and this is sufficient to burn through any concrete, steel reinforcement or prestressing cables. The lances are threaded at each end so that new sections may be added as required.

The burning process produces molten slag which limits the depth of cut if it cannot escape. Therefore, for holes cut vertically downwards the depth is no more than 1 m, but for horizontal holes it is possible to cut holes up to 2 m in length without impairing efficiency. If holes are required vertically upwards the slag will fall back on the operator, but this can be overcome by bending the lance. The burning process is rapid and it is possible to bore a 40 mm diameter hole to a depth of 300 mm in only one minute. A horizontal cutting operation is being demonstrated on a mass concrete block in Fig. 1.

The equipment is simple, easy to operate and can be made highly mobile with an appropriate size of oxygen cylinder. The capital cost of the equipment is negligible compared to pneumatic drills, but full asbestos clothing is required for the operators and it may be necessary to provide breathing apparatus in confined conditions. It may even be necessary to provide extractor fans to remove fumes, and there is always a risk of fire from the molten slag which can fly back in large showers and cover a wide area.

Figure 1.   Thermic boring in a mass concrete block

Figure 2. Mass concrete block split by a hydraulic burster

Hydraulic Burster

Hydraulic bursting is used extensively to split mass concrete and large areas of reinforced concrete walls and slabs into convenient sizes. Fig. 2 shows a concrete block that has been split by a burster, which has several rows of pistons that expand perpendicularly to the axis of the cylinder. This type of burster is used for heavy concrete breaking, and there is a wide choice of burster available with splitting forces ranging typically from 800 kN. A second type of burster is made which operates on the wedge-and-feather principle. It consists of a hydraulic cylinder with a piston rod having a wedge shaped end arranged to expand two steel feathers. This type is employed where finer control is required in the splitting process.

Pre-drilling of holes is required for both types of burster, the diameter and spacing of the holes depending upon the size of cylinder in use and the density of reinforcement in the section. The drilling is often carried out by a compressed air powered drill, which is relatively quick but creates considerable noise and dust. A pneumatic powered drill can produce 50 mm diameter holes, 400 mm deep at 1 m centres, at a rate of approximately 300 mm/min.

A hydraulic burster can be handpumped but this is a slow and laborious operation. Electrically driven pumps and air/hydro pumps working from an air compresser are much quicker, although more noise is generated. The process of breaking concrete with a burster produces no noise or vibration. There is some sound arising from the concrete as it splits, but is is mainly the associated equipment and drilling that creates a disturbance.

The initial capital cost of hydraulic bursters and the supporting equipment is relatively high, although it varies according to the size and type. The equipment is portable, and bursting can be a far quicker method of fragmentation for large areas of concrete than conventional pneumatic breakers. Therefore, in spite of the high initial costs, the speed of operation and the small labour content involved make this type of demolition extremely attractive for particular applications. In general, the movement produced in the splitting operation is insufficient to break the reinforcement. However, the steel can be quickly cut with a small diameter thermic lance or an oxy-acetylene torch if it is easily accessible.

Cardox

The cardox process is a method based upon the sudden release of high pressure carbon dioxide gas from an indestructible shell of high grade alloy steel. The charge is set off by the electrical firing of a non-explosive chemical mixture, which thermally increases the gas pressure until it ruptures a steel diaphragm. This allows the carbon dioxide to escape at a pressure of approximately 20 N/mm^2.

A force is developed in the form of a heaving thrust rather than a violent explosion, and it is accompanied by a sudden drop in temperature. Therefore, the method does not cause any shattering effect or shock vibration which could damage nearby buildings. It is also particularly suitable for demolition work in dangerous situations such as petrochemical plants, where there is a high risk of an explosion.

Holes must be pre-drilled in concrete members by conventional methods, the charged shell inserted and weights placed over the top to prevent it being expelled. The initial cost of a Cardox shell is very small and it can be re-charged on site as many times as required. Consequently, both the man-power requirements and the operating costs are minimal. There is considerable scope for the use of this technique in built-up areas where the use of normal explosives is prohibited, and it is necessary to create minimum ground vibration and noise. The technique is subject to a similar disadvantage as in the case of hydraulic bursters, since the splitting operation is unlikely to break the reinforcement and this will need cutting separately.

Diamond Saw

Diamond sawing is often used where precision cutting is required in a concrete member during partial demolition or structural alterations. Complete panels can be removed in wall or floor slabs with little dust or vibration. There is some small hand-held equipment available, but it is safer and more accurate to mount the saw on a track which is in turn rigidly bolted onto the surface being cut. The rig holding the saw has to be securely braced in order to counteract the high forces exerted at the cutting face.

The conventional circular saw consists of a metal blade with a cutting edge containing industrial diamonds. The diamonds are surface set or bonded into a matrix, and are very effective in cutting even the hardest aggregates and high tensile reinforcement. The matrix is designed to wear away during the cutting operation, so that fresh diamonds are constantly being exposed. The efficiency and cost of cutting largely depends upon the hardness of the aggregate.

The effective depth of cut obtained from a circular saw is little more than one third of the blade diameter. Consequently a massive blade diameter of 900 mm is needed to cut a groove only 375 mm deep and the power input required is substantial. An alternative type of saw is illustrated in Fig. 3, where the concrete slab is being cut with a vertical reciprocating blade in a similar manner to a hand-

Figure 3. Diamond saw cutting through a concrete slab.

Figure 4.   Diamond core drill cutting a concrete wall panel.

held sawing movement.

Diamond sawing is a slow process and the blade is always run
with a supply of cooling water, which also serves to lubricate the
blade, remove the debris and reduce the dust. The cooling water
presents a spillage problem since the demand for water can be as high
as 1500 litres per hour. The initial capital cost of a diamond saw,
the ancillary power unit and supporting rig is very high. The saws
are susceptible to breakage so that the operating costs can be
substantial.

Diamond Core Drill

Diamond core drilling is frequently employed in a similar
situation to diamond sawing. Panels can be removed from reinforced
concrete walls and floors by cutting a continuous line of holes, but
the resulting edge needs trimming with conventional tools. The
drilling is carried out with barrel bits constructed of thin steel
tube with a cutting edge of exposed diamonds or a diamond impregnated
matrix. A typical example is included in the foreground of Fig. 3.
The annular cutting surface is perforated, so that cooling water and
debris can easily escape between the slots. The consumption of
cooling water is far less than for diamond sawing, and is typically
in the range from 50 to 70 litres per hour.

There are a wide variety of tools available with maximum drilling
depths reaching several metres. The largest core drills available go
up to 1 m in diameter and require a massive hydraulic motor to power
the drilling operation. Electrically driven rigs are usually used
for the small power range, but pneumatic motors are widely employed
for the mid-range rigs because of their power-to-weight advantage. A
pneumatic motor requires a large volume of compressed air, and this
requires a high capacity mobile compressor within 30 m of the drilling
operation to avoid large pressure drops in the air supply.

A diamond drill can be operated in any position between horizontal
and vertical. However, the drilling rig has to be securely clamped to
the surface in order to generate the high contact pressure at the
drilling face. Providing the rig is rigidly braced, it is possible
to produce deep holes with great accuracy and this facility is vital
for some types of remedial work. The cutting of a series of holes in
a concrete slab is illustrated in Fig. 4, where the drill is clamped
in a horizontal position.

Concrete Nibbler

This name is given to a concrete breaking device that is
operated by a normal hydraulic excavator. The jaws of the tool are

designed to grip the free edges of a concrete slab and snap off
sections by applying a bending moment. The "Nibbler" was developed
at the Building Research Station[8] in order to create a quieter
technique for the demolition of concrete road slabs. It can also
break up low rise concrete walls, foundation slabs and runways.

The Nibbler has a high work rate, and the overall noise level
is no greater than that of the excavator itself since the breaker
does not impart any impact to the concrete. The absence of vibration
during the fragmentation of a concrete road slab reduces the risk of
damage to the underground services. Mesh-reinforced concrete slabs
up to 375 mm thick can be broken up by this machine, and the
shattered debris is readily picked up and transported from site in
a normal haulage lorry.

Special Techniques

A number of demolition methods have been developed for specific
applications, where high speed, accuracy or noise levels are of prime
importance in the cutting process. Some of the techniques are still
in the development and research stage, and include methods such as
water jetting, powder cutting, microwaves, plasma jets, laser beams,
electron beam drilling, electro thermic-lance, oxy/kerosene rocket
jet burner and expansive cements.

A very fine water jet at an extremely high pressure is used to
cut holes and slots in concrete members. The volume of water
consumed is relatively low, but the water pumps have to be highly
specialised in order to develop the necessary pressure. A major
disadvantage is that the power output is extremely low compared to
the power input required to generate the very high pressures.
Consequently, the water jet is far less efficient compared to a
conventional pneumatic breaker.

Water jets do not cut the reinforcement or the aggregate, but
produce the cutting action by eroding the softer cement matrix.
Trials have been carried out with high frequency pulsed jets to
improve the cutting performance and this system uses even less water.

Powder cutting of concrete is a burning technique which
elevates the temperature of an oxy-acetylene flame by introducing
very fine iron powder. The resulting flame temperature is 2800°C,
so that it operates in a similar manner to a thermic lance but can
be used in more restricted spaces. Although the powder cutting
process is less expensive in materials consumed than the conventional
thermic lance, it is slower and more difficult to operate in damp
conditions where the flow of powder becomes a problem.

When sufficient power is transmitted into a material containing water, the water will absorb enough energy to convert to steam. The material itself will also get hot and expand throughout its depth when high radiated power is used. Therefore, the internal forces produced in concrete when irradiated by microwaves can exceed the tensile strength of the material.

A mobile concrete breaker was developed at the Building Research Station[9] using microwaves to split concrete slabs up to 200 mm thick. The method of crack propagation is to heat a spot or series of spots with the microwaves. Each heated spot is filled with sand before the material cools down and this induces further cracking along the line of the heated spots. The process is extremely quiet apart from the noise resulting from the splitting of the concrete and is virtually dust-free. Consequently, it has been used for cutting a concrete slab in a hospital where noise, vibration and dust had to be strictly controlled. High energy radiation is a potential danger to personnel, so that the apparatus had to be designed to prevent the accidental leakage of radiation.

The remaining specialized techniques for cutting and splitting concrete have not yet achieved acceptance in the demolition industry in Britain. There could well be a place for new forms of burning equipment to improve on the conventional thermic lance, and the use of expansive cements could compete favourably with hydraulic bursters for controlled splitting of large concrete slabs in the near future. At the moment, there seems little scope for the introduction of cutting techniques by laser beams or electron beam drilling until the technology has developed further to make these methods competative in practical situations.

CONCLUSIONS

Traditional methods evolved in the demolition industry for use on brick and masonary buildings are now being widely employed on the dismantling of modern structures in reinforced and even prestressed concrete. However, they are not particularly efficient for concrete structures and would be highly dangerous if used indescriminately on some types of prestressed concrete. An exception here is the use of explosives, which have met with notable success and good economy in the demolition of tall concrete buildings. Explosives have not been fully investigated in the area of partial or selective demolition in reinforced concrete and have only been used on a few occasions for prestressed concrete demolition. There appears to be a considerable potential here for savings in both cost and time, if explosives could be more widely employed.

The fragmentation of reinforced concrete can now be approached using a number of efficient splitting techniques such as hydraulic bursters and cardox shells. In the case of concrete road slabs, the

demolition is considerably quieter and more effective with the
Concrete Nibbler. There is a good possibility that bursters and
cardox shells could be used as an aid in prestressed concrete
demolition and this is a field that should be investigated by a
series of trial tests.

Accurate cutting and boring of reinforced concrete sections is
expensive and tedious, but it can be reliably performed by diamond
tipped saws and core drills. The splitting technique using
microwaves has already been developed to a practical stage that has
allowed concrete slabs up to 200 mm thick to be accurately split.
This method is particularly suitable for partial demolition where
vibration, noise and dust are unacceptable.

Further development work and proving trials will be necessary,
before the more recent techniques and ideas for concrete demolition
can become widely established in demolition practice.

REFERENCES

1. Factories Act 1961 Part IV, "Protection of eyes (Section 65),
        and removal of dust and fumes (Section 63)", H.M.S.O.,
        London (1961).
2. Health and Safety at Work Act 1974, Chapter 37, H.M.S.O.,
        London (1974).
3. The control of Pollution Act 1974, Chapter 40, H.M.S.O.,
        London (1974).
4. British Standards Institution B.S.5228, "Code of Practice for
        noise control on construction and demolition sites", B.S.I.
        London (1975).
5. Department of Employment, "Code of Practice for reducing the
        exposure of employed persons to noise", H.M.S.O., London (1972).
6. A.A.B. Musannif, "Thermic Boring", Building Research Establishment,
        Current Paper CP 58/75, Watford (1975).
7. P. Lindsell, "Demolition of post-tensioned concrete", Concrete,
        Vol. 9, No.1 (1975).
8. A.A.B. Musannif, "The Nibbler – a new concept in concrete
        breaking", Building Research Establishment, Current Paper
        CP83/74, Watford (1974).
9. J.L. Smith, "Removing concrete-cutting by microwaves", Concrete
        Society Symposium, "Advances in Concrete", Birmingham (1971).

1.2.2 OBSERVED ENERGY-DISSIPATIVE FEATURES OF CRACK PROPAGATION

IN MORTAR

Sidney Mindess   Sidney Diamond

Dept. of Civil Engineering   School of Civil Engineering
University of British Columbia Purdue University
Vancouver, B.C.   West Lafayette, Indiana

INTRODUCTION

The propagation of cracks in hardened cement paste, mortar and concrete has been studied extensively for more than fifty years, both theoretically and experimentally. However, details of the nature of the origin and development of cracks in these systems are still uncertain. In hardened cements and concretes, crack development and propagation at stresses far below those required to induce structural failure are common. However, attempts at applications of linear elastic fracture mechanics and its extensions to cement and concrete have not been very successful. This may, in part, be due to the fact that there have been only limited.observations on the way in which small cracks are formed and then propagate in these materials. The present work is aimed at providing more detailed information on the microstructural characteristics of crack processes in cementitious systems, through use of a device which permits the testing of cement and mortar specimens directly in the specimen chamber of a scanning electron microscope (SEM). The objective is to provide information about crack processes on a fine enough level so that the observations can be incorporated into the assumptions and simplifications that are necessary for the application of fracture mechanics to cementitious systems.

DESCRIPTION OF LOADING DEVICE AND SPECIMEN GEOMETRY

Both the specimen and loading device have been described elsewhere[1,2]. Briefly, the specimen configuration is that of a compact tension specimen, 31.8 mm long, 24,0 mm wide, and 12.7 mm thick, as shown in Fig. 1(a). A 13 mm notch, about 0.6 mm wide, was cast into the specimen, and 3 mm steel posts were cast into the specimen

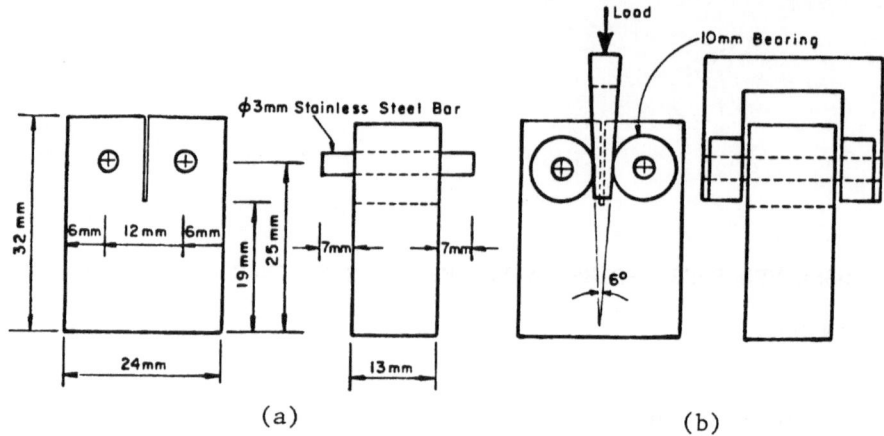

Fig. 1.   (a) Compact tension specimen;
          (b) Method of loading

on either side of the notch.   Steel bearings placed on the posts were
used to prevent substantial frictional forces when the specimens were
wedge-loaded as in Fig. 1(b).   The test frame was specially designed
to be accommodated within the specimen chamber of an ISI Super III-A
SEM; its overall dimensions were 83 mm x 54 mm x 41 mm.

LOADING HISTORY

        The specimen described in this study was a w/c=0.4 mortar,
made with quartz sand passing the 300 μm sieve and retained on the
150 μm sieve, with a sand:cement weight ratio of 1.09.   The specimen
was cured for 3 days in water and 46 days in air before testing.   It
was then polished with a fine grit to reveal the sand grains.   A very
light conductive coating was applied before loading.

        The assemblage and specimen were placed in the SEM, and the
tip of the preformed notch was located.   The specimen was then loaded
until visible cracking occurred, in this instance at a load of 43.6N.
The motor was immediately turned off, and the crack pattern was
"mapped" by taking sequential micrographs along the crack path.   The
specimen was then removed from the SEM, a heavier conductive coating
was applied, and the specimen was replaced in the SEM and examined at
higher magnifications to study finer microstructural details.

OBSERVATIONS

        The extent and pattern of cracking that occurred on loading
(i.e. the crack "map" referred to above) have been described in
detail elsewhere[3].   It was noted that, while the main crack proceeds

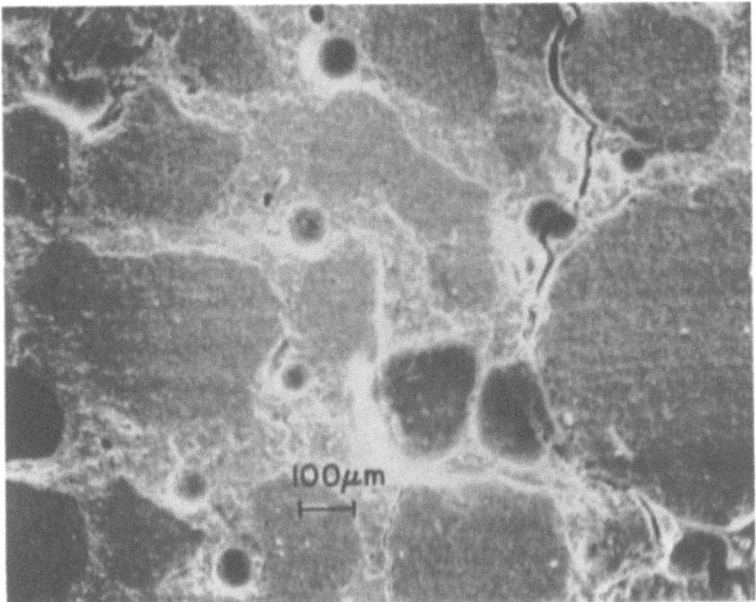

Fig. 2.   Typical crack propagation along the
sand-hardened cement interfaces.

generally in a forward direction, there is considerable tortuosity in
the crack path.   In addition, a considerable amount of crack branching occurs.   The micrographs show also that the cracks (at least on
the surface) appear to propagate preferentially along the interfaces
between the sand grains and the hardened cement paste, as shown in
Fig. 2.   As has been pointed out by many investigators[4], the interfacial regional appears to be the "weak link" in mortars and concretes.

     The crack appeared to terminate in the large surface void
shown in Fig. 3.   Beyond that point, a definite crack trace could
not be identified, even after a careful search carried out by
traversing back and forth across the specimen.   The crack simply
seemed to blend into the other surface discontinuities of the specimen.   The zone of disturbance above the void in Fig. 3 did not appear
to be a continuation of the crack, being no different from similar
surface discontinuities present in other (uncracked) areas of the
specimen.

     On observing the specimen at higher magnifications, a number of
other features were revealed.   These indicate a variety of energy
dissipative processes that can take place when a crack propagates in
mortar.   Fig. 4 shows a shattered sand grain, indicating that the
crack does not always grow around sand grains, even in this relatively weak material.   A considerable amount of crack branching is

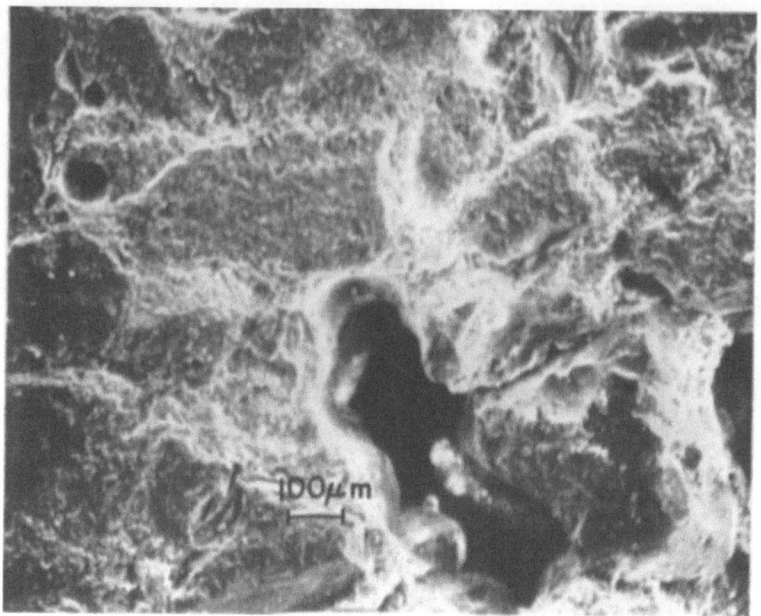

Fig. 3.   Apparent end of crack formed on loading.

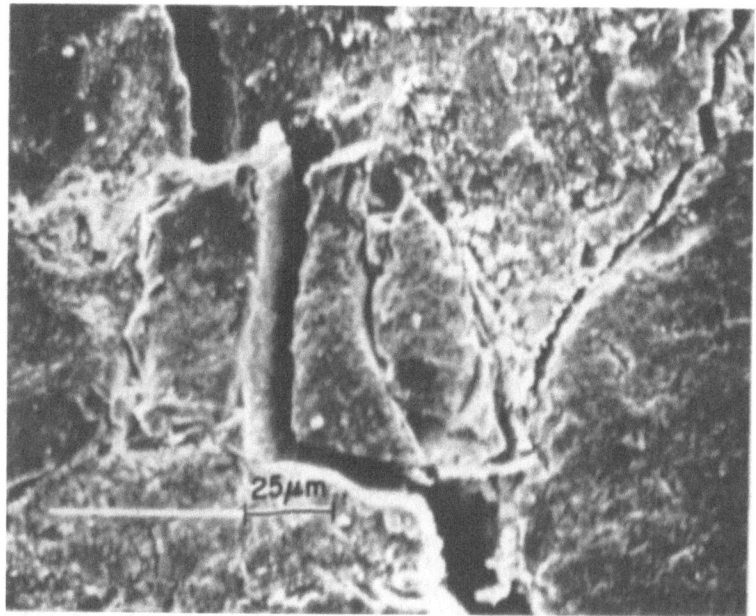

Fig. 4.   A shattered sand grain.
           Note the amount of crack branching.

also shown in this microphotograph.

As stated earlier, the cracks generally propagate along cement-sand interfaces. However, even before loading, as shown in Fig. 5, (taken on a companion specimen), some incipient cracks at the interface appear, presumably due to drying shrinkage. Fig. 6. shows clearly the typical debonding that takes place on loading. It also shows that separate cracking may occur around two adjacent aggregate particles. Fig. 7 reveals the very clean separation that takes place at the interface.

Fig. 8 shows the crack pattern around a large surface void; the multiple cracking in this region is very extensive. Finally, Fig. 9 reveals the presence of discontinuous cracking. That is, the trace of the crack that appears on the surface is not always continuous. Instead, the crack appears to terminate and an adjacent crack, parallel to the original crack but not obviously connected to it, is seen to have developed, and proceeds in the forward direction. This phenomenon is not yet understood, but it may be related to the 3-dimensional nature of the system, and the complicated stress distribution near the crack tip as the crack propagates.

In general, it should be stated that the crack described is perfectly stable. Upon loading, the crack appeared and extended for

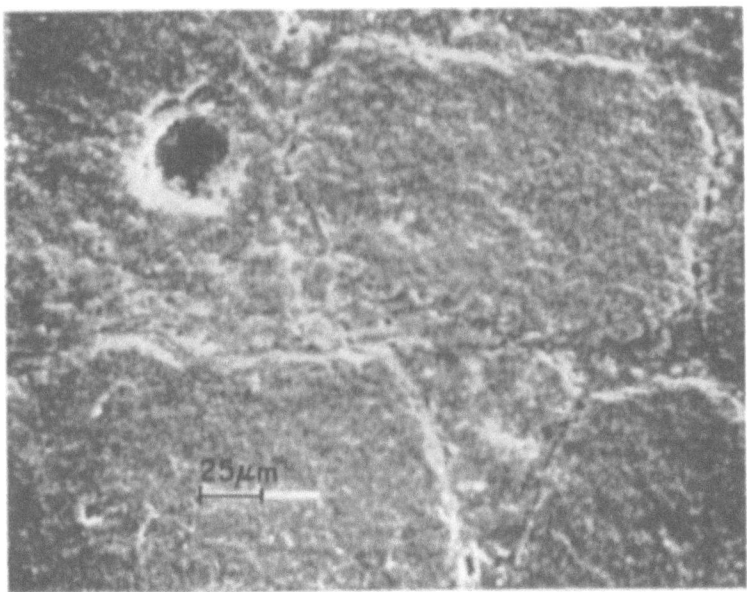

Fig. 5. Bond cracks prior to loading on
companion specimen, probably due
to drying shrinkage.

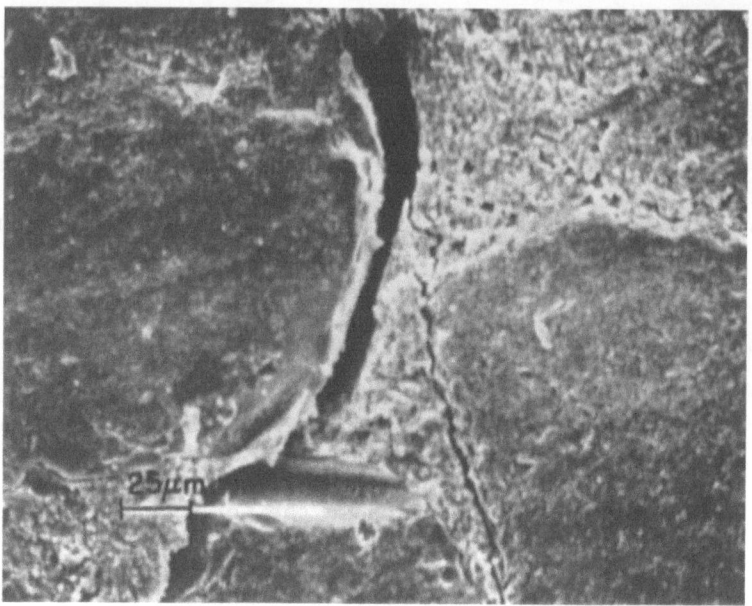

Fig. 6.   Bond cracks around adjacent aggregate
          particles.

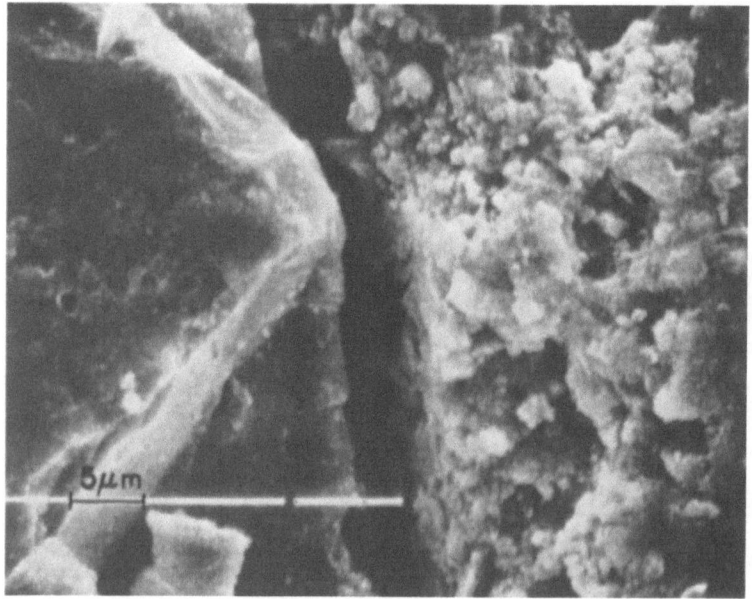

Fig. 7.   Bond crack, showing clean separation
          at the interface.

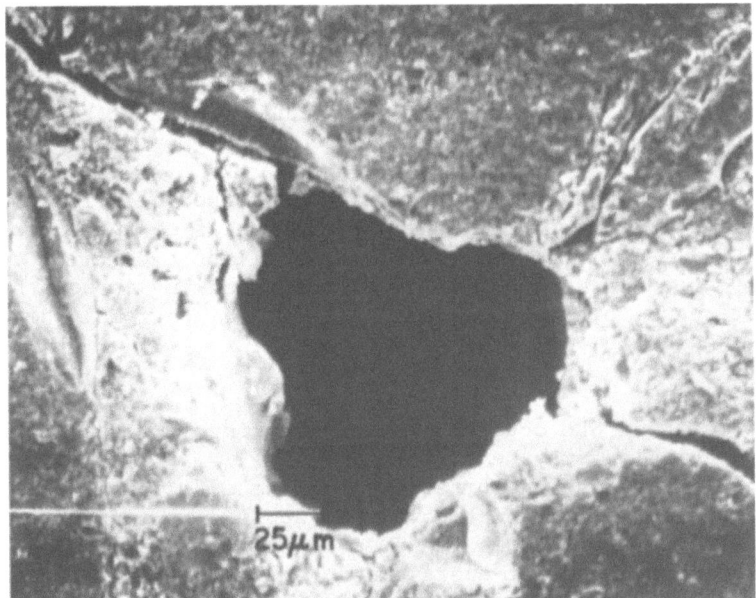

Fig. 8. Crack pattern around a surface void.

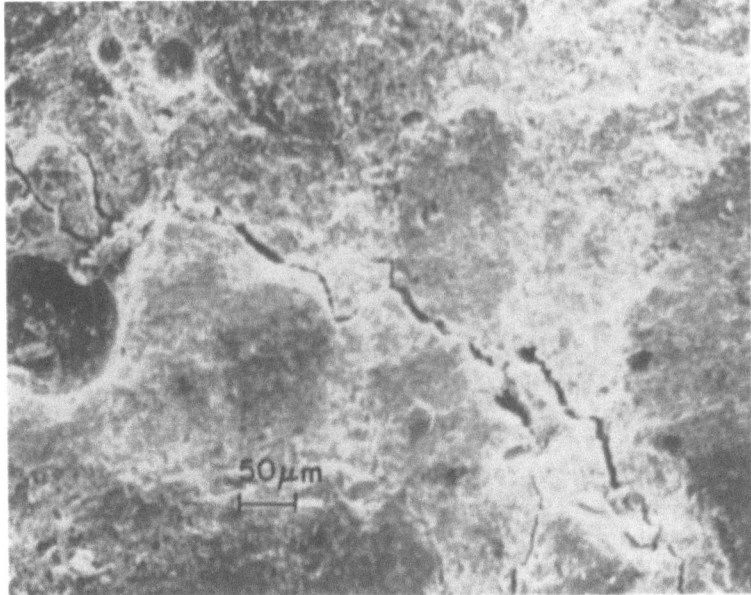

Fig. 9. Discontinuous cracking.

approximately 12 mm and then stopped. It did not grow measurably during the two hours in which the specimen remained in the SEM under load. On initial cracking, the load fell immediately from 43.6N to 17.8N; it decreased slightly during the next two hours, probably due to creep. The specimen could easily be handled without breaking when it was removed from the SEM.

It should be emphasized that using the method described here to study cracking in mortar, only the surface traces of the cracks can be observed. However, subsequent examination of the fracture surfaces that are produced makes clear the fact that cracking in both hardened cement pastes and mortars is a complicated, 3-dimensional phenomenon.

CONCLUSIONS

Based on these SEM observations of cracking in mortar, the following conclusions may be drawn:
1. The cracking process in mortar is complicated. The simple fracture mechanics models which have been applied to mortars greatly oversimplify the geometric features of the crack extension process.
2. A considerable amount of energy must be dissipated in the large amount of branch cracking that is seen to occur, as well as in creating the tortuous crack surfaces. These energy dissipative processes should be taken into account in developing fracture models for the material.
3. Loading cracks develop preferentially along the interfaces between sand and cement, but are not limited to these regions.

ACKNOWLEDGEMENTS

This paper represents a contribution from the Joint Highway Research Project, Purdue University, in co-operation with the Department of Civil Engineering, University of British Columbia. Support was also provided by the National Sciences and Engineering Research Council Canada. We thank Janet Lovell for preparing the specimens, and technical assistance.

REFERENCES

1. S. Diamond and S. Mindess, Scanning electron microscopic observations of cracking in portland cement paste, in: "Proceedings of the Seventh International Congress on the Chemistry of Cements," Vol. III, Paris (1980).
2. S. Mindess and S. Diamond, A preliminary SEM study of crack propagation in mortar, Cem. Concr. Res. 10:509 (1980).
3. S. Mindess and S. Diamond, The cracking and fracture of mortar, in: "Fracture in Concrete", W.F. Chen and E.C. Ting, eds., American Society of Civil Engineers, New York (1980).
4. L. Struble, J. Skalny and S. Mindess, A review of the cement-aggregate bond, Cem. Concr. Res. 10:277 (1980).

2. WORKSHOP 4 - FRAGMENTATION OF PRESTRESSED CONCRETE

Chairman: H.K. Hilsdorf,   secretary M. Geudelin
Attendance: R. Brepson, P. Cormon, C. De Pauw, M. Geudelin,
                H.K. Hilsdorf, O. Ishai, H. Lambotte, P. Lindsell,
                S. Mindess, C. Molin, U. Neck, A.T.F. Neerhoff,
                D. Tabor, J.W. Weber, J.F. Young, S. Ziegeldorf.

2.1   Introductory lecture - Demolition of prestressed
                concrete structures - P. Lindsell

2.2   Minutes of workshop 4 - M. Geudelin

2.3   Contributions

2.3.1   Explosives for localized cutting in concrete -
                M. Molin

2.1 DEMOLITION OF PRESTRESSED CONCRETE STRUCTURES

Peter Lindsell

Department of Civil Engineering
University of Surrey
Guildford, Surrey, GU2 5XH

SYNOPSIS

Controlled dismantling of prestressed concrete structures requires adequate knowledge of the design calculations and construction details. The sophisticated techniques now being employed in prestressed concrete construction have created structures which will require elaborate planning and detailed calculations when the time arrives for demolition. For complex structures, it is suggested that designers simultaneously prepare a "demolition sequence" for the benefit of future generations.

Various techniques may be used for relieving the tension in prestressed concrete structures and the suitability of these methods is discussed with reference to the basic forms of prestressing. Investigation of alternative demolition techniques and systematic recording of case histories are suggested as a means of improving the present lack of general knowledge in this field.

INTRODUCTION

Over the last ten years there has been a sharp increase in the number of prestressed concrete structures requiring demolition. One reason for this is due to the fact that many early prestressed structures have now become obsolete. In addition, alterations are often required to change the use of an existing building and this may demand the removal of a prestressed concrete member. There has also been a number of instances where accidents, corrosion, chemical conversion and structural failure have prompted premature dismantling of a complete section of a prestressed concrete structure.

The condition of the concrete, prestressing tendons and the state of stress in a structure after many years in service, create uncertainties which must be allowed for during demolition. Consequently, the potential problems facing a structural engineer or a specialist contractor concerned with the demolition of pre-stressed concrete can be just as varied as the structures themselves.

Prestressed Concrete Construction

The application of prestressed concrete in all branches of civil and structural engineering has flourished dramatically over the last 30 years. Detailed development has taken place in the design procedures and construction techniques, but there remain only two basic types of prestressed concrete:-

a)   Pretensioned. Mainly used for mass production of precast beam units.

b)   Post-tensioned. Widely employed for both precast and in-situ concrete structures.

The majority of the bridges built in Britain from the mid-1960's contain either pretensioned units for the smaller spans or post-tensioned concrete for long spans and multi-span continuous decks. Applications in high-rise buildings and multi-storey car parks have become common, and prestressed lightweight concrete has achieved popularity in roofs for sports halls and grandstands. In the early years of prestressed concrete, stressing in one direction was the original theme, but now designers are fully accustomed to stressing structures in two or three dimensions.

The variety and type of prestressing systems has kept pace with demands for more applications and larger prestressing forces. Plain high tensile wires have generally been superseded by low relaxation high quality strands and high tensile bars. The prestressed concrete designer is constantly replacing "obsolete systems" with the latest catalogue supplied by the firms specializing in prestressing equipment.

This perpetual change only makes the future tasks facing a demolition contractor even more bewildering. Not only is a condemned concrete structure likely to contain some form of prestressing, but the type of system, strand pattern and standard precast unit may all be long forgotten by future structural engineers.

Construction Details

A demolition contractor cannot be expected to dismantle any

structure without adequate details of the original construction.
This statement is particularly relevant to a prestressed concrete
structure where, apart from the type of stressing used, a precise
knowledge is needed of the location of all prestressing strands and
anchorages. This information may be sufficient for simple structures,
but in cases where incremental stressing was applied during
construction some knowledge of the design calculations is essential
if the demolition is to be adequately controlled.

Structures are seldom built exactly to the design drawings.
Changes occur by mistake or modifications are introduced as
construction progresses. The actual materials used for the concrete,
grout and prestressing steel, and the degree of exposure to the
atmosphere or contaminating substances are all factors that need to
be recorded. Therefore, a complete set of design calculations and
"as-built" record drawings is vital information for the future
generation of demolition contractors.

Where and how this information is to be stored is a matter that
needs planning at the highest level. In Britain The National
Federation of Demolition Contractors[1] has initiated proposals with
these objectives in mind, and have made general recommendations on
the information and precautions needed when demolishing prestressed
concrete.

Many international consulting engineers have designed prestressed concrete structures in foreign countries. One can only
speculate at the difficulties that may be experienced in years to
come when a "local" demolition contractor is invited to demolish a
complex structure that was designed long ago by a foreign design
team and constructed by a consortium from various other countries.
Clearly, an international policy is needed in the immediate future
to avoid the possibility of serious accidents occurring.

DEMOLITION

In theory, the demolition of a prestressed concrete member is a
simple operation. Immediately the cables or stressed bars are cut,
then the corresponding prestress in the concrete is lost and the
member degenerates to a reinforced concrete fragmentation problem.

However, the safe demolition of a prestressed concrete structure
depends upon a sound knowledge of prestressing theory and structural
design principles. Herein lies the main danger, because many
existing demolition contractors have no experience of prestressed
concrete and at present do not employ any personnel qualified to deal
with this form of construction. Consequently, education and training
in this topic is one of the key requirements for the future.

Suitable techniques for prestressed concrete demolition depend primarily upon whether the entire structure is to be removed or whether selective demolition is required for structural alterations. The location of a structure is also of fundamental importance regarding the methods that can be employed. For example, the demolition of a structure adjacent to a hospital requires strict control over dust, noise, fumes and vibration. These restrictions will effectively dictate the techniques that can be used.

Techniques

Structures in congested city centres demand special treatment where demolition of prestressed concrete is concerned. Safety requirements are likely to be stringent and the use of techniques such as controlled explosives or a traditional demolition ball would certainly meet with considerable opposition. More time consuming procedures are then inevitable, and a combination of drilling and cutting of individual prestressing tendons is the alternative procedure.

Simple structures on industrial sites away from residential areas or bridges that can be isolated from the public may be safely demolished with nothing more than the conventional "ball and chain" and burning gear[2]. Heavy structures in this situation may be more effectively disposed of by controlled explosives, but insufficient evidence is available and research in this area is needed to establish the relative merits of alternative methods.

Thermic lances and pneumatic concrete breakers provide a crude but controlled method of demolition for structures open to the atmosphere[3,4]. However, in confined conditions the regulations for noise, ventilation and fire risks are difficult to satisfy. This can be overcome by using percussive/rotary drills in combination with electric burning gear to melt the prestressed tendons. This technique is also ideal for selective demolition work, where strict control is required and adjacent parts of the structure must not be made unserviceable by the demolition process.

Demolition Sequence

Controlled dismantling of any structure requires a carefully planned sequence of operations. In the case of prestressed structures, it is imperative that the workmen are thoroughly briefed on this aspect and constantly supervised by a qualified engineer. Mistakes have already occurred in practice, and they can lead to hazardous conditions and structural instability.

For simple prestressed beams it is often possible to remove the superimposed dead load as a first stage, and then lift out the entire beam unit. Prestressed beams are often relatively thin and special attention should be paid to bracing the section to prevent lateral instability. Final demolition can then be performed with the unit on the ground and full safety precautions can be easily provided.

Some prestressed beams are built into a structure with the superimposed dead load added in stages and the total prestressing force applied in increments. In this situation the demolition sequence must be planned in such a manner that the dead load and pre-stress can both be safely removed, maintaining lateral and vertical stability at all times. Substantial propping and even temporary pre-stressing can be used to control the release of the tension forces.

The more complex structures that contain post-tensioned cables over many spans and stressing in several directions, demand very careful consideration of the demolition procedure. Breaking up such a structure into manageable units will certainly require extensive propping and temporary works. Box girder bridges are a typical example of structures in this category. Detailed examination of the design calculations and original erection procedure will be vital information for a future demolition sequence.

Pretensioned Concrete

Demolition of beam units which have been pretensioned does not present any great difficulties. If the unit is to be cut up into sections in-situ then, provided the beam is fully braced, the inherent bond between the strands and the concrete will ensure adequate safety as the tension is released. Any convenient technique for cutting or sawing through the section may be used here. However, due regard to the stability of the section must be made as the strands are cut. A sequence of strand cutting must be specified and strictly followed by the site operators.

Where a pretensioned beam can be removed from the site in one section, then final fragmentation of the unit can be simply performed with a demolition ball. Again, the excellent bond between the strands and the concrete will prevent any "flying missiles", and the process should be no more explosive than a reinforced concrete beam failure.

Post-tensioned Concrete

The demolition of structures which have been post-tensioned is not difficult, providing the stressing system and structural behaviour at all stages of demolition are clearly understood. Unfortunately, to the workmen who fail to follow the demolition sequence or who cut into cables that are not in the positions originally supposed, then the potential dangers are very real.

Grouted Cables. The use of fully grouted cable ducts provides an efficient bond between the steel and the concrete member. Consequently, cables can be cut with minimum danger to the demolition team. Release of the cable tension has been successfully completed by direct cutting, burning[3,4] and heating-up of the tendons at the anchorages[5]. Tension forces can often be in the range from 3,000 to 10,000 kN per tendon, so that allowance must be made for the "impact effect" that takes place as a force of this magnitude is suddenly released.

In spite of the immense forces in a tendon, cutting of cables in a fully grouted member may be necessary at several positions. The high friction and bond forces can prevent full release of the pre-stress over the entire length of a long section[3,4].

A recent survey by the Building Research Establishment[6] in Britain has revealed that the use of calcium chloride in prestressed concrete buildings was widespread in the late 1950's, in spite of its ban in specifications. In this type of situation, individual tendons may be seriously corroded in isolated areas. Therefore, it is possible that a well-planned demolition programme may be quite unsafe if some of the tendons happen to be weakened in critical positions.

Unbonded Cables. The use of unbonded tendons in flat-slab construction has been widespread in the United States and there is now a growing interest in this technique in Britain. Initially, concern was expressed by many engineers worried about the eventual demolition of these structures and the possible sudden release of large amounts of strain energy. Experience so far has shown that there are no major hazards and several procedures have evolved.

Tendons have been cut in a flat-slab during structural alterations and far from an explosive effect, the actual energy release was scarcely noticed. The mortar plugs covering the end anchorages moved outwards by 10 mm, but apart from this, there were no other visible signs that the tension had been released.

This type of behaviour is not surprising. The force per tendon is relatively small compared to bonded tendons, and a greased tendon inside a close fitting sleeve must experience considerable friction as the tension is relieved. The "shock" produced by the tendon release is in the plane of a structurally continuous concrete slab which has high in-plane strength. It usually has adjacent tendons and ordinary reinforcement, which can easily distribute the potential tensile splitting action across a wide area of slab.

Therefore, providing sufficient concrete cover is available and

precautions are made to enclose the end anchorages, there should be
no danger in releasing the cable tension by cutting or burning.
Once the cable tension has been systematically released throughout
a slab section, then normal techniques for reinforced concrete
fragmentation can be safely employed.

Relief of the cable tension in unbonded flat slabs has also
been performed by progressive removal of the concrete slab in a
series of slices, using traditional pneumatic concrete breakers. The
prestress force gradually decreases as the slab area is removed, but
the associated problems with this technique are likely to be dust,
noise and vibration.

Occasionally, unbonded tendons may be met in other applications.
They can be de-stressed with a hydraulic jack if the tendons have
been left uncropped at the anchorages. Caution is needed since the
previous stressing operations and anchoring at the end blocks
introduces small stress raisers into the cables and corrosion may
aggravate the situation even over a short time period. This may
seem insignificant, but when a cable is in a curved profile the
reverse friction effects and inter-strand friction can be very
significant. Consequently, the applied force required to de-stress
the tendon can be over 90% of the ultimate tensile strength of the
strand[3,4].

Unbonded External Cables. External post-tensioned cables have
been used on numerous structures in Britain. At first sight, the
eventual replacement of old strands or complete demolition would
seem to be relatively simple.

A recent corrosion problem in a four-span continuous bridge deck
has demonstrated that the de-tensioning operation may not be
straightforward. In this case, secondary effects in the transverse
beams linking the longitudinal webs caused the condemned strands to
lock in position.

Detensioning of any cables that have been subject to long-term
corrosion must always present a potential hazard. Snapping of any
cable can be quite unpredictable and careful surveys must be made in
advance. Safety measures along any exposed sections or at end
anchorages must also be stringent.

FUTURE DEMOLITION PROBLEMS

Developments and extensions of the basic prestressing concepts
are continually being introduced throughout the world. These
innovations represent totally new problems for the demolition team
in the future. Two relatively unknown concepts which are gaining in
popularity are discussed here, to illustrate where research and

practical investigations might be needed in the next few years.

Stress-ribbon Concept

Numerous structures have now been constructed using the stress-ribbon approach. The technique has already been applied to off-shore conveyor belt structures, suspended pipe-line bridges, long-span roofs and highway bridges. Rigid horizontal end-anchorage points are necessary for the stability of these structures. The necessary resistance can be obtained by using further prestressing in the form of ground anchors. The tendons in the end anchorages may be synchronised with the stressing of the stress-ribbon in order to maintain horizontal equilibrium during construction.

A continuous stress-ribbon structure is commonly supported at intermediate points by a hinged pier. The pier behaves in an analogous manner to the cable saddle of a suspension bridge. The required tensile force in the stress-ribbon is initially very high, but losses due to creep and shrinkage cause a significant reduction.

Controlled demolition of this type of structure would demand progressive release of the total tension force in the stress-ribbon and the end-anchorages to maintain stability. Provision for propping the deck and support of the piers might also be needed for safety reasons. This type of demolition work is an area that needs monitoring in the future as no examples have yet been published.

Post-compression Concept

The potential advantages of using a combination of post-tensioning and post-compression have yet to be fully realised in practice. However, several years ago the world witnessed the first bridge constructed using this technique in Austria. In Britain, research has been continuing on the practical application of post-compression for the past 20 years. A trial footbridge is to be constructed near London during 1981, and the problems encountered will be closely monitored and compared with conventional prestressed concrete construction.

Demolition of members and structures containing both post-tensioned and post-compressed tendons will be an interesting problem. Tendons under compression and under tension will require de-stressing in strict sequence, and research work in this field could help to identify and solve the difficulties that might be encountered.

CONCLUSIONS

The release of the strain energy stored in the tendons of pre-stressed concrete structures can be carried out by using a variety of existing cutting and burning techniques. To perform these operations with maximum safety for the demolition team and the public does require a carefully planned and closely supervised operation.

The stress state and cable layouts in modern prestressed structures can be complex. Therefore complete sets of design calculations and record drawings should be stored for future generations of structural engineers and demolition contractors to be able to comprehend and anticipate the potential hazards. It would be a major step forward if designers of the more complex prestressed structures simultaneously proposed a safe demolition sequence. This "demolition procedure" could be filed away with the as-built construction drawings and would be particularly useful in the event of premature failure or an accident, when a rapid demolition operation might be essential.

There is little experience available on the demolition of some recent forms of prestressed concrete construction. This type of work is often unpublished for a variety of reasons, and there is a need for systematic recording of case histories by structural engineers specializing in this field. If a brief account of all prestressed concrete demolition was sent to a central organizing body, then a substantial store of useful information could be quickly compiled. This knowledge should then be made freely available to all member countries and organizations interested in the safe control of prestressed concrete demolition.

REFERENCES

1. National Federation of Demolition Contractors, "The Demolition of Prestressed Concrete Structures", Report by N.F.D.C., Leicester (1975).

2. D. A. Andrews, "Durability of Prestressed Concrete with Reference to a Structure Under Demolition", F.I.P. 6th Congress, Prague (1970).

3. P. Lindsell, "Demolition of post-tensioned concrete", Concrete, Vol. 9, No.1 (1975).

4. P. Lindsell, "Demolition of post-tensioned concrete", Concrete Construction, Vol. 21, No.6 (1976).

5. Construction in Southern Africa, "The most exacting demolition ever undertaken", Construction in Southern Africa, (1970)

6. Building Research Establishment, "The Structural Condition of Intergrid Buildings of Prestressed Concrete", H.M.S.O., London (1978).

2.2 MINUTES OF WORKSHOP 4

M. Geudelin

Direction de la Recherche UTI, Paris

Professor Hilsdorf opened the session with an introduction to
the practical problem of demolishing reinforced concrete. Increased
energy will be needed for this because in the initial building the
reinforcement was included to improve tensile strength and ductility
through optimum bond. However bars in concrete to be demolished
could be used in electric and thermal conductivity methods to obtain quicker bond failure, also there remained the possibility of
dismantling whole reinforced components for reuse elsewhere.

A list was drawn up of special demolition problems, including
the use of techniques to locate reinforcement before demolition,
drilling aids, zêtapotential and the recycling of the bars.

American participants stated that there was no recommended demolition practice in the U.S.A., whereas Lindsell was able to outline the British Code of Practice under revision which fails to
distinguish between demolition of plain and reinforced concrete,
mentions only traditional techniques and makes no distinction between
the different types of structures to demolish.

Hilsdorf seized the opportunity of immediately formulating research needs for methods based on material and safety aspects, also
on breaking bonds for example producing split by yielding of bars.

Molin was then called upon to present Swedish research involving
the use of small explosive charges for localized cutting. Explosives
take the form either of a V-shaped strip producing blasting and cutting forces on small cartridges from 5-20 g. These techniques were
illustrated by slides which emphasized the advantages of localized
blasting - controlled cracking, safety, small debris ready for re-

cycling, interest for structural alterations where noise, and dust
must be restricted. Ishai, following Molin's slides showing cracking
within the reinforced concrete slab after application of the "hole-
demolishing" technique, asked for the type of failure, mere debonding or separation of the slab. Molin answered that whatever the type
of failure, it did not seem to be too harmful for the structure.
Ishai further wondered if there is known any static demolition technique which uses the concept of an inverse loading pattern of the
concrete structure. In such a technique the destruction could, for
example, be effected by inducing tensile tresses into what originally
was designed as an unreinforced compressive zone. Molin asserted
such a technique has been developed in Japan: upward loading is applied by powerful jacks positioned between the floor and ceiling
of a building.

De Pauw's research presentation (see contribution, 5.3.2.,
workshop 2) also covered fragmentation by blasting, since by calculating charges when demolishing beams he produced rubble of reusable
size without further crushing, coupled with recovery of bars. As most
promising crushing technique he mentioned a hammermill in which the
hammers turned through adjustable grates in such a direction that
the fragments are thrown up and not as usual are pressed through
the hammers, leading to tensile loading of the concrete. In throw-
ing up, the concrete is crushed until the fragments are small enough
to fall through the grates, while bars stay above the grates and
can be removed.

Lindsell's introductory lecture outlined the reasons for the
sharp increase in the demolition of prestressed concrete:
obsolete structures, structural alteration or failure, chemical
conversion, corrosion of tendons, accidents mainly through impact.
Using two case histories of selective demolition (faulty prestressed
bridge beams with displaced cable ducts and deteriorated high alumina prestressed concrete beams in a school building) he stressed the
problems encountered: release of prestress in-situ, avoiding damage
to adjacent members and hinges, providing temporary support and
stability, imperative need for safety. His conclusions fell under
three headings - economics and the need to develop alternative demolition techniques, - safety to be ensured by training of workers
and demolition contractors, appropriate planning and supervision -
and information in the shape of long-term records of design calculations, as - built drawings, erection sequence, demolition procedure and archiving of case histories. This last point should be organized at national level by professional bodies. The associations
of demolition contractors are now trying to insist that all buildings
be marked for demolition procedures by the designer. Complete records
should be kept by engineering and local authorities of all construction details including structural function, erection procedure,
location of tendons, anchorage design, chemical details of grouting,
concrete mix, strength, composition.

Tension release is a very acute problem and needs more research to
design anchorage devices easily removable. In conclusion he proposed
ideal demolition procedures including temporary works (propping,
lateral bracings), removing dead loads, release of tension in sequence, fragmentation.

Subsequent discussion focussed on research needs, especially
work on controlled explosives (shaped charges), cutting techniques
(water jetting), de-tensioning systems and alternative grouts comprising the possibility of a phase change through heating, reduc-
cing the grout to powder easily removable by blowing.

2.3.1 EXPLOSIVES FOR LOCALIZED CUTTING IN CONCRETE

C. Molin

Swedish Cement and Concrete Research Institute

Drottning Kristinas Väg 26, Fack S 10044, Stockholm, Sweden

Full scale field tests recently carried out by the Swedish Cement and Concrete Research Institute and the Nitro Nobel group in an old concrete building indicates that blasting with explosives is technically and economically feasable even for ordinary walls and slabs thicker than 150-200 mm. Small charges 10-20 g, placed at 250-300 mm distance in both directions can be used for those thin constructions. Heavily reinforced slabs, 300 mm thick, demand bigger charges, probably three times as much. The distance can be increased to 300-350 mm. The concrete near the edge will probably get some cracks parallel and perpendicular to the wall or slab surface. These cracks will, however, in most cases probably not cause any problems. Some crushed concrete round the edge of the hole needs be cleaned. The reinforcement will not be affected very much. Vibrations, shock-wave and flying debris can be controlled in a safe manner. Photo no 1 shows a hole 1x2,5 m, big enough for a small lift, in a heavily reinforced slab. Photo no 2 shows three tests in an ordinary wall with different distances between the charges.

Debris from localized cuttings can easily be used when recycling concrete. One reason is that this debris, mostly coming from modernization sites, is not so easily mixed up with other materials, for example gypsum from partition walls, which could contaminate the recycled concrete. Another reason is that fragmentation and crushing effort will probably be decreased due to the smaller size of debris. The size of the debris is of course dependent on which method being used during the demolition work. The full scale field tests show that blasting with explosives gives the smallest size of the concrete particles, i.e most suitable for use as aggregates. Maximum size is 200x130x70 mm and median size 70x40x30 mm, which

however, indicated that some more crushing is needed. Percussion
methods give somewhat bigger particles, sawing the biggest ones.

Figure 1.   Localized cutting in heavily reinforced concrete slab
            with explosives.

Figure 2.   Three test holes with different distance
            between the charges.

3. WORKSHOP 5

Contamination effects on fragmentation,
fibre and polymer concrete

Chairman: S.P. Shah, secretary M. Geudelin

Attendance: R. Brepson, S.H. Carpenter, J.P. Collin, M. Geudelin,
              O. Ishai, J.J. Mills, S. Mindess, A.T.F. Neerhoff,
              C. De Pauw, C.D. Pomeroy, S.P. Shah, J.W. Weber,
              S. Ziegeldorf.
3.1   Introductory lecture: Materials characterization for
              fragmentation - S.P. Shah
3.2   Report on workshop 5 - S. Mindess

3.3   Contributions

3.3.1   Correlation between fracture toughness and zeta-
              potential of cementstone - A.T.F. Neerhoff.

3.1 MATERIALS CHARACTERIZATION FOR FRAGMENTATION

S. P. Shah

Department of Materials Engineering
University of Illinois at Chicago Circle
Chicago, Illinois, USA

INTRODUCTION

Demolition and fragmentation of concrete is currently being done using several techniques. Concrete has been fragmented using explosives, using hydraulic and pneumatic hammers, using a ball and a crane of using a hydraulic crane with ripping teeth and shovel. Depending on the method used in breaking concrete, concrete is subjected to different rates of straining and different modes of fracture. For the purpose of materials characterization, it may be convenient to differentiate various techniques of demolition according to the strain rate employed.

Strain-Rate Effects

The available information on the mechanical properties of concrete can be broadly divided into three ranges of strain rates as shown in the following table:

	Strain Rate	Stress Rate	Time to Fracture
Static Testing Rate	$\sim 10^{-5}$/sec.	~ 10 psi/sec.	~ 5 sec.
Impact Testing Rate	$\sim 10^{0}$/sec.	$\sim 10^{5}$ psi/sec.	~ 0.5 milisec.
Explosive Tesing Rate	> 10/sec.	$> 10^{6}$ psi/sec.	< 5 microsec.

Most of the available information on fracture and other mechanical properties of concrete is for the static testing rates, while many

of the demolition techniques employ impact or explosive testing
rates.   Very little is understood about how cracks propagate at
strain rates higher than the static rate. Material inertia near
the crack tip becomes an important consideration at explosive
rates when the time to fracture, t_f, is less than r/c where
r = crack length and c = wave velocity. This means that when the
fracture time for concrete is less than 10-26 microseconds, inertia
effects become predominant. This can be seen in Fig. 1a where the
effect of strain rate on the tensile strength of concrete is shown
(Ref. 1). It can be seen that as the strain rate approaches the
explosive testing rates, the rate of increase in strength increases.
For strain rate up to the explosive testing rates, it has been
observed for concrete, rocks and other brittle materials that the
maximum dynamic strength is about two times the static strength.
A much higher increase in dynamic strength is observed at the
strain rates corresponding to the explosive rates as can be seen
in Fig. 1b where the data for oil shale are given (Ref. 2).

The effect of the strain rate on fracture is likely to depend
on the temperature.   From testing rock specimens at various
temperatures and strain rates, Heard (Ref. 3) developed the following equation:

$$\varepsilon^\circ = \varepsilon^\circ_0 \left(\exp \frac{-U}{KT} \right) \sin h\left(\frac{\sigma_f}{\sigma_o} \right)$$ (1)

where ε° = strain rate, σ_f = fracture strength, ε°_0 and σ_o
are constants, U = activation energy and KT = Boltzman's constant.

For explosive rates, a cubic root dependence of fracture
stress on strain rate was observed by Birkimer for concrete
(Ref. 4).   Such a relationship can be derived by including the
effect of inertia on the classical linear elastic fracture
mechanics (Ref. 5):

$$\sigma_f \sim \frac{K_{IC}}{\sqrt{r}}$$ (2)

where K_{IC} = critical stress intensity factor. From this it
follows that:

$$\sigma_f \sim \frac{K_{IC}}{\sqrt{ct_f}}$$ (3)

The strain rate to fracture is of the order:

$$\varepsilon^\circ \sim \frac{\sigma_f}{\rho c^2 t_f}$$ (4)

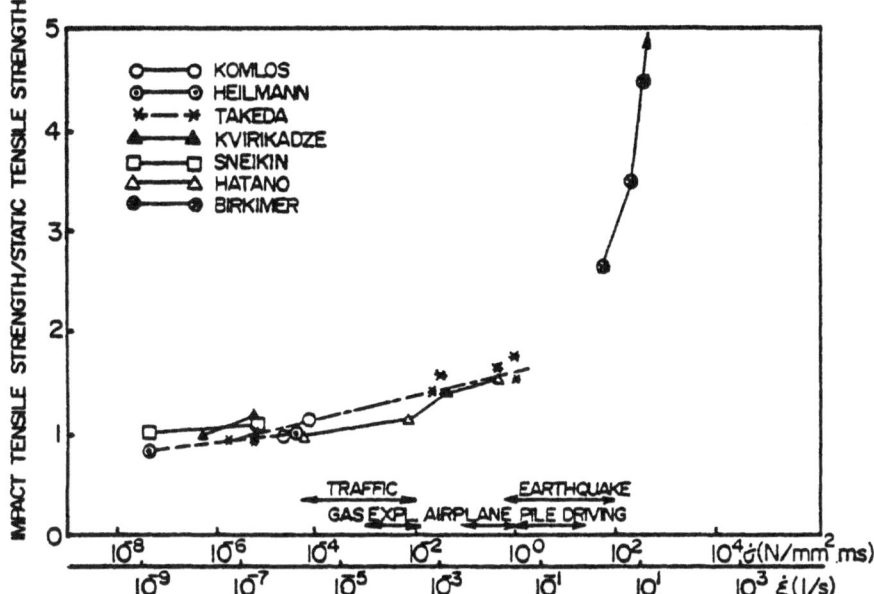

Fig. 1a.   RESULTS OF TENSILE STRENGTH VS. LOADING RATE
FOR CONCRETE (1)

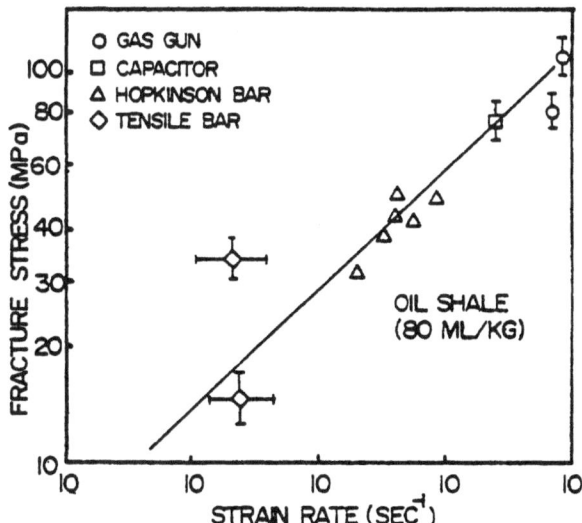

Fig. 1b.   FRACTURE STRESS DEPENDENCE ON STRAIN RATE
FOR OIL SHALE (2)

where σ = density of the material. Combining Eqs. (3) and (4), we get:

$$\sigma_f \sim \left(\rho c K_{IC}^2 \varepsilon_o\right)^{1/3} \tag{5}$$

It can be shown that for a constant strain rate loading the exact expression is (Ref. 6):

$$\sigma_f = \left(\frac{9\pi}{16} \rho c K_{IC}^2 \varepsilon^o\right)^{1/3} \tag{6}$$

Note that approaches similar to Eq. (1) and Eq. (6) may be useful in theoretically predicting the fracture strength of materials subjected to various strain rates.

Crack Propagation

The velocity of crack propagation plays an important role in determining the size of fragmentation and the energy required for fragmentation. Kumar (Ref. 7) has proposed the following equation to relate the strain rate and the velocity of crack propagation:

$$\varepsilon^o = KNCV_c \tag{7}$$

where K = orientation constant, N = N(σ) = number of microcracks, σ = stress, C = average length of microcracks and $V_c = V_c(\sigma)$ = crack propagation velocity. When the velocity of crack reaches the terminal velocity then decreasing the strain rate increases the fragment size as shown in Fig. 2a.

In blasting of rocks, it has been suggested that the knowledge of dynamic fracture toughness can be used in designing the spacing of bore holes and the amount of explosive charges in the holes (Refs. 8 and 9). Critical bore hole pressure necessary to initiate cracks has been related to the static fracture toughness K_{IC}. In order to sustain a running crack, it is necessary to maintain bore hole pressure such that a K_D value above the crack arrest value K_{Ia} is necessary. On the other hand, if the dynamic stress intensity is larger than K_{Ib}, then crack branching will occur (Fig. 2b).

If similar concepts are to be applied in breaking concrete, a knowledge of K_{Ic}, K_{Ia} and K_{Ib} are necessary. In the remaining part of this paper some work that is under progress at the University of Illinois at Chicago Circle for determining fracture mechanics parameters under static loading and under dynamic (impact) loading is summarized.

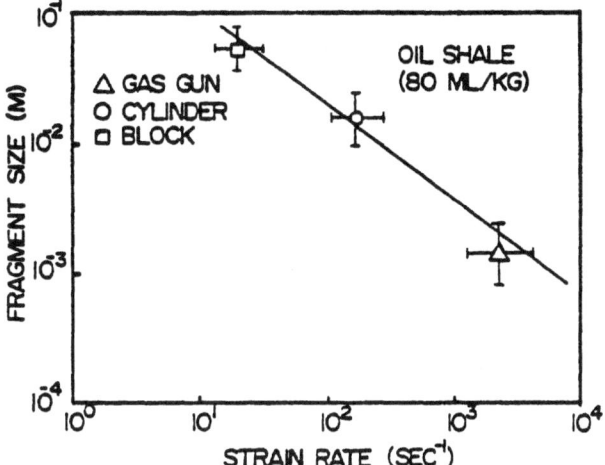

Fig. 2a.   FRAGMENT SIZE DEPENDENCE ON STRAIN RATE
            FOR OIL SHALE (2)

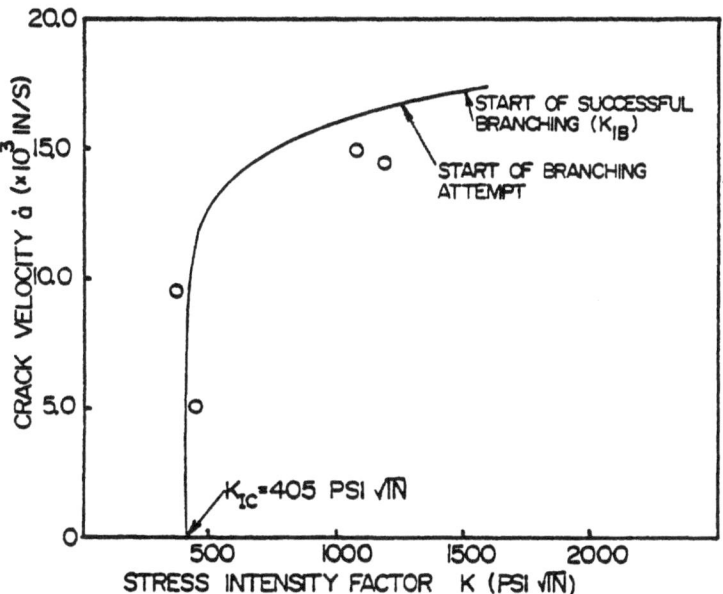

Fig. 2b.   CRACK VELOCITY VS. STRESS INTENSITY FACTOR
            IN HOMALITE (8)

STATIC FRACTURE MECHANICS PARAMETERS

Many attempts have been made to apply linear elastic fracture mechanics and elastic-plastic fracture mechanics concepts to fiber reinforced composites. No single fracture mechanics parameter which can uniquely quantify the resistance to crack propagation has been found acceptable. For example, different investigators have reported widely varying values of critical stress intensity factors (K_c) using different types of specimens and applying different techniques to measure the onset of unstable crack propagation of otherwise similar materials. One of the primary reasons for this discrepancy is the slow crack growth that precedes the unstable crack propagation. Unless the size of the uncracked ligament is substantially larger than the slow crack growth (also often called fracture process zone), a size independent fracture parameter may not be obtained.

To study slow crack growth, notched beams, double cantilever beams specimens and double torsion specimens are being tested in a continuing investigation at the University of Illinois at Chicago Circle. The applicability of various fracture mechanics approaches including: critical stress intensity factor, J-integral, critical crack opening displacement (COD), compliance techniques for determining the slow crack growth and R-curve analysis are being evaluated. Attempts are made to identify a fracture parameter which is independent of test-specimen geometry and which can correctly predict the effects of fiber addition.

Some of the results of our investigation are described in detail in Refs. 10-15. In this paper, our attempts to apply the R-curve (resistance curve) analysis to concrete and fiber reinforced concrete are briefly summarized.

R-curve is the relationship between the resistance in terms of energy absorbed (G_R) and the crack extension (Δ_a). It is common to calculate the strain energy release rate from the elastic stress intensity factor (K_c) and the crack extension from the compliance calibration. This approach was found not to be valid for concrete and fiber reinforced concrete.

Different methods of calculating G_I are shown in Fig. 3. The method A assumes a linear, elastic and brittle material in the Griffith's sense. If the material is linearly elastic, exhibits stable crack growth and shows no permanent deformations upon unloading, then the method B may be applicable. For materials exhibiting permanent deformations upon unloading the reloading (or unloading) compliance method (method C) is often used to calculate G_I or K_I. As can be seen from Fig. 3, this approach ignores the inelastic energy absorbed during crack growth.

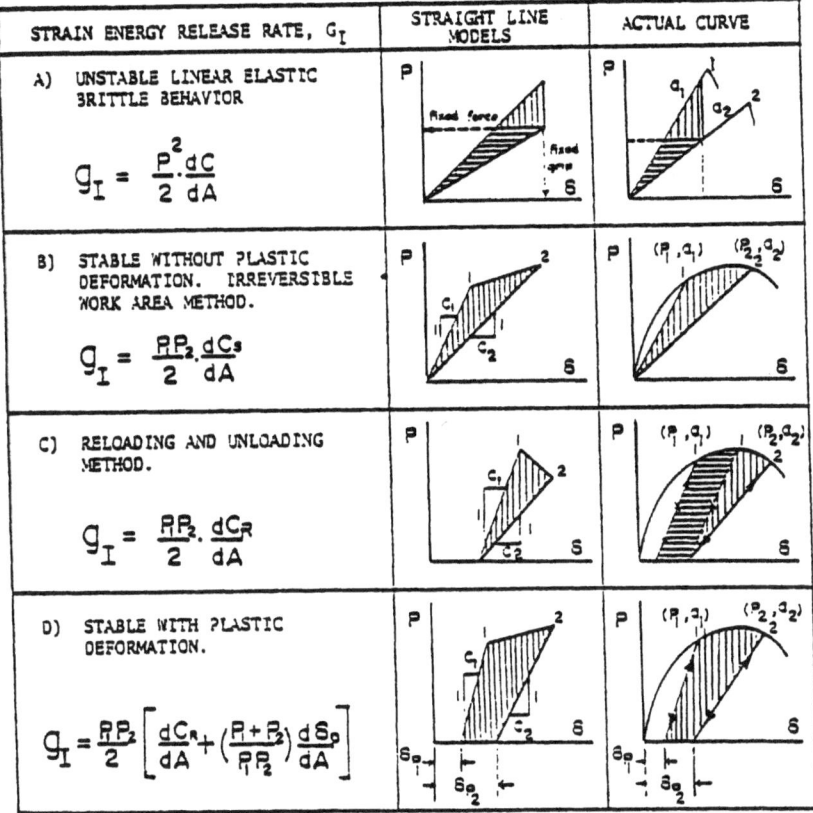

Fig. 3 Schematic Models for Different Fracture Behavior and Corresponding Strain Energy Release Rates

To include this, we have modified the definition of the strain
energy release rate as shown in method D, Fig. 3.

R-curves were experimentally determined from two types of
specimens: double cantilever and double torsion (Fig. 4). These
specimens were designed so as to study the crack growth up to about
600 mm (24 in.). The testing procedure is shown in Fig. 5. The
crack growth was measured with an accuracy of about 0.1 mm
(0.004 in.).

It was observed that only when R-curves were calculated using
the proposed definition of G_I the curves for mortar and concrete
were similar for both the double cantilever and the double torsion
specimens (Fig. 6). However, for fiber reinforced mortar the two
types of specimens did not give comparable results. This was
thought to be due to the different amount of crack opening displacement for a given crack extension for these two types of tests.
Different crack opening displacement means varying amount of
energy absorbed in fiber pulling-out processes. When the strain
energy release rates were compared for the identical crack opening
displacement, then the difference between the two types of tests
became less significant (Fig. 7).

DYNAMIC PROPERTIES

The main objective of this part of our continuing study is to
investigate the dynamic properties and fracture behavior of
concrete and fiber reinforced concrete (FRC) subjected to impact
loading using an instrumented impact system (Fig. 8). The study
consists of:

a. obtaining a valid testing procedure and specimen dimensions
    for instrumented impact study of concrete and fiber reinforced
    concrete;

b. determining the influence of the reinforcing parameters and
    matrix properties on the impact behavior of the composite;

c. formulating appropriate models to predict the impact properties
    of cementitious composites from the measured constitutive
    relationship at high strain rates of the matrix, fiber and the
    fiber-matrix interface; and

d. studying the applicability of various dynamic fracture criteria
    for fiber reinforced concrete.

During instrumented impact testing, a record of impact-load
vs. time is obtained. From this trace other quantities such as
energy vs. time and the load vs. deflection can be calculated.
Such analysis of the data presupposes that the load recorded by

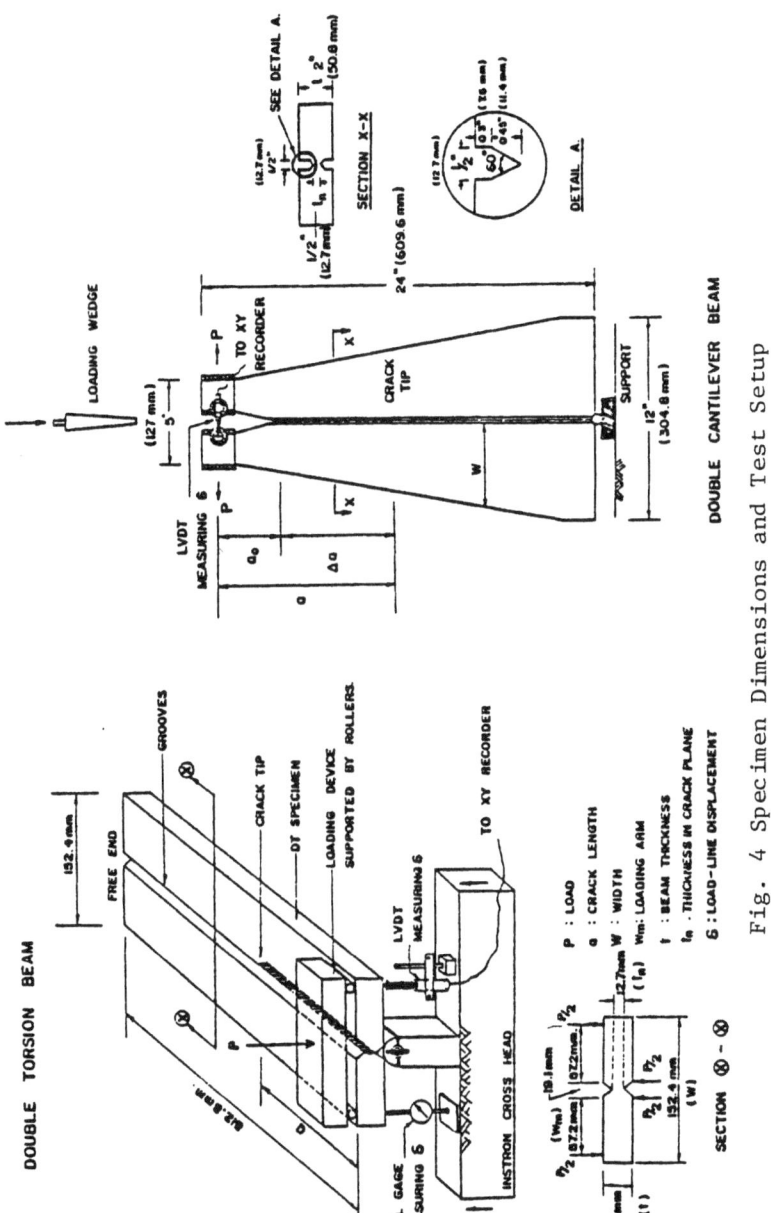

Fig. 4 Specimen Dimensions and Test Setup

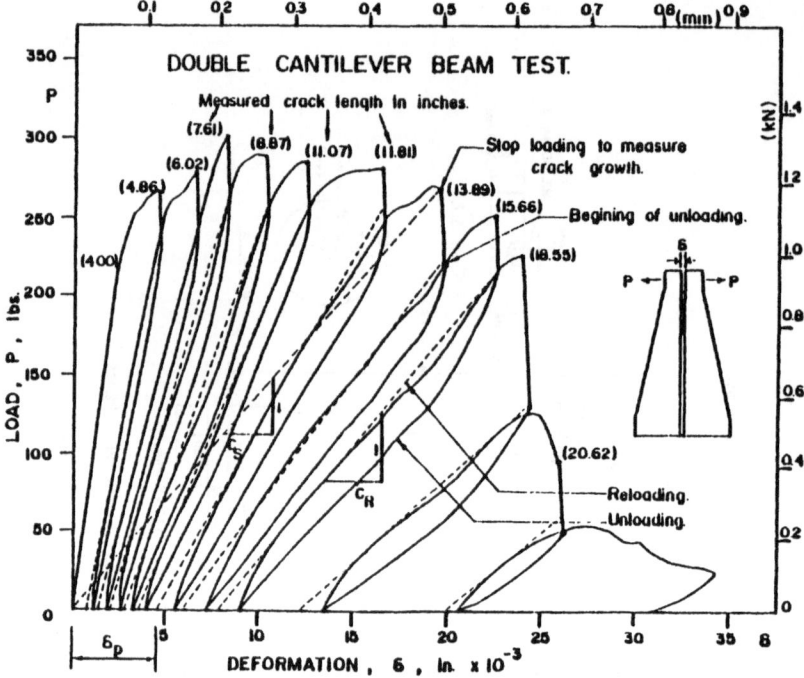

Fig. 5 Typical Load Deformation Curve

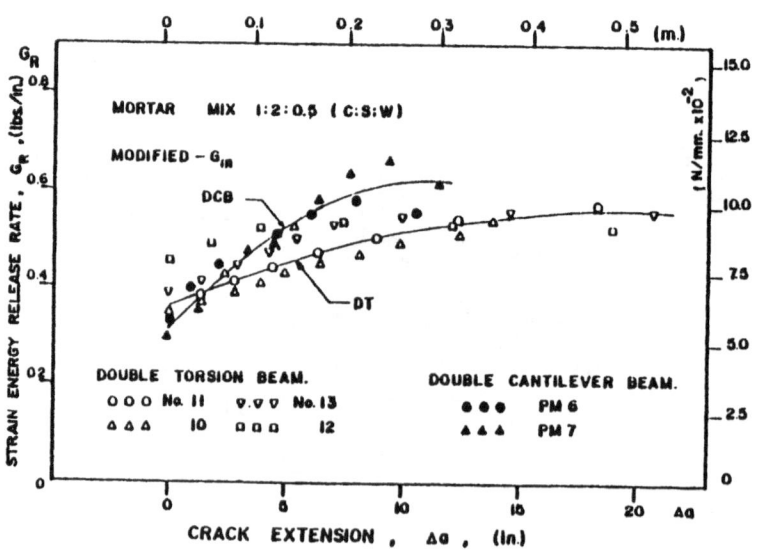

Fig. 6 Strain Energy Release Rate and Crack Extension Relation

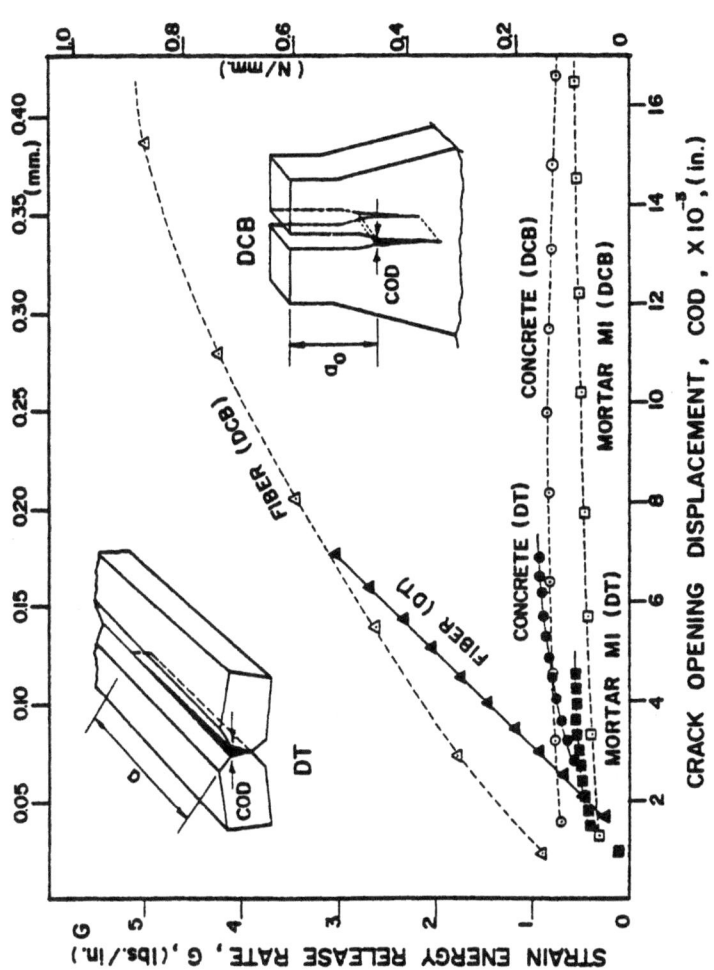

Fig. 7 Strain Energy Release Rate and Crack Opening Displacement Relation

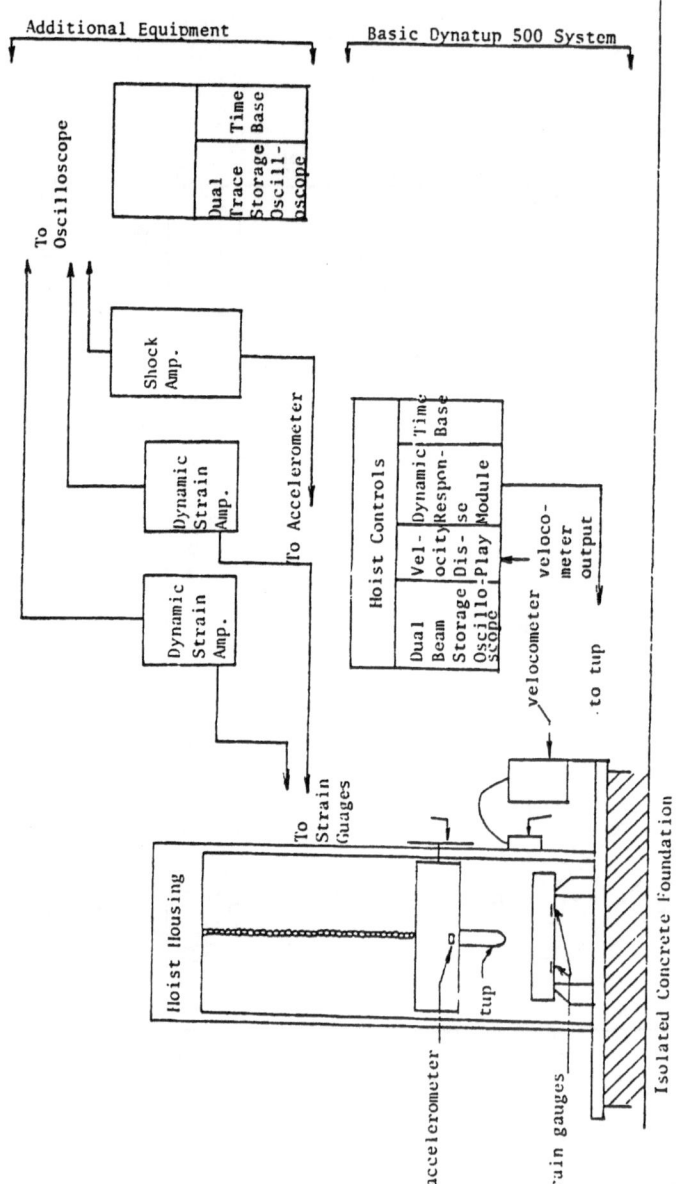

Fig. 8 Complete Block Diagram of Instrumented Impact System

the instrumented tup is the same as the bending load on the beam.
It has been shown that for metal specimens the load recorded by
the tup does not correspond to the flexural load during an initial
period of time. By assuring that the fracture time is substantially greater than this initial period, one can minimize the
so-called inertial effects. For concrete specimens, the problem
is compounded since the fracture time is substantially shorter
than metallic or polymeric specimens and the inertial effects are
relatively higher because of the relatively low strength/weight
ratio and because of the need to test large size specimens.

To better understand the interactions among the inertial
effects, the fracture time and the velocity of impact, a two
degree of freedom of vibration model was developed (Ref. 16).
This proposed model, unlike many of the previous analytical models,
is sensitive to the bending stiffness and weight of the specimen,
the effective stiffness between the tup and the specimen and the
velocity at impact. The validity of the model was verified by
testing a series of asbestos cement specimens of varying dimensions and tested at various impact velocities. The effective
stiffness between the specimen and tup was changed by introducing
a rubber pad of different thicknesses. Both the experimental and
the analytical results showed that the introduction of rubber pads
substantially reduced the inertial effects.

Results of tests on concrete and fiber reinforced concrete
showed that the inertial effects have been substantially reduced
and a valid test procedure and specimen dimensions have been
developed (Fig. 9). The analytical model correctly predicted
both the magnitude and the frequency of the inertial oscillations
for the concrete and the fiber reinforced concrete specimens.

Comparison of the results of fiber reinforced concrete
specimens subjected to impact loading with those of static loading
(Table 1) showed that the equivalent modulus of rupture is approximately twice for dynamic loading for the strain rates employed.
However, the energy absorption for both static and dynamic loading
were somewhat comparable. Since the pulling out of the fibers
from the matrix absorbs a significant amount of energy in fiber
reinforced concrete, it may seem that the debonding energy is not
very strain-rate sensitive. This observation was confirmed by the
results of pull-out tests (Fig. 10). A special device has been
built to study the load-slip relationship of the fiber-matrix
interface (Ref. 17).

The investigation on notched specimens subjected to varying
strain rates has just started. From this study we hope to obtain
some idea of dynamic fracture toughness as well as on the velocity
of crack propagation.

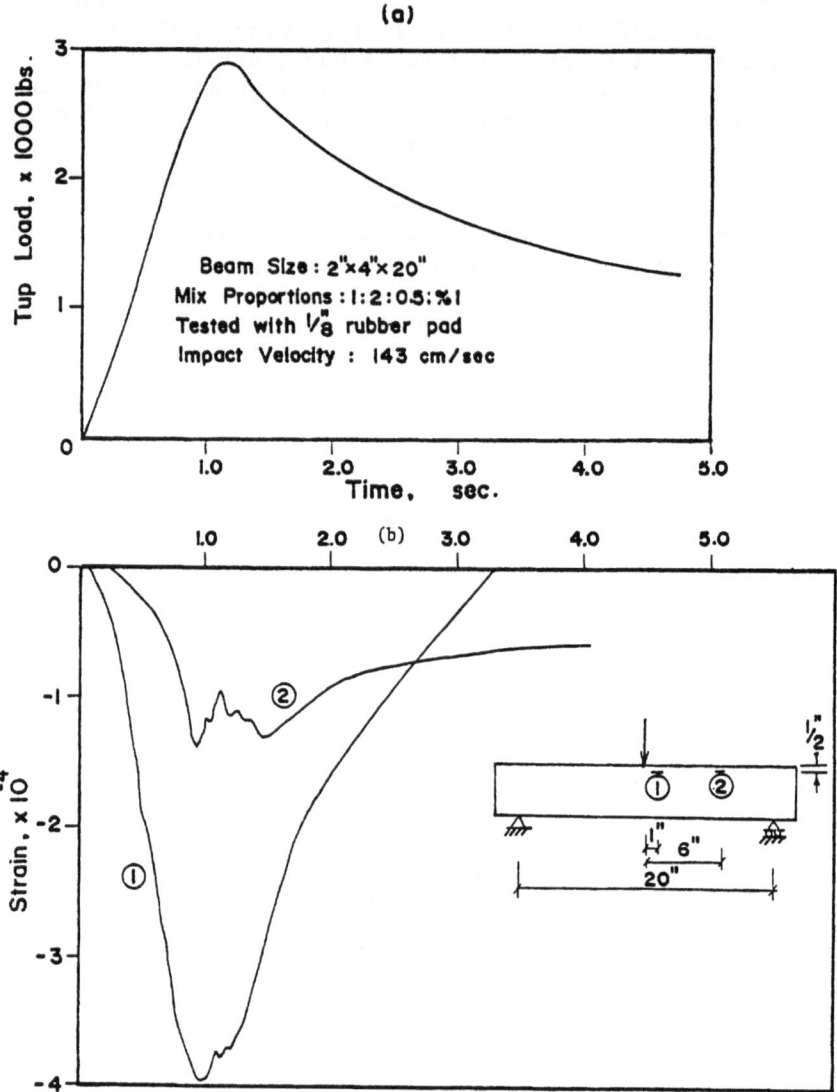

(a)

Beam Size : 2"×4"×20"
Mix Proportions : 1:2:0.5:%1
Tested with $\frac{1}{8}$" rubber pad
Impact Velocity : 143 cm/sec

(b)

Fig. 9   Typical Load-Time and Strain-Time Records Obtained
for FRC Specimens

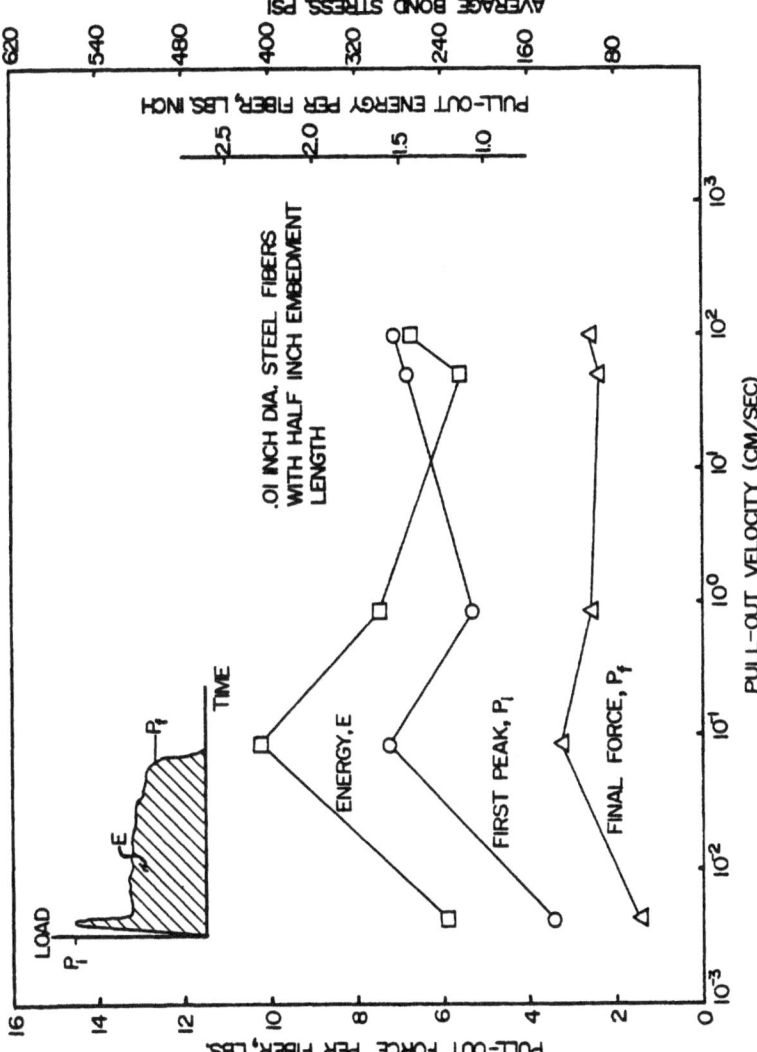

Fig. 10 Pull-Out Force, Energy and Velocity Relationships

Table 1.   Fiber Reinforced Concrete Beams Under Static
and Impact Loading

Stress Rate (psi/sec)	M.O.R. (psi)	Bending Energy (ft lbs) at Defl. of 0.30 in.
2.35×10^6	2700	43.1
2.25×10^6	2590	41.3
1.11×10^6	2440	45.5
20×10^0	1400	32.0

REFERENCES

1. H. A. Kormeling, A. J. Zielinski and H. W. Reinhardt, "Experiments on Concrete Under Single and Repeated Impact Loading," Report 5-80-3, Stevin Laboratory, Delft University of Technology, Delft, The Netherlands.
2. D. E. Grady and M. E. Kipp, "Continuum Modeling of Explosive Fracture in Oil Shale," Int. J. of Rock Mech., Minerals, Sci. & Geomech., Vol. 17, 147-157 (1980).
3. C. H. Heard, "Effect of Large Charges in Experimental Deformation of Yule Marble," J. of Geol., Vol. 7, No. 2, 163.
4. D. L. Birkemer, "A Possible Fracture Criterion for Dynamic Tensile Strength of Rock," Proc. of 12th Symp. on Rock Mech. (Edited by G. B. Clark), 573 (1971).
5. D. E. Grady and M. E. Kipp, "The Micromechanics of Impact Fracture of Rock," Int. J. of Rock Mech., Minerals, Sci. and Geomech., Vol. 16, 293-302 (1979).
6. L. B. Freund, "Crack Propagation in an Elastic Solid Subjected to Stress Wave Loading," J. of Mech., Phys. and Solids, Vol. 21, 47-61 (1973).
7. A. Kumar, "Effect of Stress Rate and Temperature on the Strength of Basalt and Granite," Geophysics, Vol. 33, No. 3, 501-510 (June 1968).
8. V. K. Der, D. C. Holloway and T. Kobayashi, "Techniques for Dynamic Fracture Toughness Measurements," Report, Mechanical Engineering Department, University of Maryland, College Park, Maryland (1980).
9. J. W. Dally and W. L. Fourney, "Fracture Control in Construction Blasting," University of Maryland Report to NSF (1976).
10. M. Wecharatana and S. P. Shah, "Double Torsion Tests for Studying Slow Crack Growth of Portland Cement Mortar," Cement and Concrete Research, Vol. 10, 333-844 (1980).

11. G. Velazco, K. Visalvanich and S. P. Shah, "Fracture Behavior and Analysis of Fiber Reinforced Concrete Beams," _Cement and Concrete Research_, Vol. 10, 41-51 (1980).

12. K. Visalvanich and S. P. Shah, "Evaluation of Fracture Techniques in Cementitious Composites," _Proc. on Fracture in Concrete_, ASCE Annual Convention, Florida (October 1980).

13. M. Wecharatana and S. P. Shah, "Resistance to Crack Growth in Portland Cement Composites," _Proc. on Fracture in Concrete_, ASCE Annual Convention, Florida (October 1980).

14. S. P. Shah, "Whither Fracture Mechanics in Concrete Design," _Proc. of the Engineering Foundation Conference on Concrete Production and Use_, New Hampshire, 175-185 (June 1979).

15. K. Visalvanich and A. E. Naaman, "Compliance Measured Fracture Toughness of Mortar and Fiber Reinforced Concrete," to be published in the _Proc. of the ASTM Conf. on Fracture Mechanics Methods for Ceramics, Rocks and Concrete_, Chicago (June 1980).

16. W. Suaris, U. Gokoz, O. Youngquist and S. P. Shah, "Analysis of Inertial Effects in the Instrumented Impact Testing of Fiber Reinforced Concrete," Progress Report, U.S. Army Research Office (1980).

17. U. N. Gokoz and A. E. Naaman, "Effect of Strain-Rate on the Pull-Out Behavior of Fibers in Mortar," Progress Report, submitted to U.S. Army Research Office (Jan. 1981).

3.2 REPORT ON WORKSHOP 5

S. Mindess

Dept. of Civil Engineering, University of British Columbia, Vancouver, B.C.

Dr. Shah began by summarizing the conclusions of the preceding workshop on fragmentation. There was produced a list of those subject areas in which more knowledge was needed before the efficiency of fragmentation could really be improved:

1. Methods of breaking
2. Modes of deformation and fracture
3. Stress analysis
4. Critical materials properties (K_{ic}, γ eff. and the static, dynamic, acoustic, thermal etc. characteristics)
5. Energy balance in demolition and fragmentation
6. Improvement to the various techniques

The questions that had been raised were:

1. Is fracture machanics a useful tool for the fragmentation of concrete? Do we know enough about fracture mechanics to influence demolition and fragmentation?
2. What else do we need to know?
3. Are other approaches, such as continuous damage theory, helpful?

Shah then reviewed some of the work that too already has been done for rock fracture. He distinguished between three testing rates: static, impact and explosive, as follows:

	strain rate	stress rate	time to fracture
Static	10^5/s	$\simeq 10$ psi/s	$\simeq 5$ s
impact	μs to 10/s	10^5 psi/s	0.5μs
explosive	> 10/s	10^6 psi/s	$< 0.5 \mu$s

For concrete, most of the available information is on static loading, some information is available for impact loading but virtually none for explosive loading. Only for explosive testing does the material inertia near the crack tip become important.

However we do know that apparent strength increases with increasing strain rate. The rate of increase also increases at very high strain rates. For relatively slow loading rates, a number of models have been used to describe this rate effect, such as the Arrhenius equation. For explosive loading rates a cube root relationship between stress at failure (σf) and strain rate ($\dot{\varepsilon}$) are observed: $\sigma_f = f(\dot{\varepsilon})^{1/3}$

Another model considers the crack orientation, the number of micro-cracks and crack velocity. At the terminal crack velocity, an increase in strain rate will increase the amount fragmentation. In rocks the following should be true:

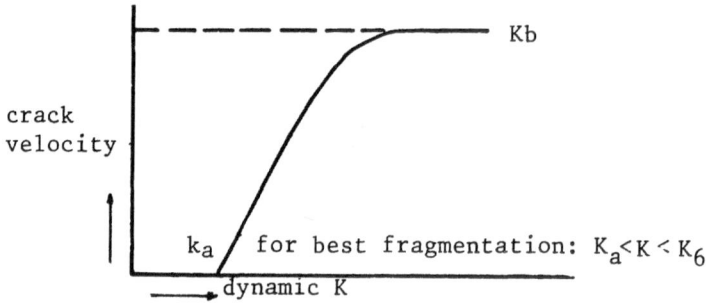

Following this general introduction, Shah then presented an extensive account of his recent research, describing quasi-static double torsion and double cantilever beam tests, whose results one best described using R-curve analysis. He also described his most recent work using an instrumental impact machine. On Ishai's question what the contribution is of elastic and plastic energy involved in the stretching of steel fibres to the total debonding energy of an FRC beam, Shah answered that this contributions is negligible compared to the energy involved in the pullout of fibres. There followed a brief presentation by Neerhoff on the correlation between zêta potential and the fracture toughness of hardened cement paste. Although his measured changes in K_{ic} were small, he claimed that K_{ic} reached a maximum at the iso-electric point of the solution in which the studies were done.

After these two presentations a general discussion followed. First there was considerable debate over whether we were able to measure a "true" K_{ic} for concrete. There seemed to be general agreement that at this time we probably could not, but that it might not make very much difference to practice even if we could. Clearly there is still no agreement on to how K_{ic} should be measured even with quasi-static loading; at high strain rates, the problem is even more confusing.

Pomeroy argued convincingly that rather than studying the propagation of single cracks, we should be looking at the relationship between the total energy input and the total damage in concrete. He went on to state that it was possible to use very simple, empirical tests to get a measure of damage. For instance, in rocks he was able to get useful correlations by simply calculating the energy required to crush a cube of rock and relating this to the amount of degradation simply by then measuring the amount of material passing some arbitrary sieve size. He suggested the desirability of some similar empirical tests for concrete; fundamental tests could be left for subsequent "fine-tuning".

Ishai argued that we must distinguish between structural and material parameters. Echoing Pomeroy he too suggested that we must measure the total energy input to the structural system in analyzing demolition techniques. We want a design tool that will enable us to select the best technology for the job at hand. He argued that, as a first approximation, the area under a "$\sigma-\epsilon$"-curve could be used to estimate the energy input to a system.

Following this general discussion, a series of questions was posed that need answers before we can systemetically improve demolition and fragmentation techniques:

1. What microscopic (empirical) techniques can we use to assess or predict the demolition efficiency? That is, how do we relate the external energy applied to the damage that results? For example ϵ (vs) fragmentation, other methods?
2. Can fracture mechanics lead to any useful results with regard to demolition and fragmentation?
3. How can we relate (1) and (2) above? At present there is a high gap between fracture mechanics and demolition techniques.
4. What correlations are there with rock mechanics?
5. How important is the fracture energy? We may wish to use the applied energy more efficient, say to improve the capacity of a recycling plant. Can we develop more efficient machines?
6. How can we improve explosive techniques to improve demolition efficiency?
7. How can we relate the energy per unit amount of damage to the various test parameters?
8. We must also consider the properties of the resulting material after fragmentation. What do we want to end up with?
9. Can we use some statistical damage theory instead of (or combined with) fracture mechanics, such as the Weibull theory or the Krachinov theory?

10. How do contaminants affect the efficiency of fragmentation?
11. How can environmental effects (e.g P_H, T, R.H., stress corrosion) facilitate degradations?
12. How can we apply energy other than by purely mechanical loading (e.g thermal shock, etc.)

3.3.1 CORRELATION BETWEEN FRACTURE TOUGHNESS

AND ZETA POTENTIAL OF CEMENTSTONE

A.T.F. Neerhoff

Eindhoven University of Technology, Department of
Architecture, Building and Planning, Group Science of
Materials, P.O. Box 513 - 5600 MB Eindhoven - Netherlands

SUMMARY

A brief account is first given of the difficulties encountered
when trying to make a proper choice of fracture-facilitating surface-
active agents for cementstone. In this context we describe the "Rebinder-effect", the notion of zêta potential, and the present knowledge about the adsorbtion behaviour of calcium alumino silicates in
alkaline aqueous environments. Comparison of the results of zêta potential measurements by means of electroosmosis, and measurements
of the fracture toughness K_{1c}, both performed on cementstone in
aqueous electrolytic solutions of varying concentration which were
kept saturated vs. calciumhydroxide, shows a distinct maximum for
K_{1c} at the so-called "Iso Electric Point". A preliminary model is
suggested to explain the observed behaviour of K_{1c} as a function of
the concentration of the electrolytic solution.

I INTRODUCTION

A vast amount of literature exists about the chemical, physical
and mechanical properties of cementstone (1,2,3,4). As far as is
known however no definite results have been published up till now on
the effects of environment on the intrinsic strength of the cement-
stone-gel during a fracture process.

Essentially cementstone is a short-range- order ionic lattice.
Its structural units are formed by hydrated calcium silicate-
and calcium alumino silicate srystallites. De to the small-
ness of these crystallites (typically 10 Å to 1 μ) a relative
large number of atoms is present at their surface. This will
make every surface effect to a volume effect 5, 6, 7). For in-

stance the strength of the crystallites will undoubtly be determined
by their adsorbate (5,7), which mainly consists of water saturated
vs. Ca(OH)$_2$. The crystallites themselves induce structural changes
in their adsorbate at distances from the interface that greatly exceed molecular dimensions (8,9,10).

Forces between crystallites which increase the strength of cementstone are (1,3):
- London – van der Waals dispersion forces across pores smaller than
  about 5 Å
- primary chemical cross-links such as Si-O-Si or Si-O-Ca-O-Si
- hydrogen bonds
- pure mechanical entanglement of crystallites

Forces which decrease the strength of cementstone are:
- double layer forces due to a net electric charge of the crystallites
- penetration forces of water between crystallites

At increased crack velocities the visco-plastic behaviour of
load-bearing water films becomes important. An important question
further is wether the crystallites themselves will break (brittle
failure) or wether they shear apart (ductile failure).

During a "fracture" process all the above mentioned forces act
together. A theoretical strength- vs. structure-relationship of cementstone, containing in addition pores and microcracks in varying
proportions, therefore seems to be impossible. As a matter of fact,
even for "well-defined" silicates such a treatment is largely qualitative, and has only been possible for a very limited number of simple cases (11,12). Most realistic values of the excess-energy of
solid fracture surfaces seem to be given by indirect measurements,
such as the heat-of-solution method of Lipsett (13) which was applied on tobermorite by S. Brunauer (14). We shall therefore not attempt
here to estimate a quantity such as the "surface energy of cementstone", whatever its microscopic definition may be.

In view of the expected difficulties in making a proper choice
of fracture-facilitating agents for cementstone, the author felt it
as a most valuable approach to investigate the applicability of the
so-called "Rebinder-effect" (15) (A recent review of adsorbtion-sen-
sitive fracture phenomena including the "Rebinder-effect" can be
found in ref. 16, or see the paper by Dr. J.J. Mills, this conference). We shall now briefly discuss this effect as well as the notion
of zêta potential with which it is closely related.

When a solid is immersed in an electrolytic solution it may obtain a net electric surface charge due to the preferental adsorbtion

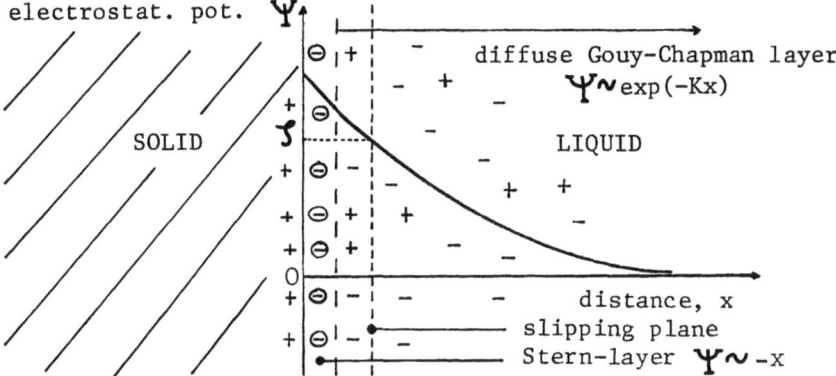

Fig. 1 Electric double-layer at the interface between a solid and
       an electrolytic solution.

of (potential-determining) ions from the solution (10). In fig. 1
these ions are supposed to have a positive charge. A thin (Stern-)
layer of hydrated counter-ions also strongly adsorbed, stays behind
in the liquid, just at the solid-liquid interface. After applying an
electric field parallel to the interface the diffuse layer of electrolyte outside the so-called slipping plane (bearing a negative
charge in fig. 1) will move. The liquid flow is proportional to the
value of the electrostatic potential ψ at the slipping plane which
is known as zeta potential (ζ), and which can be measured by the
method of electroomosis. The concentration of electrolyte for which
$\zeta = 0$ is called the Iso Electric Point (IEP). Now for most inorganic materials - wether crystalline or amorphous - there is a definite correlation between its plastic deformation characteristics
and zêta potential, with the restriction that enough time is allowed
for the adsorbtion equilibrium to be established (16). In fig.2
such a correlation is depicted between microhardness and zeta potential as a function of concentration of electrolyte (i.e. the actual
"Rebinder effect") (16).

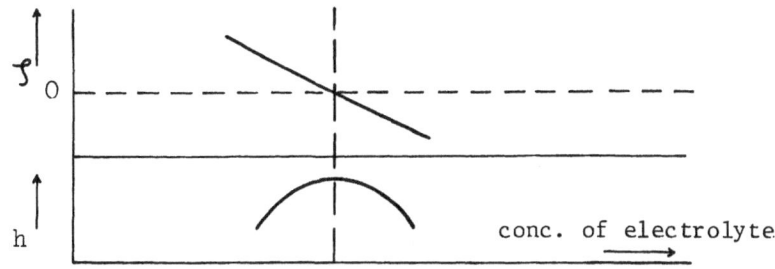

Fig. 2 Correlation between microhardness (h) and zeta potential (ζ)

A striking feature of this type of correlations in fracture-experi-
ments is the optimum value of the strength-parameter at the IEP. A
speculative explanation states that the near-surface (1 - 10 μ m
deep) electronic properties, and therefore near-surface mechanical
properties (such as mobility of dislocations and point-defects (17))
are influenced by a surface-charge, and therefore the solid is in
its most stable bonded state at the IEP (18).

As for the adsorbtion behaviour of cementstone in electrolytic
solutions no definite data are available in literature. From measurements of electromechanical bending (20) performed on cementstone one
can deduce that the crystallites in saturated cementstone specimens
bear a positive electric charge. Siskens (21) measured the zeta potential of various calcium silicates and calcium alumino silicates
at a constant pH = 12.0 as a function of concentration of $CaCl_2$, by
 he method of electroosmosis. He found a sign seversal of zeta potential from a minus to a plus at concentrations of $CaCl_2$ varying
from 1 to 10 mmole/litre.He also found that calcium- and hydroxyl-
adsorbtions mutually stimulate each other, resulting in a small net
electric charge behind the slipping plane for $\zeta \neq 0$. With calcium
alumino silicates extra adsorbtion sites for calcium ions occured,
which was ascribed to adsorbed aluminate (Al $(OH)_4^-$) ions. By means
of electrophoresis Stein (22) found for hydrated tobermorite that at
the high ambient pH = 12.5 of saturated $Ca(OH)_2$ - solution most of
the surface silanol-groups will be dissociated, and due to the excess
of calcium ions in the solution tobermorite has a positive surface
charge. Hydrated C_3S (23) and Ca $(OH)_2$ have a positive zeta potential
in saturated $Ca(OH)_2$ - solution. Spierings (25) found a positive zeta
potential for C_3A when hydrating in a 0.1 M solution of NaOH, and
ascribed it to preferential adsorbtion of calcium ions, not fully
compensated for by hydroxyl- or aluminate-ions. Cementstone, due to
its inhomogeneous structure and chemical composition, probably will
show an at random distribution of many different types of adsorbtion
sites.

In the following selected measurements of zeta potential on cementstone and quartz first are described, using electroosmosis in
aqueous electrolytic solutions, kept saturated vs $Ca(OH)_2$. The results and discussion will show amongst others the predominant role
of calcium ions. Hereafter K_{1c} - measurements on double cantilever
specimens of cementstone, when immersed in an electrolytic solution
are described. The paper ends with an explanation of the observed
correlation between K_{1c} and the log concentration of electrolytic
solution.

II ZETA POTENTIAL OF CEMENTSTONE AND QUARTZ

1. Experimental

   Electroosmosis apparatus

        The electroosmosis apparatus which was constructed from Pyrex
glass is shown schematically in fig. 3. Grains of the solid which is
to be examined (see below) fill the lower part of compartment (a)
as a porous plug. They are surrounded by the electrolytic solution
which also fills the remaining part of compartment (a). A dc current
I is applied via non-gassing electrodes which consist of zinc rods
(b) in saturated zinc sulphate solution (c). Compartment (d) contains
a 0.5 M KNO$_3$ solution which seperates the zinc sulphate solution
from the solution in compartment (a). Glass balls (e) prevent the
mixing of the liquids. Liquid flow is observed in precision bore
tubes (f) by means of a travelling microscope. The temperature of
the apparatus is kept constant at (25.0 ± 0.1) ºC. Zeta potential is
calculated from the Smoluchowski-equation (26):

$$\zeta = \frac{\phi \; \eta_\omega \; \sigma}{\varepsilon_w \cdot I}$$

where:

ϕ = liquid flow (m^3 s^{-1})
η_w= viscosity of water at 25 ºC = 8.904.10^{-3} kg m^{-1} s^{-1}
σ =   specific electric conductivity of the electrolytic solution
      (Ω^{-1} m^{-1})
ε_w= permittivity of water at 25ºC = 6.629.10^{-12} F m^{-1}
I = dc current trough the porous plug (A)

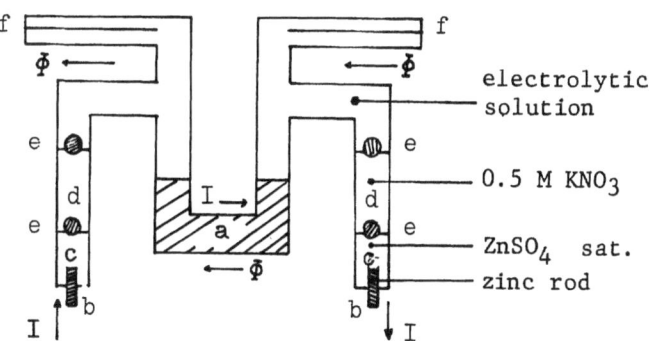

Fig. 3 Electroosmosis apparatus

Samples

Cementstone grains were obtained by grinding a 28 days old sample which had a W/C - ratio of 0.30 in an agate ball-mill. The fraction with diameters smaller than 45 μ m was sieved off wet with a small amount of water, and kept in a polythylene bottle with a magnetic stirrer. For each measurement a small amount of this suspension was repeatedly mixed (more than 10 times) with fresh electrolytic solution, and decanted after 15 minutes. Equilibrium was observed by measuring the electric conductivity (HACH cond. meter, type DR/2) and the pH value (Beckmann pH meter type 123300 with blue glass combination electrode type 39501) of each decantate. With quartz the same procedure was followed as with cementstone.

Reagents used

The chemical compositions of the Dutch commercial cements we used are given in table I below. Fused quartz of pro analysi grade was obtained from Merck (Germany) as grains with a mean size of about 0.2 mm.

Table I   Chemical compositions of cements (wt %)

cement	CaO	SiO$_2$	Al$_2$O$_3$	Fe$_2$O$_3$	SO$_2$
Portland (Encilite-B)	64.9	20.7	5.1	2.3	2.6
Portland-blastfurnace (Robur-B)	50.6	26.0	11.0	1.5	2.6

Water used was twice destilled and boiled shortly before use (conductivity< 2 μ mho/cm   All preparations were done in a glove-box under nitrogen atmosphere. The chemicals used were of pro analysi grade. Zeta potentials of portland cementstone, portland blastfurnace cementstone and quartz were measured with the following aqueous electrolytic solutions:

1. 0---0.100 M K$_3$Fe(CN)$_6$ sat. vs. Ca(OH)$_2$
2. 0---0.300 M K$_4$Fe(CN)$_6$ sat. vs. Ca(OH)$_2$
3. 0.250 M KCl, KBr, KI, KC 10$_3$, KBrO$_3$ and KNO$_3$ sat. vs.   Ca(OH)$_2$.

Ca(OH)$_2$ was heated at 1100 °C for 24 hrs and ground in an agate mortar before use.

2. Results and discussion

In saturated Ca(OH)$_2$-solution (i.e. no second electrolyte added) zeta potentials ζ_o of portland cementstone, portland blastfurnace cementstone and quartz have a positive value (see figs. 1,5
and table II below). This can most probably be ascribed to preferential adsorbtion of calcium ions on solid surfaces which bear a negative charge of their own (see also Introducion). Quartz is known
to be covered by a thin layer of calcium silicate hydrate in the ambient medium (27). Its smaller value of ζ_o as compared with that of
cementstone might be explained by the larger number of strong adsorption sites for calcium ions ("holes") of the latter, which has an
intrinsic calcium- and alumina content. The larger value of ζ_o for
portland blastfurnace cementstone as compared with that of portland
cementstone might be ascribed to the larger alumina content of the
former (see table I and also Introdution).

The porosity of cementstone and quartz (28) has as a consequence
that calcium ions which adsorb on a pore wall pull their counter ions
(which may not enter the pore) strongly against the outer solid
surface. This reduces both the effective adsorbtion energy for calcium ions and the number of counter ions outside the slipping plane
(i.e. zeta potential) (28). Grinding which transforms the solid
surface from a crystalline to a glassy state (29) has the same effect
as porosity (21); sharp protuberances on the solid surface increase
zeta potential (39). The relative magnitudes of these effects with
regard to our measurements are not yet clear.

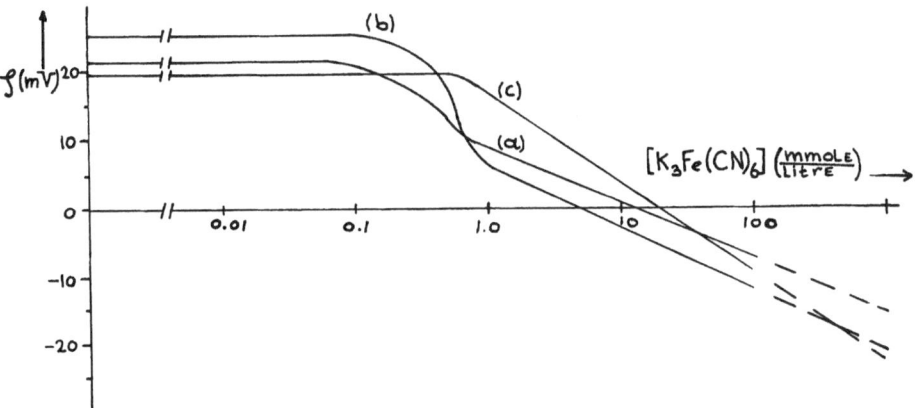

Fig. 4 Zeta potential of portland cementstone (a), portland blastfurnace cementstone (b) and quartz (c) in an aqueous solution of K$_3$Fe(CN)$_6$ kept saturated vs. Ca(OH)$_2$

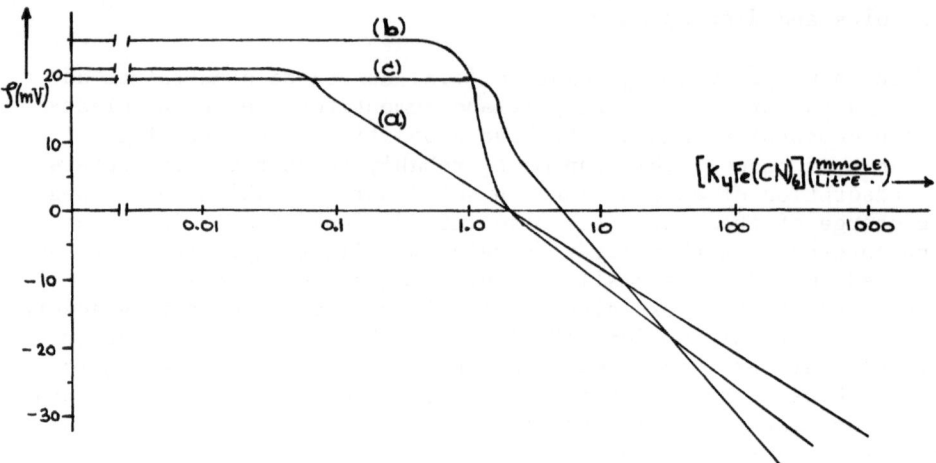

Fig.5 Zeta potential of portland cementstone (a), portland blast-
      furnace cementstone (b) and quartz (c) in an aqueous soluti-
      on of $K_4Fe(CN)_6$ kept saturated vs. $Ca(OH)_2$

Table II Zeta potential values ζ_o in sat. $Ca(OH)_2$ - solution and
          slopes S of the straight-line parts of the curves of
          figs. 4 and 5.

solid	ζ_o (mV)	S (mV/decade)	
		$K_3Fe(CN)_6$	$K_4Fe(CN)_6$
portland cementstone	+ 21.0	− 8	− 12
portland blastf. cement-stone	+ 25.0	− 9	− 15
Quartz	+ 19.3	− 13	− 21

    With increasing concentration of $K_3Fe(CN)_6$ (fig.4) or
$K_4Fe(CN)_6$ (fig.5) zeta potential curves of cementstone and quartz
bend towards the log concentration axis, gradually passing into
straight lines with slopes S as given in Table II above. The larger
charge of ferricyanide ions is clearly reflected into larger S-va-
lues. The straight line behaviour indicates - amongst others - that
for each of the solid - liquid combinations of figs. 4 and 5:

a. There are no distinct (spatial seperated) groups of sites with different adsorbtion behaviour (31)
b. There are no saturation effects such as were observed with anionic superplasticizers (to be published: see also ref. 24).

As potassium ions do not adsorb specifically on oxidic surfaces, the main mechanisms which lower zeta potential and change its sign in figs. 4 and 5 probably will be:

1. super - equivalent adsorbtion (30,32) of ferro- or ferricyanide ions on surfaces which are positive of their own due to adsorbed calcium ions
2. Desorbtion of calcium ions from these surfaces

Both mechanisms are strongly suggested by the observed raise of pH with increasing ferro- or ferricyanide concentration (see figs. 6 and 7 below), which indicates a (weak) complex formation between these anions and calcium ions. The solubility of $Ca(OH)_2$ was also found to increase with increasing concentration of potassium-ferrocyanide or potassiumferricyanide. Especially mechanism 2. above can account for the larger slope S for portland blastfurnace cement-stone as compared with that for portland cementstone (see table II). The largest S-value for quartz can be explained by its lower bond energy towards calcium ions, due to which desorption of the latter is enhanced.

In order to obtain additional information about the adsorbtion behaviour of cementstone and quartz we measured their zeta potenti-als in a 0.250 M solution of KCl, KBr, KI, $KClO_3$, $KBrO_3$ and KNO_3 respectively, which were kept saturated vs. $Ca(OH)_2$.

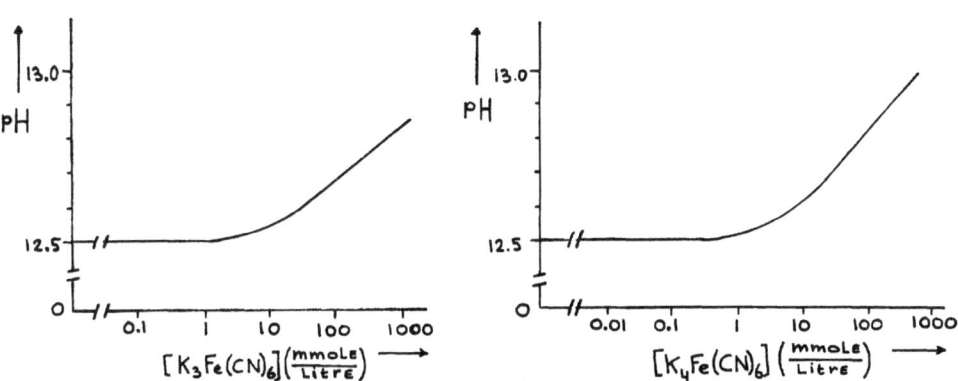

Fig. 6 pH of $K_3Fe(CN)_6$-solution,   Fig. 7 pH of $K_4Fe(CN)_6$-solution
       kept saturated vs. $Ca(OH)_2$ kept saturated vs. $Ca(OH)_2$

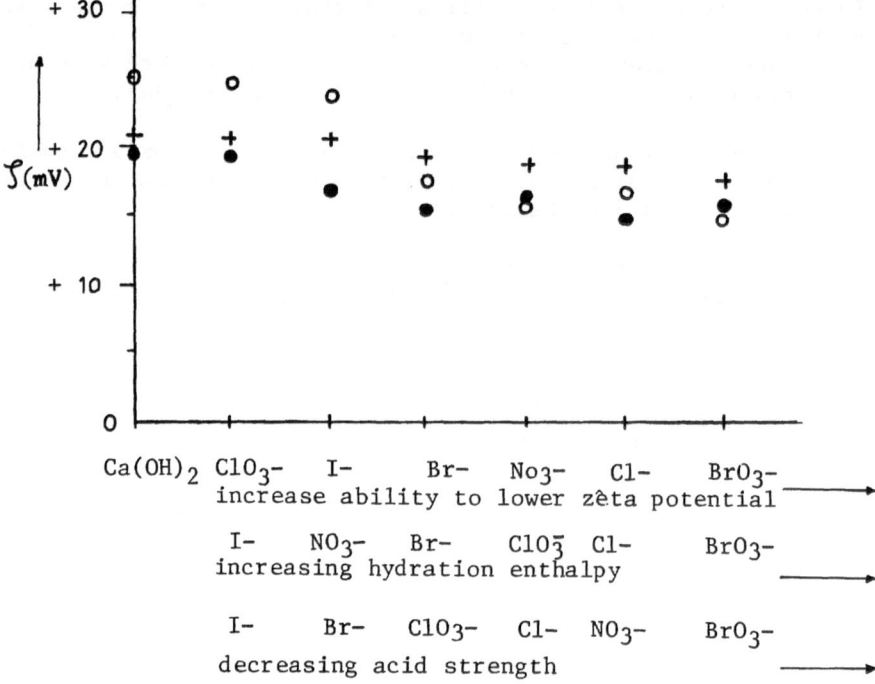

Fig. 8 Zeta potential of portland cementstone (+), portland blastfurnace cementstone (o) and quartz (.) in various 0.250 M
solutions of potassium salts, kept saturated vs. $Ca(OH)_2$

The results are shown in fig. 8 above, where we have arranged the
anions in the order of their increasing ability to lower zeta potential. No specific reactions (colour changes, phase separations, cry-
stallite growth etc.) were observed between cementstone and the solutions of figs. 4, 5 and 8. The order of the anions in fig. 8 reasonably agrees with the order of their increasing nucleophility,
such as is reflected by their increasing hydration enthalphy (33),
or the decreasing strength of their respective acids (34). Deviations
may be due to both sterical factors and partly dehydration of adsorb-
bed anions. No differences could be detected (i.e. δ pH < 0.02)
between the pH-values of the solutions of fig. 6 and a saturated
$Ca(OH)_2$-solution. Probably only weak coulombic adsorbtion of the anions occurs on the positive surfaces at the present concentration.
This effect is the strongest for portland blastfurnace cementstone,
as can be expected from its largest value of ζ_0 in a saturated
$Ca(OH)_2$-solution (see Table II). The results obtained with the anions
of fig. 8 again emphasize the importance of the adsorbtion mechanisms
1. and 2. above.

III FRACTURE TOUGHNESS OF CEMENTSTONE IN AN ELECTROLYTIC SOLUTION

1. Experimental

Fracture toughness K_{1c} of portland cementstone and portland blastfurnace cementstone was measured as a function of the concentration of an aqueous $K_3Fe(CN)_6$ -solution which was kept saturated vs. $Ca(OH)_2$. We used double cantilever beam specimens, whose webs were immersed in the solution (see fig. 9). The web was made as to conform such a profile that its increasing width exactly compensates for the effect of increasing crack length upon the critical load F_c. The relationship between K_{1c} and F_c is given by (35):

$$K_{1c}^2 = \frac{12\ F_c^2}{b.h^3.k}$$

where:

K_{1c} = fracture toughness $(N\ m^{-3/2})$

F_c = critical load (N)

h   = cantilever beam height (m)

b   = cantilever beam width (m)

k   = constant (m^{-1})

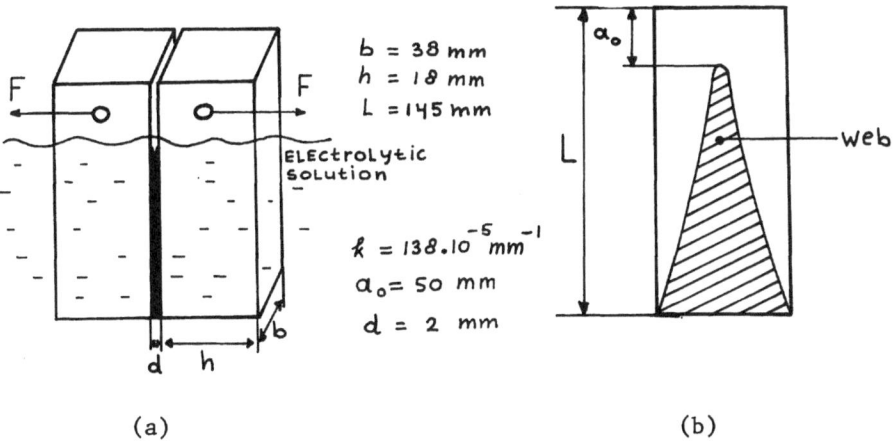

b = 38 mm
h = 18 mm
L = 145 mm

ELEctroLytic soLution

$k = 138.10^{-6}\ mm^{-1}$

a_o= 50 mm

d = 2 mm

(a)   (b)

Fig. 9 Double cantilever specimen, fractured in direct tension while immersed in an electrolytic solution (a), and cross-section of the specimen (b).

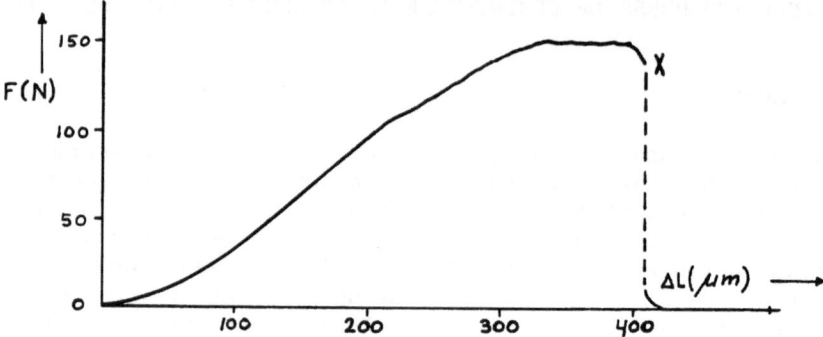

Fig. 10 Load (F) versus desplacement(Δ L) curve of double
       cantilever beam

As K_{1c} is a constant the crack will propagate at a constant load F_c
(see fig. 10). Slow stable crack growth was allowed by loading the
specimen in direct tension at a constant displacement rate of 0.30
μ m/s. Our tensile machine, which was especially constructed for this
purpose, was provided with adjustable springs to compensate for the
specimens weight, as well as pendulous grips. The load on the specimen could be detected with an accuracy of 1 % (Inductive transducer
HBM type Q3 and oscillator/demodulator HBM type MC1-A.

    The specimens, which had a W/C - ratio of 0.30, were cast in
stainless steel molds. The web was formed by a thin polished and ra-
zor-edged steel plate of the appropriate profile, which slid into
the grooved sides of the molds. After curing for 1 day at a relative humidity of 90 % the specimens were left to hydrate in a saturated $Ca(OH)_2$-solution untill they were tested after 28 days. All preparations and measurements were performed in a climatized room at a
temperature of (25.5 $\pm$ 0.2) $^{\circ}$C. All reagents used were of the same
quality as with the electroosmosis experiments (see paragraph II),
and precautions were taken to avoid contamination by CO_2 from the
air.

2.   Results and discussion

    Fig. 11 shows the behaviour of K_{1c} which we measured for port-
land cementstone and portland blastfurnace cementstone as a function
of log concentration of $K_3Fe(CN)_6$, and which we shall call the "K_{1c}-
log c correlation". Each point of the curves of fig. 11 corresponds
with the average K_{1c} - value of six samples which we took from six
consecutive casts by means of a permutation procedure (36 samples
were cast and tested for both curves).

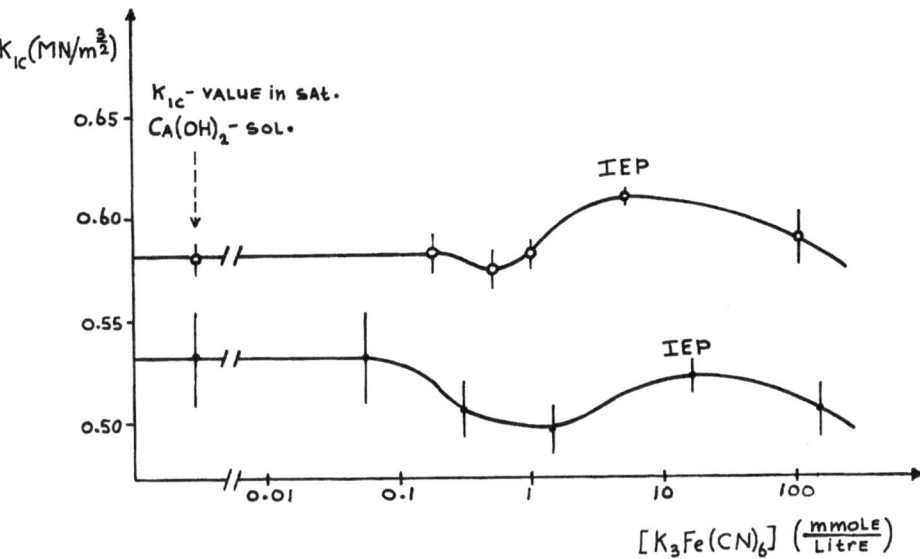

Fig. 11 Fracture toughness K_{1c} of portland cementstone (.) and portland blastfurnace cementstone (o) in a solution of $K_3Fe(CN)_6$
kept staturated vs. $Ca(OH)_2$.

For both types of cementstone there is a distinct maximum value of
K_{1c} at concentrations, which correspond with their respective iso
electric points (IEP) such as measured by means of electroosmosis
(see paragraph II). The overall variation of K_{1c} amounts to about
6 %. From the average time $\Delta t \simeq 100$ sec. we measured between the onset of stable crack propagation and final failure, and the length of
the web (see fig. 10) we estimate as an upper limit for crack velocity a value of about 100 μ m/s. This should make continious diffusion of the electrolytic solution to the crack tip possible (36).
The observed "K_{1c} - log c correlation" proves - amongst others- that
in the accessible part of the microfractured zone of cementstone a
non-neglible amount of bonds probably are present whose strengths
influence K_{1c}.

IV A MODEL FOR THE "K_{1c} - LOG C CORRELATION"

As explanation for the observed behaviour of K_{1c} as a function
of ferrocyanide concentration a model is proposed in which two elementary mechanisms are thought to influence strength - determining
bonds of cementstone. These mechanisms will be defined as Type I and
Type II respectively (see fig. 12).

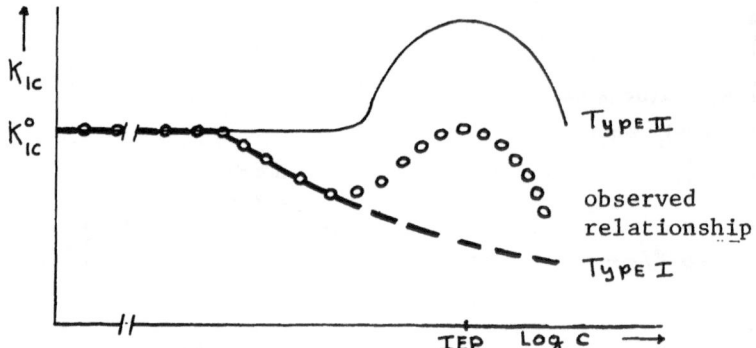

Fig. 12 Combined effect of short-range chemomechanical (Type I)
        and long-range electromechanical (Type II) processes on
        K_{lc}.

Type I mechanism causes K_{lc} to decrease with increasing log c,
and includes short-range chemomechanical processes which occur on
the plane of direct contact between the crack-tip material and the
electrolytic solution. Such processes might be for instance (see
Fig. 13):

1. Stress corrosion of surface Si-O-Si-bonds due to the increase of
   the pH - value of the solution with increasing ferrocyanide concentration (see paragraph II.2). This process is well-known
   in glass-science (36,37,38).
2. Dissolution of calcium ions under stress from surface Si-O-Ca-O-
   Si-bonds due to complex-formation with ferrocyanide ions from the
   solution (see paragraph II.2).
3, Lowering of the surface-energy of calciumhydroxide-crystals present in cementstone.

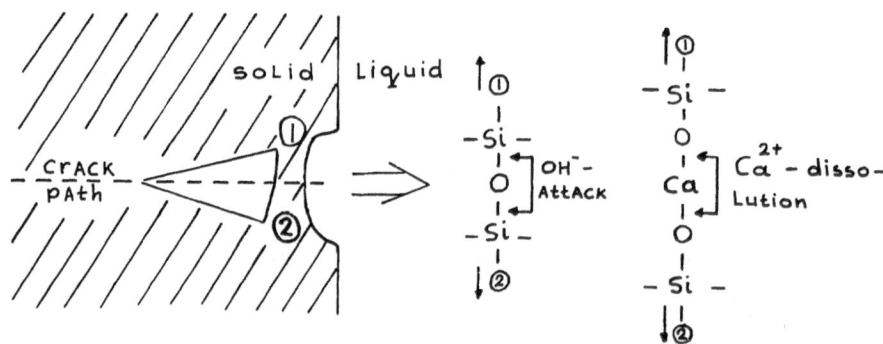

Fig. 13 Chemomechanical attack on surface bonds (Type I)

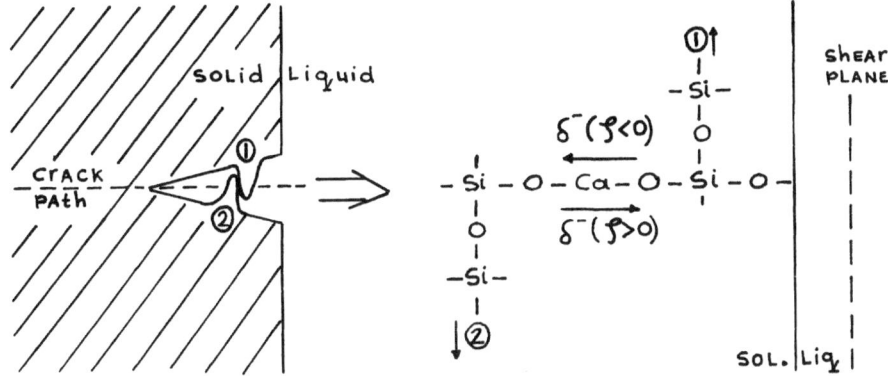

Fig. 14 Electromechanical attack on near-surface bonds (Type II)

The negative charge acquired by the solid surface with the processes
1. and 2. above can (partly) be compensated for by means of adsorbtion of calcium ions from the solution.

Type II mechanism is a long-range electrostatic process which
causes K_{1c} to have a maximum at the IEP. In order to make this process plausible we must assume that:

1. Part of the fracture process just beyond the crack-tip occurs by
   shearing-off of anionic silicate complexes (29,40,41,42,43, and
   see also paragraphs I and II) which are linked together by means
   of relative weak and polarizable bonds such as Si-O-Ca-O-Si-
   bridges (see fig. 14).
2. Due to a non-zero charge of the compact part of the electrical
   double layer in the liquid (i.e. $\zeta \neq 0$) these bonds become polarized to such an extent that the mechanical energy which is
   needed to break them is reduced (37,38). For large values of $|\zeta|$
   the polarization probably reaches a saturation value (see fig.12)

Now the extension of a diffuse charge layer is proportional to the
inverse square of the electrical carrier concentration (10,44,45).
The electric conductivity of the solid material in the present case
is smaller by several orders to magnitude (19,46) than that of the
electrolytic solution with which it is brought into contact. This
will result in a penetration-depth of an electrostatic field into
the solid from the order of 1 micron, which is a typical value for
a semiconductor in a 0.01 M electrolytic solution (45). As compared
with the submicron-dimensions of the silicates-complexes (see paragraph I) this penetration-depth is very large. However, when esti-
mating to which degree the interlinking bonds become polarized,
there is the magnitude of the electrostatic field in the solid as a
missing factor, since we cannot measure the electrostatic potential
at the solid-liquid interface.

Concluding remarks

When proposing the above model the author was aware if its preliminary character. However, it indicates some paths on should follow in selecting experiments which support it, and prepare the way to the optimization of the observed strength-concentration relationship.

Acknowlegdements

The author is indebted to Prof. Ir. P.C. Kreijger for promoting his research, to Drs. J.J. Hardon and especially to Mr. A.W.B. Theuws for his skillful technical assistance.

References

1. F. Lea, "The Chemistry of Cement and Concrete", Edward Arnold Ltd. (Publ.), London (1970)

2. H. Taylor, "The Chemistry of Cements", Academic Press, London (1964)

3. T.C. Powers, "The Properties of fresh Concrete", John Wiley & Sons, New York (1968)

4. A.M. Neville, "Properties of Concrete", Pitman Publ. Corp., New York (1972)

5. W.A. Weyl, Advances in Chemistry Series no. 33, Am. Chem. Soc., Washington, D.C. (1961), 72

6. R. Feldman and R.F. Sereda, J. Appl. Chem. 14(1964)87

7. F.H. Wittmann, Materials and Structures 1(1968)547

8. V.V. Strelko, Theoretical Experimental Chemistry (USSR) 3(1967) 263

9. D.A. Cadenhead, Progr. in Surface and Membrane Sci., Ac. Press, New York (1978), 336

10. H.R. Kruyt, "Colloid Science", Vol. 2, Elsevier Publ. Cy., Amsterdam (1952)

11. A. Dietzel, Sprechsaal (1942), 82-85

12. K. Fajans, Ceramic Age 126(1959)288

13. S.G. Lipsett, J. Am. Chem. Soc. 49(1927)925, 49(1927)1940, 50(1928)2701

14. S. Brunauer, Advances in Chemistry Series no. 33, Am. Chem. Soc., Washington, D.C. (1961), 5

15. P.A. Rebinder, Proc. 6th Phys. Conf., Moscow, 29, 1928

16. R.M. Latanision and J.T.Fourie, "Surface Effects in Crystal Plasticity", Noordhoff, Leiden (1977)

17. A.R.C. Westwood and J.J. Mills in ref. 16, p. 835

18. A.R.C. Westwood, C.M. Preece and D.L. Goldheim in: "Molecular Processes on Solid Surfaces", E. Drauglis (ed.), Mc. Graw-Hill, New York (1968), 591

19. Ch. Schulte, H. Mader and F.H. Wittmann, Cement and Concrete Research 8(1978)359

20. Chr. Hollenz and F.H. Wittmann, Cement and Concrete Research 4(1974)389

21. C.A.M. Siskens, "The interface of calcium silicates and calcium alumino silicates in an alkaline aqueous environment", Eindhoven University of Technology (Thesis), The Netherlands (1975)

22. H.N. Stein, Report CL 60/35, T.N.O. Delft, The Netherlands (1960), p. 19

23. M.E. Tadros, J. Am. Ceram. Soc. 59(1976)344

24. D.M. Roy, M. Daimon and K. Asaga, 7th Int. Congr. Chem. Cement, Paris (1980), Vol. II, p. II-242

25. G.A.C.M. Spierings, "The influence of Na_2O on the formation and colloidchemical properties of calcium aluminate hydrates", Eindhoven University of Technology (Thesis), The Netherlands (1977)

26. M. von Smoluchowski, Bull. Intern. Acad. Polon. Sci., Classe Sci. Math. Nat. 1903, 182

27. H.F.W. Taylor, J. Chem. Soc. (1950) 3682

28. J. Lyklema, J. Electroanal. Chem. 18(1968)341

29. R. Koopmans and G.D. Rieck, Brit. J. Appl. Phys. 16(1965)1913

30. E. Matjevic, "Surface and Colloid Science", Vol. 7, John Wiley & Sons, London (1974), pp. 29-31

31. J.T.A.M. Welzen, "The influence of surface-active agents on ka-
    olinite", Eindhoven University of Technology (Thesis), The
    Netherlands (1979)

32. J. Perin, J. Chem. Phys. 3 (1905)30

33. C.J.M. Houtepen, "The dehydration of some calcium aluminate
    hydrates", Eindhoven University of Technology (Thesis), The
    Netherlands (1975)

34. Gmelin, "Handbuch der anorganischen Chemie", Verlag Chemie
    Gmbh, Berlin (1927), Syst. nr. 6 ("Chlor") 315-320, Syst.
    nr. 7 ("Brom") 233, 307-310

35. J.H. Brown, Mag. of Concrete Research 24(1972)185

36. S.M. Wiederhorn and H. Johnson, J. Am. Ceram. Soc. 56(1973)192

37. S.M. Budd, Physics and Chemistry of Glasses, 2(1961)111,
    2(1961)115

38. W. Hinz, "Silikate", Vol. I, VEB verlag für Bauwesen, Berlin
    (1970)

39. F. Schröder, Zement-Kalk-Gips 9(1969)423

40. C.W. Lentz, Spec. Rep. 90, Highway Research Board (1966)296

41. E.E. Lachowski, Cement and Concrete Research 9(1979)337

42. A.K. Sarkar and D.M. Roy, Cement and Concrete Research 9(1979)
    343

43. S. Mindess and S. Diamond, 7th Int. Congr. Chem. Cement, Paris
    (1980), Vol. III, p. VI-114

44. M.J. Sparnaay, Surface Science 1(1964)213

45. H. Gerrischer, "Physical Chemistry" Vol. IX-A, Ac. Press, New
    York (1970), p. 463

46. A.K. Chatterji and T.C. Phatak, Nature 197(1963)656

4. WORKSHOP 8 – <u>Fragmentation of all types of concrete – research</u>
<u>needs</u>

Chairman: S.P. Shah, secretary M. Geudelin

Attendance: R. Brepson, M. Geudelin, O. Ishai, J.J. Mills, S. Min-
             dess, A.T.F. Neerhoff, S.P. Shah, D. Tabor, S. Ziegel-
             dorf.

4.1   Minutes of workshop 8 – M. Geudelin
4.2   Contributions
4.2.1 Recommended flow chart for future research activities concer-
      ning demolition and fragmentation of concrete – O. Ishai

4.1 MINUTES OF WORKSHOP 8

M. Geudelin

Direction de la Recherche, UTI, Paris

The group began the session by drafting a statement to define a goal: "Establish structural and material parameters relating applied energy to damage with the aim of optimizing fragmentation and demolition processes and the final recycled product".

Shah underlined the importance of distinguishing between demolition which involves structures, and fragmentation connected with materials. The workshop should emphasize research needed to improve the efficiency of both, this implied defining structural and material parameters along with an engineering and a fundamental approach.

Once the basic pattern of thought was established and approved, intense use was made of the blackboard to draft flow carts which were modified as each participant contributed with knowledge from his particular competence and experience.

The principle theme in the fundamental approach to energy/ damage was fracture mechanics, already widely debated in preceding workshop 5, 4 and 1. It appeared essential to develop valid fracture mechanics parameters for concrete (e.g. K_{ic}, K_d, fracture energy and fracture toughness, K-V curves). Several damage theories were referred to (Mines, Weibull, Kachanov) but further work was necessary to determine the best applicable to concrete. A third element was proposed for the fundamental approach, namely surface physics and chemistry of debonding and fracture, although it was agreed that many aspects of engineering approach would probably take priority.

Indeed the aim of this topic was to set up empirical measures relating energy to damage, for example load-deformation to fragment size. Under this heading the participants grouped several research needs, such as acceptable measure of energy and of damage, a relation between energy and damage(energy) per unit area of surface, energy per unit volume of fragments, energy per number of fragments. The next point-distribution of energy and damage in different fracture modes - implies study of interfacial debonding, matrix failure, aggregate-fibre fracture, elastic-inelastic behaviour, deformation-fracture etc.

The chapter of parametric studies combines the fundamental and engineering approaches to obtain the optimal energy-damage ratio depending on the method of breaking and the product to obtain. Such studies should concentrate on the problem of loading rate, size effects, loading modes, environmental effects, type of structural element, material composition (including contamination), time history and ageing (material history and condition, for example exposure to fire or thermal method of demolition).

It was hoped that by defining seperate experimental problems to be dealt with by separate sets of people, it would be possible to propose quidelines for the rational selection of the appropriate procedures in demolition and fragmentation.

Reference was made to other important issues such as environment, safety, economy, nevertheless covered by other workshops, and the advantage of studying progress in other disciplines - rock mechanics, mining, tunnelling - were stressed.

Table 1 gives the final scheme of proposed research needs.

Table 1.   Scheme of research needs for demolition and fragmentation

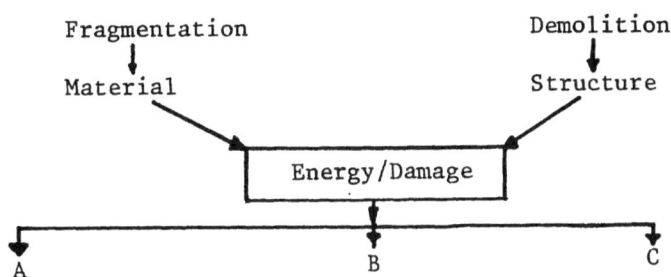

A	B	C
Engineering approach	Parametric studies	Fundamental approach

A	B	C
- empirical (measurement relating energy to damage e.g. load deformation/ /fragment size	- loading rate - size effects - loading modes - environmental effects - type of structural element - material composition (including contamination) - time history and ageing (material history and condition	- fracture mechanisms - theory of damage (e.g Mines, Weibull, Kachanov) - surface physics and chemistry of debonding and fracture

<u>Research topics and needs</u>

1.acceptable measurement of energy
2.acceptable measurement of damage
3.relation between energy and damage ⟶ e.g. energy per unit area of surface, energy per unit volume of fragments, energy per number of fragments.
4.distribution of energy and damage in different fracture modes (interfacial debonding, matrix failure, fibre fracture, elastic-inelastic behaviour, deformation-fracture)
5.influence of B on 1) to 4

<u>Research topics</u>

1.develop valid fracture mechanics parameters for concrete (K_{ic} and K_d, fracture energy, k-v curves)
2.which damage theory applies to concrete?
3.influence of B on 1) and 2)

4.2.1. <u>Recommended flow chart for future research activities con-</u>
       <u>cerning demolition and fregmentation of concrete</u>

O. Ishai, Technion Israel,
Institute of Technology, Haifa

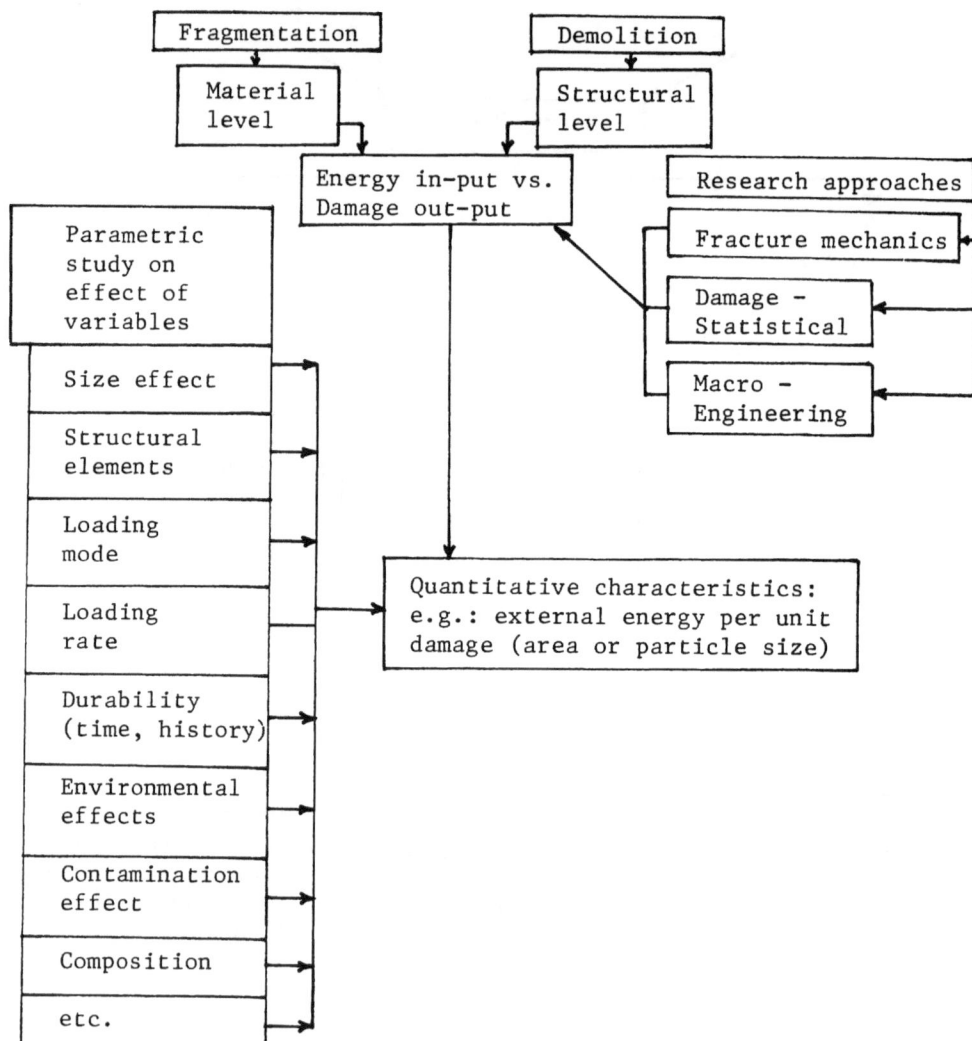

<u>Final objective</u>: To provide quantitative tools for the rational
     and optimal design of demolition and fragmentation techniques
     for a specified recycled product.

5 WORKSHOP 2 - RECYCLING OF CONCRETE
              (AGGREGATES FOR USE IN CONCRETE)

Chairman: S. Frondistou-Yannas, secretary M.J. Rubens

Attendance: B. Armbruster, Ph. Briquet, S.H. Carpenter, J-P. Col-
              lin, C. De Pauw, S. Frondistou-Yannas, T.C. Hansen,
              H.K. Hilsdorf, R. Hoffmann, O. Ishai, P.C. Kreijger,
              H. Lambotte, B. Mather, C.D. Pomeroy, V.S. Ramachandran,
              M.J. Rubens, R. Sierra, J.W. Weber, J.F. Young.

5.1   Introductory lecture: Crushed Concrete as Concrete Aggregate
                              - B. Mather

5.2   Workshop discussions

5.2.1   Minutes of workshop 2 - M. Geudelin

5.2.2   Report of workshop 2 - S. Frondistou- Yannas

5.3   Contributions

5.3.1   Classification of recycled aggregates, proposed in the
        Netherlands - P.C. Kreijger

5.3.2   Fragmentation and recycling of reinforced concrete, some
        research results - C. De Pauw

5.3.3   The role of a polymeric interlayer in improving mechanical
        characteristics of recycled concrete - O. Ishai

Bryant Mather

Chief, Structures Laboratory, U.S. Army Engineer
Waterways Experiment Station, Vicksburg, Miss. 39180
USA

INTRODUCTION

In this discussion "recycling" is used to refer to the beneficial use of a commodity or substance for more than a single cycle
of use. Hydraulic-cement concrete, a synthetic sedimentary rock,
has been being produced intentionally for beneficial use for over
2000 years. Much of this production has been recycled by entirely
similar procedures to those that have been applied for a much longer
time in the use of naturally occurring sedimentary and other types
of rocks. The classic examples of the recycling of rocks and minerals are the natural mineral, water, and the natural mineral and
rock material, soil. It is unlikely that there is any substantial
amount of water present on this planet that has not been beneficially used by people for some purpose more than once.

Turning specifically to hydraulic-cement concrete, one can
readily distinguish a variety of classes and subclasses of recycling.
If one were to start with the earliest use of rocks by people where
the rock was used--and reused--because of its shape, its mass, its
ability to resist fire and other environmental stresses such as
dissolution in water or attack by other chemical or biological
agencies; it then follows that concrete blocks or bricks, concrete
masonry units, are recycled and reused as are other masonry object
whether of natural or artificial stone without the process being

recognized as anything meriting special comment. Their re-use is
based on the obvious considerations of suitability, availability,
and economy. "Old brick," in some reuses, commands a premium price
as compared with new brick. I am not aware that "old" concrete
block or old concrete brick or any "old" precast concrete structural
elements such as bridge beams or columns or slabs has as yet reached
a degree of desirability where a premium price over comparable new
elements is commanded. Some users of hydraulic-cement concrete are
disturbed when they learn that recently placed, i.e., "new" concrete,
is a product that is still changing in levels of relevant properties
as it is in use. "Old" concrete does so much less and, as such
might be regarded as preferable for some uses. Again, so far as I
have been able to determine, this possibility has not yet emerged as
a reality. I suspect also that with the rapid increase in the use
of concrete paving elements of varied geometrically interesting
shapes, there may develop at some time in the future an antiquarian
interest in shapes that by that time are no longer made which will
be sought for reasons of nostalgia.

The nostalgia theme suggests that some mention should be made
of recycling entire structures through the method of adaptive re-
use. In the U.S.A., the National Trust for Historic Preservation,*
a national organization chartered by the U.S. Congress to encourage
public participation in preservation, works with the Heritage Conservation and Recreation Service of the U.S. Department of the
Interior and with other groups under the provisions of the National
Historic Preservation Act of 1966 and other laws and regulations to
preserve, recycle, and reuse buildings--many of which are wholly or
in part composed of hydraulic-cement concrete. In connection with
such activity there is concern about protecting hydraulic-cement
concrete in such structures from damage and about methods of resto-
ration. Even if the structure is not recognized and certified as
meriting preservation for historic reasons, many structures are recycled for purely economic reasons, i.e., it is less costly to recycle an existing structure to serve a particular purpose, different from that for which the structure was originally created, than
it is to build a new one for the current purpose. We thus have
churches, banks, armories, carriage houses, horse-car barns, railroad stations, lighthouses, used today for other purposes through
recycling.

Concrete as a Material

It is clear however that for the purposes of this particular
review, the concern is with recycling concrete as a material rather
than as structures or structural elements or masonry units. Once
a concrete structure has ceased to have utility or when a structure
must be removed to make room for some new activity at the site

*1785 Massachusetts Avenue, NW, Washington, DC 20035, USA.

which it occupies it may or must be removed. When such removal takes place, much more often than not, the rubble is regarded solely as unwanted solid waste that needs to be disposed of by transportation to a disposal site with minimum handling, processing, and transportation. Only rarely has the old concrete been regarded as useful raw material for use in new construction. The occasions when such has been the case have generally either been when the volume of rubble to be removed has been so great as to require consideration of alternatives--as after extensive bombing such as took place in parts of Europe during World War II--or when there was a shortage of granular materials for a particular need--as, for road base in portions of the middle western United States in the early stages of the construction of the U.S. Interstate Highway System.

More recently, in the United States, there has been increasing awareness that deposits of natural sand, gravel, stone, and air-cooled iron blast-furnace slag, especially those conveniently located to major urban markets, for use in producing hydraulic-cement concrete have become depleted or exhausted or have become excessively costly to procure because of environmental protection considerations and transportation costs. As a result there has been growing interest in concrete that may or must be removed as raw material to be processed as aggregate for use in new construction.

Work to see what happens when the synthetic sedimentary rock, concrete, is used as raw material for the production of crushed stone and manufactured sand as coarse and fine aggregate for hydraulic cement concrete construction has been in progress in a number of places for some time. Three major projects of the U. S. National Cooperative Highway Research Program have addressed relevant aspects of this topic:

(a) In 1972, Report 135, on "Promising Replacements for Conventional Aggregates for Highway Use" (Marek, 1972) was published. It recognized production from waste materials as a promising method. However it concluded that research was needed to characterize acceptable aggregates before one could properly assess the acceptability of unconventional materials as replacements for conventional aggregates. Another research topic that was recognized was "Use of Salvaged Structural Rubble for Aggregates." They called for a four-year study costing $300,000 to deal with the following problem: "In many instances, rubble from demolition of structures and pavements could be converted to supplemental aggregates at the site, if suitable equipment and processes were available. Characteristics of suitable equipment need to be determined. Currently available equipment needs to be evaluated to determine its adequacy. Techniques for evaluating, handling, and using the rubble should be developed. The entire system of using rubble for aggregates needs to be studied and recommendations need to be made for its economical use."; and with the objective: "To develop and

demonstrate practical methods of converting rubble obtained from
demolition of structures and pavements to useful aggregates."

(b) In 1976, Report 166, on "Waste Materials as Potential Replacements for Highway Aggregates"(Miller and Collins, 1976), was
published. Among the 31 categories of waste studied was "building
rubble" of which it was estimated that 20,000,000 tons was produced
annually in the USA. It was estimated that half of this might be
used as aggregate. The one reference given for the use of rubble as
aggregate was the paper by Buck (1972a) at HRB.

(c) In 1979, Report 207, on "Upgrading of Low-Quality Aggregates for PCC[+) and Bituminous Pavements" (P. D. Cady, et al, 1979),
was published. In the event it were to be concluded that there are
"quality" problems with aggregates produced using hydraulic-cement
concrete as the raw material, then the techniques applicable to upgrading a "low-quality" aggregate are available for use as needed.
"Adhesion" is not cited as a problem in the use of "low-quality"
aggregates in the production of hydraulic-cement concrete; only in
bituminous concrete is this cited, as "stupping." Phenomena somewhat vaguely related to adhesion are mentioned, for example:
"Objectionable Coatings" which feature is dealt with as follows:
"Surface deposits or alteration of aggregate particle surfaces may
result in reduced bond strength between the aggregate and the cement
paste. The extent of problems with aggregates in PCC owing to the
presence of surface coatings is apparently minor, as indicated by
the rare treatment of this topic in the technical literature."

The U.S. Army Corps of Engineers has, for its major civil works
construction in connection with flood control and navigation, required a quality of aggregate appropriate for the specific work in
a given contract by including in the contract a list of sources
from which, in its opinion, an adequate amount of material of satisfactory quality can be obtained. Specifically, for any given proposed job, the locally available sources of aggregates are examined,
test data on representative samples are assembled on all the various
sources, these are examined, and the sources are rated in terms of
relative quality and in most cases a line is drawn somewhere down
the list so that all sources above the line are declared to be
sources from which satisfactory material may be obtained. The
successful bidder then has the option of proposing to furnish aggregates from any one of these sources as he may elect or, if he pre-
fers, from some other one source. If he elects to propose a source
not on the list, a representative sample of the material available
in that source will be taken and tested according to the same procedures that were used to develop the data in the original study
and the results obtained will be compared with those on file for
the listed sources. If the results indicate that this material is
no worse than the poorest of those on the approved list, permission
will be given for that source to be used. However, if the results
+) PCC = Portland Cement Concrete

indicate that the material is of lower quality, the contractor will
be required to go to one of the listed sources; he only has one
chance to propose a nonlisted source. Once the question of aggregate source is settled, then it becomes incumbent upon the contractor's quality control organizations and the Government's quality
assurance people to insure that the material actually produced and
used in the work is not inferior to that upon which the acceptance
was based. And, in addition, specifically to insure that the specification requirements for grading and moisture content are met. In
some cases there will be an additional requirement pertaining to
particle shape which also must be met. The Corps of Engineers in
its civil works operations, does not require numerical limits on
aggregate quality during construction. The Corps Guide Specifications for Concrete for Civil Works Construction (OCE, 1978) specifically state (Section 8.5.1) that "Coarse aggregate shall consist
of gravel, crushed gravel, crushed stone, blast-furnace slag, recycled hydraulic-cement concrete, or a combination thereof." Fine
aggregate is required to consist of "natural sand, manufactured
sand, or a combination of natural and manufactured sands." Since
the Corps of Engineers uses the ASTM definitions of terms, it is
clear that "manufactured sand" includes crushed hydraulic-cement
concrete.

For Corps of Engineers military building construction, things
are somewhat different. There the aggregates are handled as a
specification item and reference is made to ASTM Designation: C 33
for normal weight aggregates or ASTM C 330 for structural lightweight aggregates. These ASTM specifications have specific numerical limits on a number of characteristics that relate to quality
specification limits on various types of concrete surfaces and on
the results of such tests as sulfate soundness and abrasive resistance. ASTM Designation: C 33-78 states that definitions of terms
used therein are found in Designation: C 125, which is the ASTM
Standard Definitions of Terms Relating to Concrete and Concrete
Aggregates (Designation: C 125-79a). The definition of the term
"Manufactured Sand" reads "fine aggregate produced by crushing rock,
gravel, iron blast-furnace slag, or hydraulic-cement concrete."
Thus, it is clear that so far as the ASTM Standard Specifications
for fine aggregate for concrete are concerned, if the material proposed for use is crushed hydraulic-cement concrete and if it meets
the requirements of C 33 it should be acceptable for use.

At the Seventh Congress on Large Dams, Kennedy (1961) discussed
the Processing of Aggregates for Corps of Engineers Dams. He noted
that for a series of structures the materials used included river
sand and gravel, river terrace gravel, crushed gravel, crushed limestone, crushed granite, crushed diorite, and crushed quartzite. So
far as I am aware, hydraulic-cement concrete has never been available in sufficient quantity within economical haul distance of a
Corps of Engineers civil works project to have as yet received
consideration for use as aggregate.

As noted by Marek, et al (1972), unconventional aggregates will not get fair consideration as replacements for conventional aggregates until there is a proper assessment of acceptability of conventional materials.

Typically, as indicated for example by Section 3.3 of the ACI Building Code, aggregates for concrete for reinforced concrete buildings are required to conform to a general specification, in the example ASTM C 33, Specifications for Concrete Aggregates. However, Section 3.3.2 provides an escape clause reading "Aggregates failing to meet these specifications but which have been shown by special test or actual service to produce concrete of adequate strength and durability may be used...where authorized by the Building Official."

ASTM C 33 includes a note to its scope, which note reads "This specification is regarded as adequate to ensure satisfactory materials for most concrete. It is recognized that, for certain work or in certain regions, they may be either more or less restrictive than needed." Much of the concrete in buildings does not need to resist the action of freezing and thawing when in a wet condition; in fact, relatively very little concrete in buildings must be able to resist such an exposure in order to provide satisfactory performance. Nevertheless, ASTM C 33 requires that fine aggregate must be tested by the sulfate soundness procedure and not exhibit more than a stipulated amount of degradation--or have a satisfactory service record--or perform satisfactorily in a freezing and thawing test. No option is given, other than through the general note to the scope, for waiving the soundness provisions when they are inapplicable.

The criteria for selection of aggregates for concrete need to be revised so that the levels of "quality" as measured by standard tests and the tests selected for such measurement are varied to relate to the performance required of the concrete in the service environment of the building in which the concrete is used. This concept was stated by ACI Committee 621 in 1961 in these words: "In selecting an aggregate it is economical to require only those properties pertinent to its use in a particular project." This is not the practice today.

In 1971 at the Concrete Laboratory of the U.S. Army Engineer Waterways Experiment Station, in Vicksburg, Mississippi, under the In-House Laboratory Independent Research Program, a small study of "Recycled Concrete" was begun by Alan D. Buck. It was completed and reported in 1972 (Buck, 1972). It involved crushing two kinds of discarded portland-cement concrete: (a) fragments of a driveway made using local natural siliceous sand and gravel and (b) laboratory test specimens made using a crushed limestone coarse aggregate and natural siliceous sand fine aggregate. Samples of these aggregate materials were also available and used in the study. A control was made using materials similar to those used in the driveway; a second control was made using materials from the same lots as used in the discarded test specimens. Two test mixtures used crushed, processed driveway concrete, one used it as coarse aggregate with natural siliceous sand fine aggregate; the other used recycled material as both coarse and fine. One test mixture included recycled test specimen concrete as coarse aggregate with natural sand as fine aggregate.

The use of recycled concrete as coarse aggregate at given water-cement ratio and slump required no additional cement than the control mixtures; its use as fine aggregate required a slight increase in cement content. The compressive strengths were from 300 to 1100 psi (2 to 8 MPa) lower than the corresponding controls. The resistance to freezing and thawing of concrete made with recycled driveway concrete was very much improved as compared with the control (durability factor increased from 3 to 25) but with the test specimen concrete it was reduced from 62 to 45.

In 1973 an extension of this work was begun. It was reported in 1976 (Buck, 1976). It investigated the claim that low strength concrete could not be recycled as aggregate in high-strength concrete and disproved it. Also investigated, and disproved at least for the cases studied, was the claim that unhydrated cement in recycled material stored damp as fine aggregate would cause set and lumping. It was concluded that unsatisfactory performance might be expected if recycled concrete was contaminated by gypsum from sources such as plaster if the amount exceeded 5 percent gypsum in the aggregate or 20 percent as sulfate by weight of cement in the concrete made using the recycled aggregate. The tendency of concrete containing recycled concrete as aggregate to be at lower strength for equal water-cement ratio than controls made using similar aggregate can be overcome with no increase in cement content or sacrifice of workability, by use of a water-reducing admixture; as in the case with, for example, an aggregate of less than optimum particle shape or surface texture or both.

The results of these tests, and the associated review of the literature, have been summarized in papers presented to ASTM (Buck, 1976a) and elsewhere. However, the Corps of Engineers has not authorized additional work. As part of the literature review a translation of Graf, 1948, was prepared and published (Van Tienhoven (Translator), 1973).

In July 1980 an Engineer Technical Letter (OCE, 1980) was published for the guidance of all field operating agencies having military or civil works design responsibility stating that: "Judicious use of substitute, recycled, and upgraded marginal quality materials can reduce the cost of pavements and conserve scarce materials. Use of such alternate materials may also enhance the

environment and save energy."; and listing among recycled materials:
"Recycled portland cement concrete in cement treated base, in new
portland cement concrete, in econocrete,* in trench drains, in base
course, and in shoulders."; and "Recycled rubble from buildings in
road base and in railroad ballast."

In May 1980 there was an announcement (Anon, 1980) of a concrete recycler in Detroit, Michigan, who has been in business for
six years and who has expanded his capacity to 165 tons per hour.
In July 1980, one of the construction magazines (Anon, 1980a) was
asked if concrete that had been damaged by D-cracking*could be recycled as aggregate for new concrete. The answer given was as
follows: "D-cracked pavements have been recycled but it may be too
early to know how successful the experience has been. If there is
any doubt about future susceptibility to D-cracking it would be well
to test the materials concerned. The test method is given in a
report by Paul Klieger, Gervaise Monfore, David Stark, and Wilmer
Teske, 'D-Cracking of Concrete Pavements in Ohio.' Report Number
OHIO-DOT-11-74, National Technical Information Service (NTIS),
U. S. Department of Commerce, Springfield, Virginia. (Soft cover,
201 pages, $7.25 from NTIS.)

"The first major concrete recycling project in this country to
reclaim pavement susceptible to D-cracking is a 16-mile project on
U. S. 59 between Worthington and Fulda, Minnesota. Recycling is
under way and paving with new concrete made with recycled material
is expected to begin about August 1."

Concluding Statement

Naturally occurring sedimentary rocks have been quarried,
crushed, processed, and used as aggregate in the production of
hydraulic-cement concrete for over two thousand years. Such natural
sedimentary rocks occur in a wide range of levels of properties
relevant to use in hydraulic-cement concrete. Some, such as rock
salt, have been used very successfully but only for rather limited
applications. Others such as dense relatively unweathered sand-
stones, quartzites, and limestones have been used wherever they
occur with generally quite satisfactory results. The artificial
sedimentary rock, concrete, is available under some conditions,
usually in rather limited quantity, to be quarried, crushed, processed, and used as aggregate in the production of hydraulic-cement
concrete. Use has been made of this material. Barriers to its use,
such as failure to mention it as a category of materials along with
gravel, crushed stone, and crushed slag, have been—or are being—
removed as their existence and their lack of justification for per-
petuation is called to the attention of those concerned. Probably
the potential range of levels of properties relevant to use as
aggregate that may be encountered in deposits of old concrete is
significantly less than in deposits of sedimentary rocks, e.g., few
*) see P.S. at the end of the paper.

deposits of old concrete will be found that are composed essentially of sodium chloride or calcium sulfate; however, all the relevant physical and chemical considerations that are involved in deciding to use or not to use a deposit of natural rock as concrete aggregate are relevant to making the same decision relative to a quantity of surplus hydraulic cement concrete. No more, no less.

REFERENCES

Anonymous, 1980, Detroit Concrete Recycler Expands Crushing Operation, Pit and Quarry, May, p. 17.

Anonymous, 1980a, Recycling D-Cracked Pavements, Problem Clinic, Concrete Construction, Vol 25, No. 7, July, p. 560.

Buck, A. D., 1972, Recycled Concrete, USAEWES Misc. Paper C-72-14, Vicksburg, 19 pp, 6 tables, 2 figures (NTIS AD 743-460).

Buck, A. D., 1972a, Use of Recycled Concrete as Aggregate, USAEWES Misc. Paper C-72-73, Vicksburg, (NTIS A029 832); CTIAC Report No. 9; in Highway Research Record 430, 1973, pp 1-8.

Buck, A. D., 1972b, "Recycling of Waste Concrete," Engineering and Scientific Research at WES, Misc. Paper O-72-2, October 1972, pp 3-4.

Buck, A. D., 1976, Recycled Concrete, Report 2, Additional Investigations, USAEWES Misc. Paper C-72-14(2), Vicksburg, 20 pp, 6 tables.

Buck, A. D., 1976a, Recycled Concrete as a Source of Aggregate, USAEWES Misc. Paper C-76-2, Vicksburg, 17 pp, 5 tables (NTIS AD AO 24 055); CTIAC Report No. 19. Jour. Amer. Conc. Inst., Proceedings, Vol 74, pp 212-219 (1977), also in Proceedings of a Seminar on Energy and Resource Conservation in the Cement and Concrete Industry, Canada Dept. of Energy, Mines, and Resources, Nov 1976, Paper No. 2.5A, 21 pp.

Buck, A. D., 1976b, "Recycled Concrete," The Military Engineer, Vol 68, No. 442, Mar-Apr, p 99.

Cady, P. D., P. R. Blankenhorn, D. E. Kline, and D. A. Anderson, 1979, Upgrading of Low-Quality Aggregates for PCC and Bituminous Pavements, Nat. Coop. Hwy. Res. Prog., Report 207, Transp. Res. Bd., Washington, D. C., 91 pp.

Graf, Otto, 1948, Über Ziegelsplittbeton, Sandsteinbeton und Trümmerschuttbeton. Die Bauwirtschaft, No. 2, pp 6-8, No. 3, pp 9-12, No. 4, pp 15-16 (see Van Tienhoven, 1973).

Kennedy, Thomas B., 1961, "Processing Aggregates for Corps of Engineers Dams," Proc. Seventh Congress on Large Dams, Rome, Report 52, 16 pp.

Marek, Charles R., Moreland Herrin, Clyde E. Kesler, and
        Ernest J. Barenberg, 1972, Promising Replacements for
        Conventional Aggregates for Highway Use, Nat. Coop.
        Hwy. Res. Prog., Report 135, Transp. Res. Bd., Washing-
        ton, D.C., 53 pp.

Miller, Richard H. and Robert J. Collins, 1976, Waste Materials
        as Potential Replacements for Highway Aggregates, Nat.
        Coop. Hwy. Res. Prog., Report 166, Transp. Res. Bd.,
        Washington, D.C., 94 pp.

Office, Chief of Engineers, 1978, "Concrete - Civil Works
        Construction" - Guide Specification - CW 03305, U. S.
        Government Printing Office, Washington, DC.

Office, Chief of Engineers, 1980, "Alternate Materials in
        Pavements," Engineer Technical Letter 1110-1-107,
        Washington, D.C.

Van Tienhoven, Jan C. (Translator), 1973, Crushed-Brick Con-
        crete, Sandstone Concrete, Rubble Concrete by Otto
        Graf, Stuttgart, USAEWES Translation No. 73-1,
        Vicksburg, 21 pp, 8 figures.

"P.S. Subsequent to the completion of this manuscript the July-
August 1980 issue of Transportation Research News (No. 89) was
received. It contains (pp 6-10) an article by H. J. Halm of the
American Concrete Paving Assn. on "Concrete Recycling" in which
the work by Buck is reviewed and examples of recycling of pavement
concrete are cited."

*) Econocrete = a form of hydraulic-cement concrete developed by
                the American Concrete Paving Association containing,
                typically, less completely processed aggregates
                and less cement than would be used in concrete for
                paving but more cement than used in cement - treated
                base course that can be slip form paved by normal
                equipment.

*) "D"-cracking= a form of deterioration of hydraulic-cement concrete
                resulting from the interaction of freezing and
                thawing cycles with water-saturated porous coarse
                aggregate, manifested by fine cracking parallel to
                joints, cracks and edges of slabs.

5.2.1 MINUTES OF WORKSHOP 2

M. Geudelin

Direction de la Recherche, UTI, Paris

Mather opened his introductory lecture by commenting that concrete is a synthetic sedimentary rock, most of which has already
been recycled in natural processes, as is the case for many minerals.
Viewed from this angle, recycling is not a new problem.

He then traced the background of research in the U.S.A. where
no advantage has been drawn from natural disaster debris (earth-
quakes), neither had the country been faced with the disposal of war
rubble. Several universities had nevertheless set up research projects for recycling for example the University of Illinois (1970)
to replace convential aggregate for highways and the University of
Pennsylvania which had concentrated on upgrading aggregate.

Mather's own agency - The U.S. Army Corps of Engineers - is a
large user of portland cement concrete and drafter of acceptable
requirements which affect the state-of-the-art. Work includes examining local aggregates to decide which source to use for any particular undertaking. This has not yet included recycled aggregate, although the ASTM C33 contains an escape clause mentioning "crushed
concrete".

A proper assessment of acceptability of conventional aggregate
is needed before tackling problems of unconventional material. At
the moment such assessment is based on what is available, rather
than required properties.

Research is still called for to establish a reasonable attitude
to natural aggregates. Mather concluded with slides illustrating
work by Buck where a comparison of virgin and recycled materials
pointed to a fall in compressive strength and an increase in frost-

thaw resistance in the latter case. In Mather's opinion all concrete
should be regarded as a quarry of sedimentary rock, which implies
that the requirements for recycled and natural aggregates should be
identical.
The ensuing discussion first centred on drafting definitions,

De Pauw: "demolition concrete" = "recycled concrete made with
demolished concrete aggregate", which are "aggregates obtained by
fragmentation of demolished concrete", then an attempt to classify
such aggregate according to performance i.e for use in high, medium
and low strength concrete. Kreijger (see 5.3.1) remarked that such
is approached in the Netherlands with an added distinction in size
(> and < 4 mm) but there was also a need to combine with exposure
conditions regarding moisture and quality control and assurance.
Classification implies criteria in Lambotte's opinion. The material is
no longer inert, creep and shrinkage are possible. Mather quoted
ASTM C33 which characterizes concrete in terms of the nature of the
structure and its probably exposure (three climatic zones). De Pauw
reported (see 5.3.2) on research underlining the influence of the
characteristics of the old concrete on the properties of the new,
as regards cement content and type, w/c-ratio, workability.

This prompted Hansen to present Danish research to produce high,
medium and low strength concrete with recycled aggregate , pointing
to the problems facing ready-mixed concrete manufacture because
of quality variation. An adsorbtion coefficient of 8.9% leads to
difficulties in maintaining w/c-ratio, also yield. Mix proportioning
must be constantly adjusted according to the aggregate density.

There was immediate consensus concerning the need to define
properties - long-term performance, creep (Frondistou-Yannas), tensile strength, dry/moist strength, fracture toughness and mechanisms
of failure (Ishai), carbonation (Kreijger), not forgetting improving
aggregate/matrix bond through research on the alkali-aggregate reaction, use of polymers (Ishai, see 5.3.3) Ishai asked if interfacial
bonding was considered to be a crucial problem since it seems that
in most cases failure of recycled concrete is within the mortar
phase rather than through the aggregate-cement interface.
Frondistou-Yannas thought the real mode of failure of the recycled
system is not known for sure. One has to consider at least two interfaces, namely that between the old aggregate and its cementive
coating, and that between the old, recycled aggregate and the new
mortar. It is reasonable to assume that a crack will initiate at
one of these interfaces, and crack propagation will certainly be
affected by interfacial bonds. Such a complex mechanism is very
difficult to detect in compressive tests, and further tests, e.g.
flexural or indirect shear and tension will have to be resorted to
for this purpose. It can be concluded that a fundamental investigation should be made on the modes of the failure predominating in,
and thus affecting the performance of, recycled concrete.

The importance of production technology was not neglected. Frondistou-Yannas emphasized the need for pilot plants as a first step to promote reuse of recovered concrete. De Pauw gave details of Belgian inventions in the field of crushers and explosives (see also 5.2.2) to improve the properties of the aggregate obtained, especially as regards shape, although there remained the problem of the influence of the characteristics of the old concrete (for example gypsum content, and the effect of recycling fire or frost damaged structures (Hansen).

5.2.2 REPORT ON WORKSHOP 2

RECYCLED CONCRETE AS AGGREGATE FOR NEW CONCRETE

S. Frondistou-Yannas

President Management and Technology Associates, Inc.

Newton, Massachussetts

In this workshop the need for research was established that would allow better decisions on concrete recycling. It was suggested that there is need for a better definition of "recovered aggregate" as well as a scheme to classify such aggregate. Moreover, it was suggested that more research is needed in the area of properties of concrete produced with pieces of old concrete as aggregate. Additionally, research is needed in the areas of production technology and quality assurance.

The need for a definition for "recovered aggregate" was felt by some members of the group. Is "recovered aggregate" pieces of stone from concrete rubble, is it pieces of mortar from such rubble, does it include fines after crushing, can it orginate with any kind of concrete debris, or, can it orginate with nonconcrete structures like highway bases and subbases?

The question of classification of "recovered aggregate" arose at that point. There is currently no concensus even on what are appropriate criteria for classification: Is it density, is it performance of the aggregate under well specified conditions, or is its suitability to produce a concrete with given performance characteristics? The group was advised that, in the Netherlands, a classification of recovered aggregate according to the performance properties of the resulting concrete is being contemplated. As an example, such a scheme (see 5.3.1) would list together all "recovered aggregate" that could potentially yield high strength concrete in wet or dry exposure conditions. The practicality of the above classification was questioned by some members of the group.

Clearly there is a need to learn more about some properties

307

of the concrete produced with "recovered aggregate". Included here
are toughness properties, creep performance and pumpability.
Moreover, there is a need to determine which of the characteristics
of concrete debris will importantly affect the properties of the
new concrete in which the aggregate comprises debris. It was suggested that the type of cement in the concrete debris as well as the
strength of such debris will significantly affect the strength of
the new product. Some members of the group felt that the aggregate-
matrix bond strength as well as the processes for manipulating such
strength through use of admixtures is an important area for study
for concrete produced with recovered aggregate.

In the area of production technology it was strongly felt
that a pilot plant that includes sorting facilities is needed at
this point. Indeed, the opinion has been expressed that the succes-
ful operation of such a pilot plant would probably be the most effective means of inducing entrepreneurs to enter the area of concrete
recycling.

Careful design of a quality assurance programm is essential
during the early steps of the "new" aggregate. Included are the
setting of tolerance limits for concrete debris contaminants as
well as tests for quality control.

5.3.1 CLASSIFICATION OF RECYCLED AGGREGATE, PROPOSED IN THE NETHERLANDS

P.C. Kreijger

University of Technology of Eindhoven

Den Dolech 2, 5600 MB Eindhoven, The Netherlands

In the Dutch Research Committee, dealing with recycled concrete made from building rubble, a proposal for preliminary classification (table 1) from Ir. J.G. Wiebenga of the Institute T.N.O. for Building Materials and Building Structures was agreed upon.

Table 1. Premilinary classification of crushed building rubble qualities for size < 4 mm and areas of application.

application — types of rubble	pre-stres-sed con-crete	reinforced concrete σ_c 1) $\leqslant$ 22.5 N/mm^2		plain con-crete, all qualities		plain con-crete, σ_c 1) $\leqslant$ 22.5N/mm^2	
	all qua-lities	exposure		exposure		exposure	
		moist	dry	moist	dry	moist	dry
concrete	yes	yes	yes	yes	yes	yes	yes
brick work: SO$_3 \leqslant$ 1%	no	yes	yes	no	no	yes	yes
SO$_3$ > 1%	no	no	yes	no	no	no	yes
sand lime brickwork	no	yes	yes	no	no	yes	yes
concrete brickwork	no	yes	yes	no	no	yes	yes
$\sigma_c = \sigma$ average - 1.64 x standard deviation							

Concrete aggregates   = ⩾ 95% of fragmented concrete debris

Brickwork aggregates = ⩾ 2/3 of fragmented brickwork rubble and ⩽ /3
                           of fragmented concrete – fragmented light-
                           weight-concrete, fragmented sandlime brick-
                           work – and fragmented natural stone rubble

sandlime brickwork
aggregates   =   ⩾ 2/3 of fragmented sandlime brickwork
                           rubble and ⩽ 1/3 fragmented concrete, frag-
                           mented light weight concrete –, fragmented
                           brickwork – and fragmented natural stone
                           rubble.

concrete brickwork
aggregates   =   ⩾ 2/3 of fragmented concrete brickwork
                           rubble and ⩽ 1/3 fragmented concrete –,
                           fragmented brickwork – and fragmented natu-
                           ral stone rubble

5.3.2   FRAGMENTATION AND RECYCLING OF REINFORCED CONCRETE

SOME RESEARCH RESULTS

IR.C. DE PAUW

C.S.T.C. - W.T.C.B.

BELGIUM

N.A.T.O. - ADVANCED RESEARCH INSTITUTE ON
ADHESION PROBLEMS IN THE RECYCLING OF CONCRETE
SAINT-REMY-lès-CHEVREUSE 25th-28th NOV. 1980

1. INTRODUCTION

What follows is a short presentation of some results of a research project on demolition, recycling and dismantling of reinforced concrete. This paper deals with some of the results obtained
with the fragmentation of 30 types of 15 years old concrete, with
separation of the reinforcement by means of explosives and with the
recycling of the fragments as aggregates for new concrete.

2. CHARACTERISTICS OF THE OLD CONCRETE

The table 1 below shows the characteristics of 12 types (12 out
of 30) of old concrete

(Characteristics of the fresh concrete 15 years ago)

Table 2 shows the characteristics of the old concrete
(hardened and at different ages) (also 12 types out of 30).

TABLE 1

CHARACTERISTICS OF THE OLD CONCRETE (FRESH CONCRETE 15 YEARS AGO)

Code	Composition $(1\ m^3)$				Consistency
	Cement type and content (kg) (a)	Type and content of coarse aggregate (kg) (b)	Type and content of fine aggregate (kg) (c)	w/c ratio	Jolting table
11	PN 350	P 1200	R 650	0,53	1.28
12	PN 350	P 1200	R 650	0,67	1.54
13	PN 250	P 1260	R 715	0,65	1.16
14	PN 250	P 1260	R 715	0,80	1.36
51	P-C_3A 350	P 1200	R 650	0,50	1.38
52	P-C_3A 350	P 1200	R 650	0,59	2.37
53	P-C_3A 250	P 1260	R 715	0,65	1.22
54	P-C_3A 250	P 1260	R 715	0,81	2.23
15	PN 350	P 1200	C 630	0,50	1.27
16	PN 350	P 1200	L 721	0,50	1.11
17	PN 350	P 1200	R 653	0,53	1.41
18	PN 350	C 1200	C 635	0,50	1.28

a) PN   : Portland Cement
   P-C_3A :   "   " "without" C_3A

b) P   : Porphyre (crushed)
   C   : Limestone (crushed)

c) R   : River sand
   C   : Limestone (crushed)
   L   : Furnace Slag (molded)

TABLE 2

CHARACTERISTICS OF THE OLD CONCRETE (HARDENED)

Code	At 28 days (15 years ago)	At 15 years age		
	Compressive strength N/mm^2	Compressive strength N/mm^2	Density	Water absorption % by weight
11	55.7	75.1	2.41	4.56
12	44.9	51.5	2.39	5.61
13	40.0	59.3	2.45	4.32
14	27.3	38.9	2.34	6.27
51	50.9	73.1	2.44	4.65
52	36.7	62.4	2.38	6.33
53	36.1	67.9	2.39	4.90
54	20.2	42.1	2.40	5.93
15	34.2	61.9	2.43	5.70
16	51.9	84.8	2.45	4.79
17	41.3	73.4	2.41	5.36
18	35.4	64.1	2.39	5.98

3. FRAGMENTATION OF THE OLD CONCRETE WITH SEPARATION OF THE
    REINFORCEMENT (research in collaboration with the laboratory
    of P.R.B. Nobel-Explosifs)

The different types of 15 years old concrete were submitted to
a treatment with explosive charges placed between two reinforced
concrete elements as in fig. 1 and fig. 2.

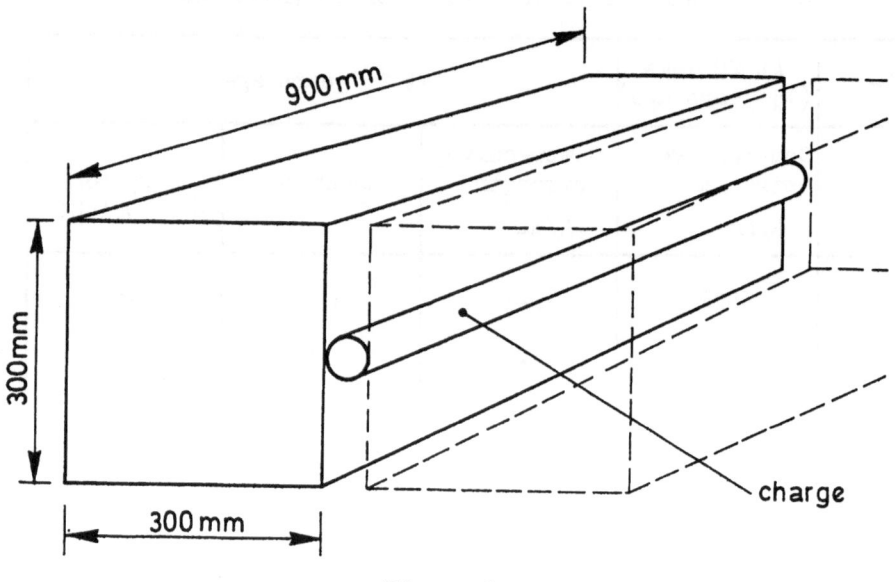

Figure 1

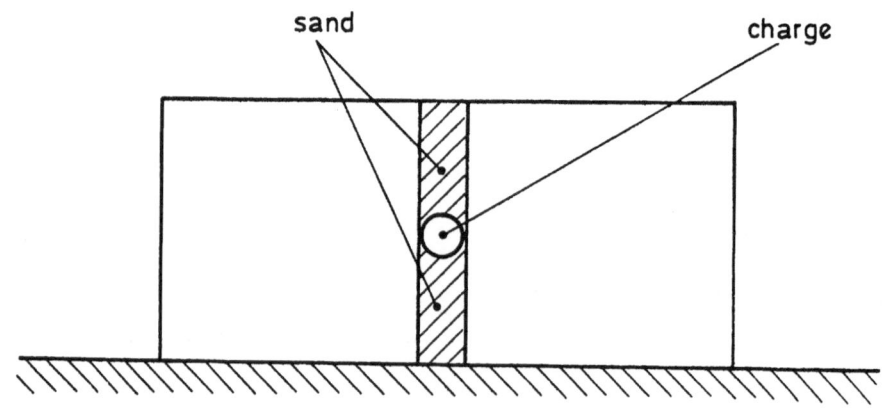

Figure 2

The concrete was in all the cases completely separated from the reinforcing bars by the explosion.

It was found that a fairly good relationship exists between the particle size (average) and the strength of the explosive charge as shown in fig. 3.

4. RECYCLING OF CONCRETE

The fragments obtained from the treatment with the explosive charges as described above were then used as aggregates for the new "recycled" concrete, without any further crushing. Before reuse, the fragments with a size under 2 mm and above 40 mm were eliminated.

One single composition was used for the manufacturing of the recycled concrete,

for 1 m^3 of recycled concrete :

- 1250 kg recycled "coarse" aggregate
- 600 kg river sand 0 - 2
- 350 kg Portland Cement P40
- w/c ratio : 0,57

The characteristics of the hardened recycled concrete are shown in table 3.

Finally the fig. 4 gives the relationship between the compressive strength of the old concrete (at the age of 28 days , 15 years ago) and the compressive strength of the recycled concrete (at the age of 28 days).

Fig. 5 on the other hand gives a more practical and useful relationship between the compressive strength of the old concrete at the age of 15 years and the compressive strength of the recycled concrete at the age of 28 days.

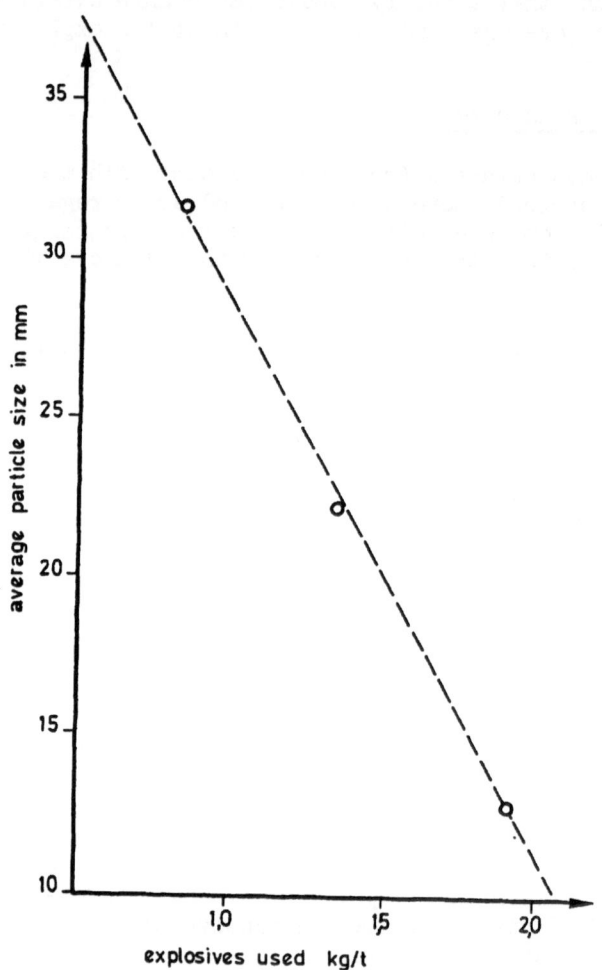

Figure 3

Influence of explosive charge on particle size

TABLE 3

CHARACTERISTICS OF THE HARDENED RECYCLED CONCRETE

Code to the old concrete	Hardened recycled concrete		
	Density	Water Absorption % by weight	Compressive strength N/mm^2 28 days
11	2.25	7.98	49.1
12	2.20	8.30	40.3
13	2.26	7.63	43.1
14	2.22	8.12	38.0
51	2.24	8.13	47.4
52	2.25	7.88	43.3
53	2.26	7.83	41.8
54	2.21	8.14	32.0
15	2.26	8.55	39.8
16	2.25	8.17	36.8
17	2.25	8.36	44.0
18	2.24	8.85	35.2

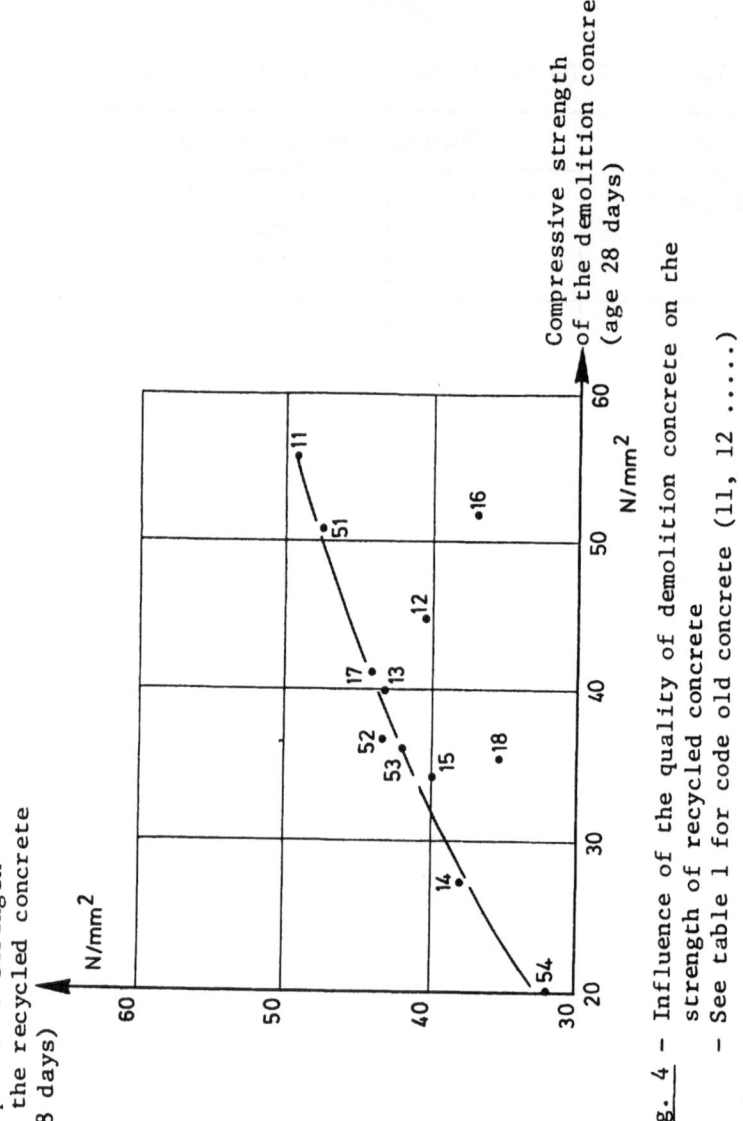

Fig. 4 — Influence of the quality of demolition concrete on the
strength of recycled concrete
— See table 1 for code old concrete (11, 12 ......)

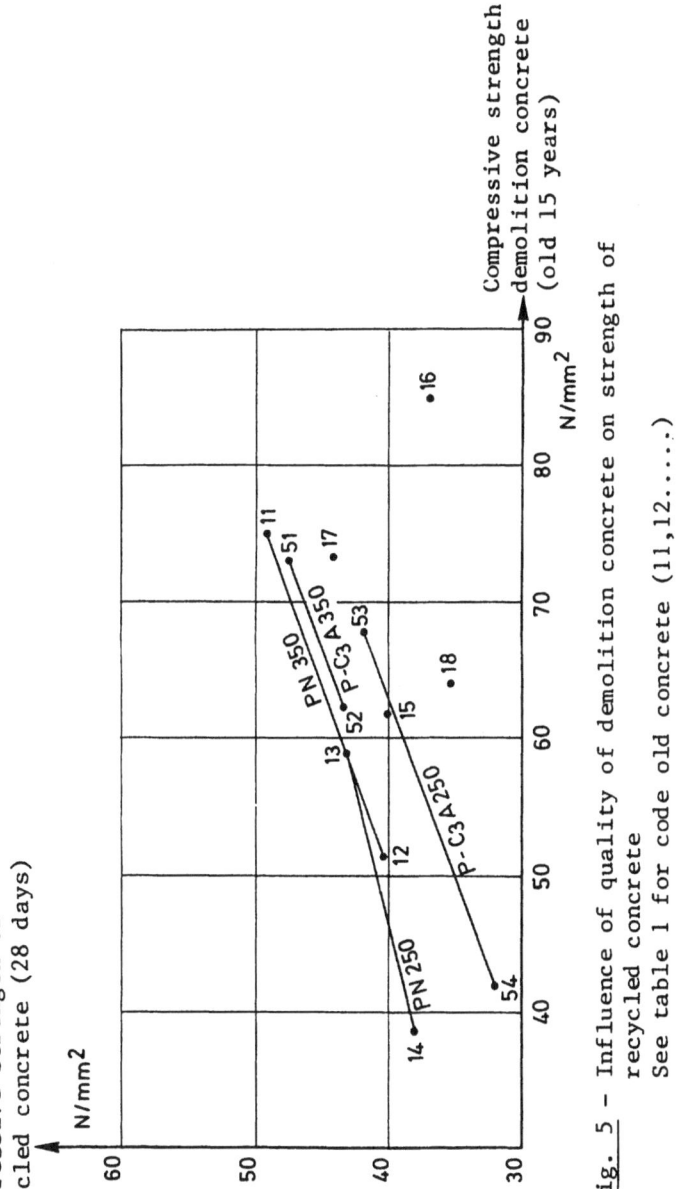

Fig. 5 – Influence of quality of demolition concrete on strength of recycled concrete
See table 1 for code old concrete (11,12.....)

5.3.3 THE ROLE OF A POLYMERIC INTERLAYER IN IMPROVING MECHANICAL

CHARACTERISTICS OF RECYCLED CONCRETE

O. Ishai

Technion Israel

Institute of Technology, Haifa

It is suggested to investigate the feasibility of introducing a thin layer of a thermosetting adhesive (epoxy) between the coarse concrete rubble (recycled aggregate) and the fresh cementive mortar matrix. Such an interlayer may improve the mechanical performance of the recycled concrete in three ways, as follows:

a. Increasing interfacial bonding at the crucial interfaces between the old and the new concrete phases;

b. Providing ductile phases, which reduce stress concentration and contribute to an increase of the fracture toughness of the whole brittle concrete system.

c. Polymeric coatings of the old - probably contaminated - concrete rubble provide insulation, which may protect the new mortar and reinforcement from the deleterious effects of chemicals and moisture diffused outward from the recycled concrete parts.

An investigation of the role or polymeric interfacial interlayers may be conducted at three levels:

- Basic research into their effects on mechanical characteristics - strength, ductility, fracture toughness, and on durability under different hygrothermal conditions.

- A study of the technical feasibility of introducing a thin polymeric layer into the fresh concrete mix, and of adequate curing processes.

- Economic feasibility study comprising the cost-effectiveness of

introducing relatively expensive polymeric constituents and special processes into the production sequence of recycled concrete.

6 WORKSHOP 3 - RE-USE OF CONCRETE (OTHER THAN AS AGGREGATE FOR
         CONCRETE)

Chairman: C.D. Pomeroy, secretary M.J. Rubens

Attendance: S.H. Carpenter, J-P. Collin, G. Frohnsdorff, S. Fron-
              distou-Yannas, T.C. Hansen, P.C. Kreijger, B. Mather,
              D.M. Roy, M.J. Rubens, S.P. Shah, P. Sierra.

6.1   Introductory lecture: Recycling of concrete into new
           applications - R.L. Berger, S.H. Carpenter

6.2   Workshop discussions

6.2.1   Minutes of the presentation of the introductory lecture
           - M. Geudelin

6.2.2   Report of workshop 3 - C.D. Pomeroy

6.3   Contributions

6.3.1   Example of re-use of concrete in France - P. Sierra

6.3.2   Reflecting properties of a road surface - J.H. de Boer

6.1 RECYCLING OF CONCRETE INTO NEW APPLICATIONS

R. L. Berger
S. H. Carpenter
Department of Civil Engineering
University of Illinois
Urbana, Illinois U.S.A.   61801

INTRODUCTION

The recycling of materials used in outdated construction into new construction is probably as old as civilization itself. Although documentation is limited, certainly materials from various existing Roman, Egyptian, Greecian, etc. structures had been cannibalized for use in other construction. The advent of a highly mechanized society with low energy costs and the abundance of raw materials resulted in low cost construction materials. However, as labor costs escalated more rapidly in this environment than material or energy costs the rapid disposal, rather than recycling of construction materials was encouraged. To compound the overall situation, the invention of the rotary kiln led to the widespread production of virtually unlimited quantities of inexpensive portland cement. This along with the expanded use of steel reinforcement resulted in widespread use of concrete to make monolithic structures. The monolithic nature of concrete which makes it so unique and useful also has made its reclamation difficult and costly compared to the cost of using virgin materials in new concrete construction. However, with the continuing rise of energy costs and depletion of quality aggregate, particularly in areas of high concrete usage, it has become necessary to consider the reutilization of the quality materials in existing concrete construction scheduled for demolition. A number of recent studies have focused on the reutilization of concrete as aggregate in new concrete pavement construction.[1-7] Although the recycling of concrete as aggregate in new concrete construction is most promising, there are other potential uses to which recycled concrete can be put. These uses may depend on type of aggregate present in the concrete as well as the use of special methods of demolition.

Demolition of concrete is a subject of another session in the current workshop however, it is necessary to consider the current methods of demolition when considering utilization of concrete rubble. Due to the monolithic nature of concrete and the presence of reinforcement steel it is necessary to crush the concrete to a relatively small size to disengage it from the reinforcing steel. This results in a granulated concrete of varying size depending on the original reinforcement spacing. Granulated concrete can be used in a number of applications some of which are independent of the type of aggregate present and other applications which may depend strongly on the aggregate type. The current paper discusses applications to which recycled concrete has been put as well as speculates on other possible applications.

PAVEMENTS

Pavement construction requires a large amount of aggregate both during construction and for rehabilitation. Crushed concrete can be used in nearly any instance where a normal aggregate would be used, although some uses may pose quality problems that would preclude their economical use. To date, concrete recycling has been involved generally with the use of crushed pavement concrete as aggregate in new pavements. Several reasons may be put forth for the preferential processing of crushed concrete aggregate from pavements into new pavements rather than using crushed concrete from, and consequently into, other construction.

First, pavement concrete to be crushed is usually in the exact place where the new pavement will be constructed. This makes the logistics problem easier, and the contractor can develop bid strategies easier and perhaps cheaper. Secondly, concrete from a given pavement tends to be more or less constant in quality, thickness, and reinforcement placement. In that reinforcement removal has been a major factor in reclaiming concrete, the consistent pattern of reinforcement placement in pavements makes it easier to reclaim than structural concrete where the amount and arrangement of reinforcement may result in difficulties in the separation of concrete and steel. In addition, concrete pavements appear to be easier to remove and transport than structural concrete from other sources. This may be due partially to the familiarity of the material processors with concrete pavements as they may be the same contractors who construct pavements. Thirdly, since concrete pavements do not contain the contamination of other building materials commonly present in, e.g., demolished concrete buildings, the crushed concrete pavements do not require sophisticated separation of "contaminates" before use in new construction. Finally, failure of a concrete pavement which utilized crushed concrete aggregate would not have the same potential consequence as the failure of such concrete used in a building.

All of these factors cannot be addressed here. With time
some of them have begun to change. One item that maintains its'
grip on the public and the highway industry is that salvaged
concrete must be placed as recycled concrete to provide the most
cost effective product. One fact that cannot be ignored is that
pavement structures will continue to require large amounts of
aggregate. Therefore, pavements will continue to provide a source
for recycled concrete. Several of the more noteworthy concrete
pavement recycling projects completed and their important features
are given below.

1. 1964 - Love Field, Dallas - Six inch cement treated sub-
    base utilizing crushed concrete aggregates.[8]
2. 1974 - California - Crushed pavement concrete as aggregate
    in econocrete subbase. First econocrete in U.S.[9]
3. 1976-77 - France - Crushed pavement concrete as aggregate
    in econocrete base, subbase and new surface.[8]
4. 1976 - Iowa - Crushed pavement concrete aggregate in
    econocrete subbase and new surface, placed as a monolithic
    composite pavement. Success led to inclusion of concrete
    recycling as a rehabilitation alternative.[10]
5. 1977 - Iowa - Two major contracts for concrete recycling
    were let.[11]
6. 1980 - Minnesota - A 16 mile concrete pavement project
    was constructed using crushed pavement aggregate from a
    "D" cracked concrete pavement.[12]
7. 1980 - Conneticut - 1000 ft. test section of concrete
    was constructed using crushed pavement concrete as
    aggregate.[13,14]
8. 1980 - Illinois - Crushed concrete aggregate used as sub-
    base material for a 15 mile 6 lane expressway.

The use of crushed concrete as aggregate in new pavement construction is accelerating rapidly. However, as can be seen in the above
list, crushed concrete can also be used in subbase construction
as well as in new pavement concrete. As a previous workshop has
considered the use of crushed concrete as aggregate in new concrete
we will consider the other aspects of use of recycled concrete in
pavement construction.

Subbase or Base Course

There have been several installations that have used crushed
concrete as a straight granular material.[15-22] The largest project
of this kind was the rehabilitation of the Edens expressway in
Chicago, Illinois.[23,24] The 15 mile six lane freeway had carried
a projected 45 years of traffic in its 29 years of service. The
total rehabilitation cost of $113.5 million is the largest roadway
type highway contract ever awarded in the United States.

The concrete surface was broken by diesel pile drivers and hauled to the crushing site where it was processed into two sizes, plus 3 1/2 inches and minus 3 1/2 inches. These two sizes were used to form two gradations of aggregate typically used in Illinois for base courses. The crushed material was hauled to the project and placed on the specially prepared subgrade and compacted as a subbase. A bituminous base, also utilizing recycled bituminous materials from the old bituminous overlay from the Edens, was placed, over which a continuously reinforced concrete pavement was then placed.

Earlier installations also have utilized crushed concrete in this manner. Love Field, Dallas, Texas utilized old runway concrete as a stabilized subbase. One of the earliest uses was in Illinois during the mid 1940's. An old concrete pavement was crushed and used as aggregate base for post-war construction.[8]

The requirements for using crushed concrete as an aggregate base pose no problem other than obtaining the proper gradation. Contaminants other than gypsum, should pose minor problems and the quality of the aggregate should be more than satisfactory. The fines produced during the crushing operation will be non-plastic and may require the addition of fines from the subgrade or elsewhere to provide the needed plasticity for good compaction and density.

Bituminous Mixes

From a quality consideration there is no reason why crushed concrete cannot be used in bituminous mixes. The aggregate quality is excellent to begin with. The gradation requirements for most bituminous mixes requires crushing of the concrete to minus 3/4 inch size. This removes a very large portion of the cement paste on the aggregate surfaces and provides fresh faces for the asphalt. This crushing also increases the angularity of the particles which in the initial concrete may have been a more rounded material than is commonly used in bituminous mixes.

Much concrete contains siliceous or gravel aggregates. These type of aggregates have a very high potential to develop stripping in bituminous mixtures. Stripping is an adhesion problem at the aggregate-asphalt interface. The presence of moisture enhances this separation. Stripping produces an unstable mixture requiring complete removal and replacement. One additive which reduces the stripping potential is hydrated lime, another is portland cement. Thus, the use of crushed portland cement concrete may lower the stripping potential through the presence of cement paste in the finer material sizes, and/or the coating of carbonated cement paste on the surface part of the aggregate, which may increase the microtexture of the aggregate surface.

The influence of other contaminants, such as the deicing salts used on pavements, on the development of stripping would have to be evaluated.

One problem may arise because of the porous nature of the crushed concrete, mainly in the cement paste structure. Initial tests indicate that the absorption of bitumen by crushed concrete aggregate is high compared to that of a normal bituminous aggregate. This increased absorption requires more bitumen for a proper mixture. Although this does not affect the quality of the mixture, it is an economic consideration in that some of the cost savings produced by using crushed concrete may be lost due to the increased amount of asphalt cement required, as illustrated in Figure 1.

Figure 1 is constructed using current hot mix cost data:
$$\text{Aggregate} \qquad \$3.50 \text{ per ton}$$
$$\text{Asphalt Cement} \quad \$160 \text{ per ton}$$
At a cost of $1.67 per ton, crushed portland cement concrete aggregate, requiring an increase of slightly over 1 percent additional asphalt cement would result in the loss of any cost savings. With the increased emphasis on bituminous recycling which uses less asphalt cement than new construction, crushed concrete aggregate may be more economical when combined with bituminous recycling. In bituminous recycling, new aggregate typically amount to only 30 to 50 percent of the total mixture. A typical series of mix design curves for a good quality mix using crushed concrete aggregate are shown in Figure 2. A more common asphalt content would be nearer 5 to 5.5 percent for the gradation used.

Filter Materials

Highway rehabilitation requires a large amount of subdrainage to relieve the moisture buildup in the pavement that is producing the distress. This subdrainage typically takes the form of edge drains placed under the pavement shoulder joint. An all too common practice is to fill this edge drain trench with a concrete sand which is too fine to function as an effective drainage material.[25] Crushed concrete could be used as a one sized material for excellent drainage. Quality and contamination problems would be minimal. A gradation produced by a laboratory crusher is shown in Figure 3 and is compared with that of a concrete sand.

In 1977 a new keel strip was needed in the runway in Jacksonville, Florida International airport.[26] The two twenty five foot wide center strips were crushed on site and used in the new keel section. The new section needed positive drainage because moisture damage had produced the present damage. The crushed concrete was graded and placed as a filter blanket under the new keel strip, an econocrete base and a new concrete surface. The openness of this gradation, has produced an excellent drainage material.

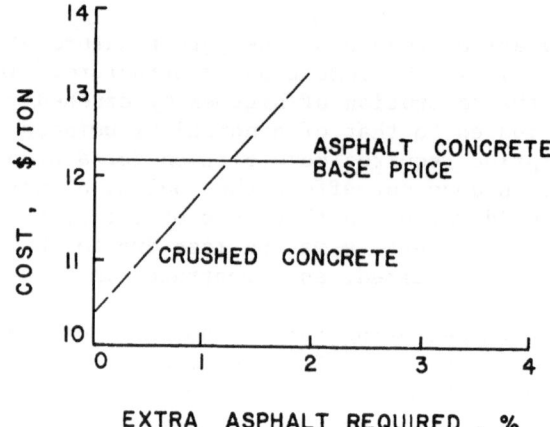

Figure 1.   Increase in Costs for Hot Mix Asphalt
            Concrete using Crushed Concrete of
            Varying Absorption

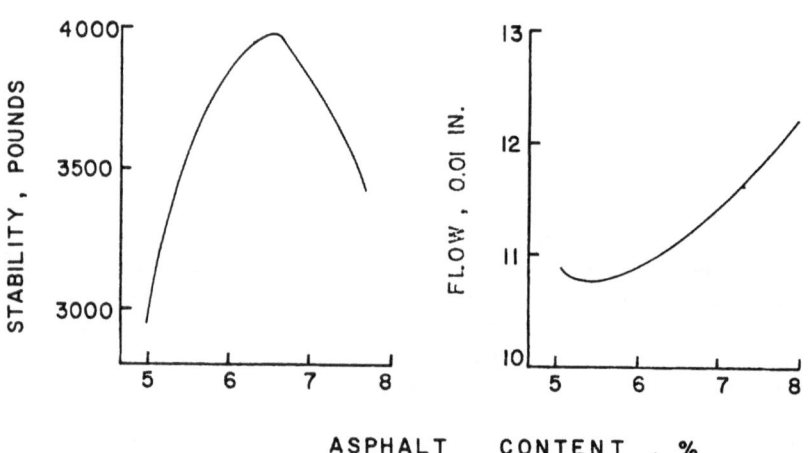

Figure 2.   Stability and Flow Curves for an Asphalt
            Concrete Made from Crushed Concrete and
            Normal Mineral Filler

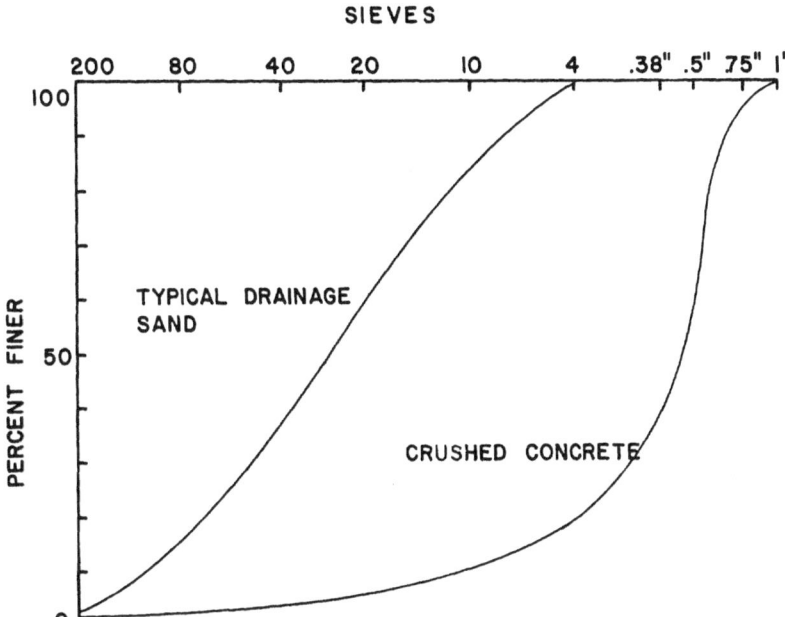

Figure 3.   Comparison of Sand Commonly Used in
            Subdrainage With the Gradation Produced
            From One Pass of Concrete Through a
            Jaw Crusher

Use of crushed concrete as a pavement filter material may require removal and disposal of the excess fines which are generated in the crushing operation.

Low Quality Stabilization

The fines from crushed concrete may contain a high percentage of cement paste. If this is the case, the fines may be ground and possibly reused as a low quality cement. It has been shown that cement paste ground to cement fineness after 28 days of hydration and mixed to form a 1:3 mortar resulted in 28 day mortar strengths[27] which were about 60% less than the original 28 day mortar strength. The fines from crushed concrete ground to cement fineness would be substantially less reactive than the case cited, since they would contain inert aggregate material and the degree of hydration of the cement would be higher. However, even a low quality cement can improve the behavior of fine grained highly plastic subgrade soils. Normally cement stabilization is most effective in granular soils lacking fines. Because of the low amount of cement being added, the main result obtained in high plasticity soils will be to lower the plasticity index. This is a major benefit in construction areas where heavy traffic must be carried directly on the subgrade. When the construction is completed the structure will benefit from a stronger, less moisture-susceptible subgrade.

Grinding the crushed concrete pavement fines to the required size will require a large expenditure of energy. If grinding is not economical, the fines may be used without grinding. However, by mixing the fines directly into the subgrade the improvement in plasticity will not be as large as if the fines had been ground. In this case, improved performance will probably result from the alteration in gradation due to the material addition rather than substantial cementation.

Reinforcement

Recycled concrete also provides another potential resource. Namely, the steel that is removed during aggregate production. At present, steel removal is the major roadblock to concrete recycling. Recent projects have shown that steel removal has not been as difficult as first thought, but the concept is still prevalent that steel removal is very difficult. Different machines have been used which fracture the concrete and debond the steel and concrete to the greatest extent possible. Ultrasonic vibrations are being investigated. To date, the reclaimed steel has been sold as scrap for $40 per ton.[8]

It is unlikely that this steel could be reused in a structural application due to its deformed condition and the possible presence of corrosion. However reuse should not be blatantly disregarded.

The development of equipment to straighten and clean the steel or new methods of demolition which result in less deformations of the reinforcement may well make the reuse of reinforcement possible.

NON-PAVEMENT USES

Buildings

Aggregates are used in buildings to provide support to foundations and drainage for water as well as in structural concrete. In expansive clay areas or frost heave areas the crushed concrete aggregate could be utilized to separate the foundation from the expansive material for structures the size of homes or apartments. In larger structures free draining backfill is required under footings and behind retaining walls to prevent pore pressure buildup. Use of crushed concrete aggregate for structural concrete would probably require thicker members or richer concrete mixes due to the reduction in strength when using crushed concrete as aggregate.[1] In addition the quality of the concrete produced with crushed concrete aggregate would be a function of the quality and variability of the rubble concrete. This increases the uncertainty of the safety factors used in current building codes and design procedures.

Fill

Utilization of concrete for fill is, of course, the predominant method of disposal for waste concrete.[28,29] This is due to the low cost of disposal, and also, concrete provides, except in extreme cases, an environmentally inert material. Concrete fill has been used effectively to reclaim or upgrade low lands, fill quarries, build bases for break waters and levees, to fill underground caverns which may otherwise fail in subsidence, etc. The list of potential uses are almost endless as the waste concrete can be substituted for practically any application that uses crushed rock. In specialized cases such as utilization of recycled concrete for fill in areas with an acid environment, concrete with limestone aggregate may not be suitable as it would react with the environment.

Rip Rap

Utilization of concrete in shore areas to prevent erosion or to form breakwaters is certainly not new. However it can provide an important use of large slabs of concrete which are better suited for this application than aggregate size material. This precludes the need to remove the reinforcement which will make the crushed concrete more cost effective compared to an aggregate which requires removal of the reinforcement.

Ballast - Subballast

Crushed concrete with the proper gradation[30,31] can be successfully used for railroad ballast and should perform as well as other
crushed stone ballast material. From an angularity standpoint the
crushed concrete will be similar to the crushed stone. The degradation properties of a highway aggregate may not be as good as the
high quality ballast normally used. If this is true the crushed
concrete could be utilized as a subballast, removing it from the
high stress areas directly under the tie-track structure.

Roofing Granules

Graded crushed concrete can be used for roofing granules as
long as it meets ASTM specifications.[32] It would be anticipated
that structural crushed lightweight concrete should perform similar
to expanded slag type materials and crushed normal weight concrete
should perform as well as normal crushed rock.

Neutralizing Beds

Concrete containing limestone aggregate could be utilized as
a neutralizing agent in acidic environments. Such concrete could,
for example, be pulverized and then substituted for ground limestone
in SO_2 scrubber systems currently used in coal burning power
plants[33]. Other applications could include the neutralization of
acid mine drainage or treatment of acidic industrial waste water.

Filtration Beds

Crushed concrete graded to the desired particle size could
be utilized as filtration beds in a number of industrial applications. Perhaps the most appropriate would be as 2-10 cm diameter
particles used in trickling filters in waste water treatment
plants.[34] The surface area of the hydrated cement may be of some
advantage in providing sites for the biological growth needed in
the removal of organics from the waste water.

Agricultural

Concrete with limestone aggregates could be reacted with nitric
acid to produce calcium nitrate for use as a fertilizer.[33] Crushed
concrete containing limestone could act as a neutralizer for acidic
agricultural soils. In specialized applications the readily available silica in hydrated portland cement could be beneficial. For
example, in highly acidic lateritic soils with low silica contents,
silica supplied by the hydrated portland cement could provide a
valuable source of plant silica for certain grains.[35-37] In Hawaii
the application of wollastonite ($CaSiO_3$) to highly leached soils
deficient in available silica resulted in a significant increase

in the production of sugar cane. The increased production was not due solely to the liming or neutralizing influence of the calcium in the wollastonite but by increased uptake of silica which helped provide a more rigid plant structure that allows an increased uptake of other nutrients.

A lesser agricultural application would be the use of finely crushed concrete containing limestone as a mineral food, (i.e. bonemeal or poultry grits).

Thermal Reservoirs

The heat capacity of concrete although influenced by the moisture content is approximately that of the aggregate contained in the concrete. Solar heating units rely on the storage of heat in reservoirs of various materials such as a one-sized aggregate. Crushed concrete as well as other demolition materials could be used as a material in such thermal reservoirs. The lack of need for separation of other building materials may make the use of inorganic building rubble an efficient source of thermal ballast.

Lowgrade Cement and Lime

Pulverizing the concrete to cement fineness may produce a very lowgrade cementitous material. A cement made from finely ground concrete would produce only very low strength concrete and would probably be utilized more readily in soil stabilization, rather than for structural purposes. (See the Section on Low-Quality Stabilization). Although impractical because of the quantities available, crushed concrete containing limestone aggregate could be calcined to produce lime.

Masonry

Concrete slabs in which the rebar was sparse or the spacing such that the slab could be prenotched (grooved) prior to demolition, a uniform fracture pattern could be produced. This could result in building stone for masonary construction. The fractured surfaces of such building stones would have asthetic surfaces composed of fractured or exposed aggregate depending on the strength of the concrete.

RESEARCH NEEDS

Pavements

Portland Cement Concrete. Fresh concrete can be made from crushed concrete and acceptable strengths obtained. The main concern when reusing pavement concrete is the damage which may result from use of "D" cracked concrete. If a "D" cracked concrete

is used, what can be done to the crushed concrete to prevent "D"
cracking in the recycled concrete mix? This may include blending
with new aggregate or crushing to a specified maximum size. It
may be that crushed "D" cracked concrete can only be used in subbase
applications due to the durability problems. The effect of contaminants on strength and durability must be examined. The levels
of deicing chemical which are acceptable must be examined.

Bituminous Mixes. The use of crushed concrete aggregate in
bituminous mixes will require study to illustrate the behavior of
the mix produced. Does the increased angularity and micro-texture
of the crushed concrete produce a mixture that will be difficult
to compact to the proper density? Will the standard criteria
applied to normal bituminous mixes apply to crushed concrete aggregate mixes? Does the condition of the concrete being crushed for
the aggregate source influence the resulting mix, e.g. "D" cracked
vs. non "D" cracked concrete? Does the amount of cement in the
original concrete mixture affect the performance of the bituminous
mixture using the crushed concrete aggregate?

These questions relate to the structural adequacy of the
bituminous mixture. There are long term considerations which must
also be considered. Factors influencing asphalt stripping must be
investigated. For example, the presence of deicing chemicals and
their concentration in the concrete may increase the stripping
potential. Likewise the influence of the fine cement paste on
stripping will need to be studied. This material may reduce
stripping and could actually reduce and detrimental effect produced
by chemical contaminants.

Low Quality Stabilization. The use of finely pulverized concrete as a low quality stabilizer additive needs investigation.
The distribution of the cement paste in crushed concrete needs
clarification. If the paste is beneficiated in the finer fractions
of the crushed concrete then the separation and use of these fractions either untreated or ground to cement fineness may be beneficial in soil stabilization. The quality of the material would have
to be related to engineering performance and its effectiveness in
the stabilization of various soil types. Standard performance
tests should be adequate for these evaluations. However, the
reactivity which may be a function of unhydrated cement content,
calcium hydroxide content, amount of carbonated cement or any
combination of these or other parameters also would warrant investigation.

Filtration Beds. The effectiveness of crushed concrete aggregate in trickling filters used for waste water treatment needs
investigation. This useage, although small, would be of value if
the crushed concrete proved superior to the other materials currently used.

Ballast. The performance of crushed concrete as railroad ballast should be evaluated. Concrete used in this application may not need to be as free from other materials as crushed concrete used in new concrete construction. The resilient and permanent deformation characteristics would need to be characterized.

Agricultural. The use of crushed or ground limestone containing concrete for acid soil neutralization, and as a source of available silica (from the cement paste) in highly leached lateritic soils could be easily evaluated.

SUMMARY

Crushed concrete aggregate has been limited to a large extent to being used in producing structural quality concrete. This use has captured the widest recognition, especially in the paving industry where the recycling concept in general has flourished in recent years. There are other areas where crushed concrete aggregate can be used directly as a raw aggregate. The quantity, location, transportation costs, and contamination are all factors that will influence the use the crushed concrete is put to. At present they favor the recycling of concrete pavements into new pavements. However, even the reuse of crushed concrete in payements presents some unanswered questions. At present, reuse of crushed concrete in the more esoteric applications discussed would appear to be minor due to problems of contamination with other materials during demolition. These applications could benefit from research, however, from a viable economic standpoint the major research efforts should concentrate on the performance of reused concrete in pavement applications.

REFERENCES

1. A. D. Buck, "Recycled Concrete," Highway Research Record No. 430, pp. 1-8, 1973.
2. S. Frondistou-Yannas, "Waste Concrete as Aggregate for New Concrete," American Concrete Institute Journal, 74:8, pp. 373-376, August 1977.
3. A. D. Buck, "Recycled Concrete as a Source of Aggregate," American Concrete Institute Journal, 74:5, pp. 212-219, May 1977.
4. P. J. Nixon, "Recycled Concrete as an Aggregate for Concrete-A Review," Materials and Structures, 11:65, pp. 371-378, September 1978.
5. R.-R. Schulz, "Recycling on Concrete," Betonwerk Fertigteil-Technik, 44:9, pp. 492-497, September 1978.
6. D. Bernard, "The Alternate Aggregate Source - Recycled Concrete," Rural and Urban Roads, 16:10, pp. 52-53, October 1978.
7. G. K. Ray, "Concrete Recycling: An Historical Overview, Part I," Rural and Urban Roads, 18:3, pp. 70-71, March 1980.

8.   H. J. Halm, "Concrete Recycling," Transportation Research Board News, 89, pp. 6-10, August 1980.

9.   "Old Pavement Recycled into New Subbase," Concrete Construction, 20:10, pp. 441-492, October 1975.

10. J. V. Bergren and R. A. Britson, "Portland Cement Concrete Utilizing Recycled Pavement," Iowa Department of Transportation, Division of Highways, Office of Materials, January 1977.

11. "Rebarred Slab Recycled as Pavement," Highway and Heavy Construction, 122:1, pp. 62-63, January 1979.

12. "Biggest PC Recycle Job Underway in Minnesota," Rural and Urban Roads, 18:79, pg. 52, July 1980.

13. K. Lane, "Connecticut - FHWA Demo Job Evaluates PCC Recycling," Rural and Urban Roads, 18:11, pg. 31, November 1980.

14. "State Pours Recycled Pavement," Engineering News Record, 204:23, pg. 29, June 1980.

15. J. L. Robertson, "Recycled Concrete Forms Base Material for an Aggregate Short Region," Rock Products, 83:2, pp. 50-51, February 1980.

16. "Crushing Converts Rubble Into Subbase Aggregate," Roads and Streets, 114:5, pp. 44-45, May 1971.

17. C. R. Marek, B. M. Gallaway, and R. E. Long, "Look at Processed Rubble - It's a Valuable Source of Aggregates," Roads and Streets, 114:9, pp. 82-85, September 1971.

18. T. B. Sadler, "A Crushing Success: Aggregate From Concrete," Public Works, 104:4, pp. 72-73, April 1973.

19. "Recycling Roads and Buildings with Portable Plants," Pit and Quarry, 65:8, pp. 91-92, February 1973.

20. "Concrete Recycling Makes Everyone Happy," Western Construction, 50:1, pg. 64, January 1975.

21. "Less Dumping, More Stone Supply with Contractor's Slab Recycling," Rural and Urban Roads, 17:34, pp. 46, 50, March 1979.

22. "Old Concrete Recycled into Saleable Aggregate," Highway and Heavy Construction, 122:7, pg. 87, July 1979.

23. "Urban Expressway Rebuilt on Recycled Concrete Base," Engineering News Record, 203:22, pp. 24-25, November 1979.

24. "The Edens 3R Project: Showcase for Recycling," Rural and Urban Roads, 18:3, pp. 34-35, March 1980.

25. S. H. Carpenter, M. I. Darter, and B. J. Dempsey, "A Moisture Accelerated Distress (MAD) Identification Manual," Federal Highway Administration, Contract No. DOT-FH-11-9175, In Review.

26. "Recycled Slab is New Runway Base," Highway and Heavy Construction, 120:7, pp. 30-33, July 1977.

27. F. M. Lea, "The Chemistry of Cement and Concrete," 3rd Edition, Chemical Publishing Co., pg. 727, 1971.

28. "Economics Spur Pavement Recycling," Public Works, 106:6, pg. 93, June 1975.

29. "Rubble Recycling Saves Time, Energy, and the Environment," Rock Products, 80:5, pp. 107-108, May 1977.

30. R. M. Knutson and M. R. Thompson, "Resilient Response of Railway Ballast," Record 651, Transportation Research Board, 1977.

31. R. M. Knutson and M. R. Thompson, "Permanent Deformation
     Behavior of Railway Ballast," Record No. 694, Transportation
     Research Board, 1978.

32. "Standard Specification D1863-80 for Mineral Aggregates Used
     on Built-Up Roofs," American Society of Testing and Materials,
     Part 14, pp. 648-649, 1980.

33. R. S. Boynton, "Chemistry and Technology of Limestone," 2nd
     Edition, John Wiley and Sons, New York, pg. 578, 1980.

34. M. J. Hammer, Water Supply and Pollution Control, 3rd Edition,
     Harper and Row Publishers, New York, pg. 857, 1977.

35. A. S. Ayres, "Calcium Silicate Slag as a Growth Stimulant for
     Sugar Cane On Low Silicon Soils," Soil Science, 1010:3, pg.
     216, 1966.

36. H. F. Clements, "Roles of Calcium Silicate Slags in Sugar Cane
     Growth," Hawaii Agricultural Experiment Station, Technical
     Paper No. 765, 1965.

37. H. F. Clements, "Effects of Silicate on the Growth and Leaf
     Freckle of Sugar Cane in Hawaii," Hawaii Agricultural Station,
     Technical Paper No. 722, 1965.

6.2.1 MINUTES OF THE PRESENTATION OF THE INTRODUCTORY LECTURE

M. Geudelin

Direction de la Recherche

UTI, Paris

In his introductory lecture <u>Carpenter</u> first reviewed the re-
use of demolished concrete pavements, mostly for highway construction. This provides best benefit for least risk due to uniformity
of the original structure (thickness, placing of reinforcement),
lower contaminant level (mostly from de-icing salts with which engineers are now familiar), lower consequences of failure than with
other types of structures.

The demolition techniques were illustrated by slides showing
diesel pile driving hammer towed along to fracture pavement at regular intervals. After crushing the aggregate obtained is used in
the surface concrete or bases and subbases where it contributes to
solving the problem of pumping (recycled aggregate has a low fines
content). To avoid the disadvantage of harshness of mix in surface
applications, 20-25% extra sand (passing # 4 sieve) is added to obtain a workable mixture, underlining a research need to optimize
the crushing for improved shape to avoid using large quantities of
new material. Another disadvantage is the adsorbency of aggregate -
5-7% more asphalt is consumed - however the increased angularity
ensures good interlock. Anti-stripping agents (hydrated lime,
portland cement) help to promote bond between the aggregate and the
asphalt.

Non-pavement uses include backfill behind retaining walls,
fill, rip-rap in coastal areas to prevent erosion, railway ballast
(after comparison with strength of traditional ballast - if strength
is inferior, it can be used as sub-ballast), roofing granules, neutralizing beds on municipal waste, soiled water treatment (appropriate because of high specific surface, agriculture (proof of
efficiency as sugar cane fertilizer - increased milk yield when used

as cattle feed), thermal reservoirs to store heat.

   Carpenter concluded by stressing the possibility of wider use
of recycled aggregate once the crushing plant becomes fully opera-
tional, encouraging procedures to be inventive.

   After the A.R.I., the editor had an interesting talk with the
President of the International Commission on Illumination, Prof. Ir.
J.H. de Boer about reflection properties of road surfaces.
In the light of recycled aggregates used in roads, it is thought
that Prof. de Boer's contribution is worthwhile to be added to the
Proceedings (as 6.3.2)

6.2.2 REPORT ON WORKSHOP 3 - "REUSE OF CONCRETE"
       (OTHER THAN AS AGGREGATES FOR CONCRETE)

C.D. Pomeroy

Material Research Department Cement
and Concrete Association
Wetham Springs, Slough SL 3 GPL, U.K.

The topics to be discussed were: - applications of theory, use
of products (e.g. fills, roads, dams), quality requirements, reuse
of reinforcement and research needs.

Dr. S.H. Carpenter presented his paper (co-authored with R.L. Berger)
on "Recycling of Concrete into New Applications" and the following
discussion ensued.

Mather questioned the quality of the unhydrated cement in the
finest material that is recovered. Is all of the cement hydrated
or is this a variable factor that depends in part on the
manufacturing standards used with cements? For example are modern
cements more likely to hydrate completely than those obtaining some
twenty or thirty years ago ? Kreijger thought that the degree of
hydration of the grains an important parameter that merited
further research. There was considerable discussion on the use of
fine materials recovered from degradation of concrete and
Frohnsdorff suggested the possible use of these materials as
chemical scrubbers for removal of SO_2. Carpenter replied that the
potential use of waste fines in chemical scrubbers was mentioned
briefly in his paper.

With regard to the use of aggregate in pavement concrete the
presence of small quantities of asphalt was discussed with
relation to its effect on air entrainment. There may be the need
to treat the finer material to remove bituminous coatings prior to
use to enable proper air entrainment to be achieved.

Carpenter admitted that there were difficulties in controlling the
air entrainment and there was no clear answer to this question at

the present time.

Further discussion took place concerning agricultural use of
Wollastonite and Frohnsdorff reported work in which there was an
increase yield attributed to the silicate strengthening from
wollastonite from sugar cane plantations which were both more
hardy and stronger in their cane properties, and it was this latter
change which was thought to account for the increased yield of
sugar that was comparable to the use of wollastonite fines at a
loading of about 16 tons per acre. Frohnsdorff also developed
thoughts regarding the use of fines for neutralisation of acid
soils and as an adsorbent.

Subsequently a wider discussion took place concerning various
aspects of the utilisation of broken concrete. Firstly economic
theory regarding the beneficial use of broken concrete was
discussed. It was agreed that in the analysis put forward by
Mrs Frondistou-Yannas no allowance had been made for utilisation
of the fines and that if this was possible the whole opportunity
for economic recycling of material was better. Consideration was
given to the classification of broken materials in terms of size
and quality, the size being related closely to the new jobs under
consideration. For example, Kreijger said that in the Netherlands
concrete was frequently used in the bases of dykes and marine
protection barriers and in these instances very large fragments
of concrete were desirable. On the other hand, in the United
States where concrete fragments were used as aggregate in subbases for roads a different requirement was present. It was
therefore, sensible to design the demolition process after
discussion with potential users to optimise the benefit of the
waste material.

Hansen asked for a definition of sand for use in concrete as this
seemed to be a variable parameter and he tabled some information
concerning the sub-division of concrete by size in his own
experiments. (3% greater than 30 mm; 27%, 20-30 mm; 36% 10-20 mm;
15% 5-10 mm; 19% smaller than 5 mm). Mather suggested that the
tax system could be used to stimulate supply and demand for waste
concrete materials and that this could have a significant affect
on the profitability of projects. However, this was considered
to be outside the terms of reference of the discussion group.

Carpenter reported that in Illinois considerable quantities of
concrete had been dumped in quarries and that some aggregate
producers were now re-exploiting this dumped material, crushing
it and using it as a basic concrete material. Mather then
referred to the use of large pieces of concrete for artificial
reefs in marine farming provided that the material had no
deleterious ingredients such as insecticides or other pollutants.
Next Frohnsdorff introduced the idea of dry compaction of fine

material at high pressure. It was not thought likely that this
would be economically viable by Mrs. Frondistou-Yannas. Later
the possibility of steel recovery was discussed and here the
chances of reusing steel in construction were considered minimal,
particularly as reinforcing steel is bent before installation and
that it was unlikely that it would be suitable for reuse, even if
it could be extracted in a clean form. Hence, the only real use
for extracted steel was considered to be as an ingredient in
steel manufacture.

Sierra asked whether basic materials, particularly from the fine
end of the size range could be used to stabilize sludges that
were produced in water treatment plants or as coal mine waste
materials; Mather advocated studying kilndust technology.

Summing up Pomeroy said that within the economic framework the
value of the products in relation to utilisation would relate
to their size and their quality and that depending on the
geographical location of the available raw materials and the
source of use different answers would apply. The principal uses
of waste would seem to be as fill materials for sub-bases, for
lean rolled concrete and as filter materials. Further research
was certainly necessary to investigate the quality of the fine
materials as it is in the fine end of the spectrum that the most
beneficial materials would seem to be available. The specifier
should be involved in the definition of material qualities and
sizes for any particular degradation product if the old concrete
is to be used successfully by a user, hence specifier and broken
concrete supplier must work very closely together. The
agricultural uses would appear to merit further study as would
the quality testing and assurance for different grades of
broken concrete. To achieve this more may need to be known of
the concrete going into a crushing and beneficiating plant but
this is a topic for the future.

6.3.1 EXAMPLE OF RE-USE OF CONCRETE IN FRANCE

R. Sierra

Laboratoire Central des Ponts et Chaussées

Centre de Nantes, Bouguenais

The "Ponts et Chaussées" department have experience of reusing 12 year-old concrete for partial reconstruction in 1976-77 of the three lane pavement of the A 1 northbound motorway. The middle and heavy truck lanes were demolished in 6 days representing about 65 000 m2. Fragmentation was undertaken in two machines where 5 ton masses exerted a guillotine action. The recycled plain concrete was used to rebuild the foundation of these same lanes (lean concrete 25 000 m3) and the emergency stopping lanes (porous concrete).

The sand obtained from fragmentation of the concrete was not reused because of pronounced heterogeneousness in the fines content. About 30% of the crusher installation production was considered as waste.

There are several suggestions for reuse of crusher sand - one being processing of sedimentation sludge (semisolid waste) from gas scrubbing and water cleansing. The treatment aims at solidifying the sludge to retain it in place or to transport it to dumps but especially to avoid transfer of pollution by checking all the harmful and toxic elements concentrated there (heavy-metal ions, pesticides). This result is frequently obtained with Portland cement. The sand obtained from crushing recycled concrete contains large quantities of basic constituents and the grains are highly microporous. With or without cement it can contribute in an economic way to consolidating sludge and to the quasi-irreversible retaining of pollutants to avoid their transfer to the environment.

6.3.2. REFLECTING PROPERTIES OF A ROAD SURFACE

J.H. de Boer

President International Commission on Illumination

The reflection properties of a road surface determine in a given road lighting installation (characterized by the type of luminaires, their mounting heigth, spacing and position relative to the carriage way) what value of average luminance the road surface will have and what the uniformity of this luminance will be. The more favourable the reflection properties the higher the average luminance and the higher the luminance uniformity of the road surface. More favourable means two things:
1) higher average reflection factor in other words "a higher brightness" of the surface.
2) reflection as diffuse as possible in other words avoidance of specular reflection also under wet conditions.

For the purpose of designing road lighting installations the reflection properties of road surfaces are usually described by the so-called luminance coefficient which is defined as the ratio of the luminance at an element on the surface to the illuminance at the same element, as given by a single light source. The luminous coefficient of a given road surface depends upon the positions of the lightsource and the observer relative to the element under consideration. The average luminous coefficient is the luminous coefficient weighted over a solid angle from the surface element containing the directions from which in a usual road lighting installation light is incident on the element under consideration.

In a given road lighting installation determined by the light distribution of the lanterns used and the geometry of the installation (mounting height and spacing of luminaires and raod width) the average luminance of the road surface is linearly proportional to the product of the luminous flux of the bare lamps in the luminaires and the average luminous coefficient of the road surface. Therefore, the power to be installed in the lighting installation in order to obtain a certain value of the road surface luminance is inversely proportional to the average luminous coefficient of the road surface.

By applying additives to the raod surface dressing in particular when the latter is composed of the generally used, rather dark, materials such as asphalt, the average luminous coefficient can easily be increased by a factor 2 and more which thus means the possibility of reducing the energy consumed in the lighting installation to 50% and less of the value required at less favourable reflection properties of the top layer of the road surface.

An important aspect of the reflection properties of a road surface is, furthermore, the degree of diffusion of the reflected light. The more diffuse the surface reflects the easier good uniformity of luminance can be arrived at. However, even the most diffusely reflecting road surfaces are rather shiny. With a convenient light distribution of the luminaires adapted to the specularity of the road surface a satisying uniformity of the road surface luminance can, nevertheless, be obtained. This means, however, that when the shinyness of the road surface gets stronger, which is the case with most road surfaces when becoming wet, the carefully built-up uniformity of the luminance is spoiled.

So a very important property of a road surface id furthermore the degree to which its reflection properties remain unchanged when becoming wet. In general a smooth surface with fine granulation, even when it reflects diffusely when dry will almost relflect as a perfect flat mirror when wet as it then will soon be covered by a continuous flat waterskin. A surface containing coarse grains and capable of draining the water so that no inundation occurs, will have a broken waterskin and accordingly maintains diffuse reflection properties. The coarseness of a road surface is a matter which has consequence to other aspects such as resistance against wear due to traffic and the skid resistance.

Literature: 1. Calculation and measurement of the luminance and illuminance - CIE Publication, no. 30 - 1976

2. Road lighting for wet conditions — CIE Publication no. 47, 1979

(CIE= Commission Internationale de l'Eclairage= International Commission on Illumination)

7 WORKSHOP 7 - CONTAMINATION EFFECTS ON RECYCLING AND RE-USE

Chairman: G. Frohnsdorff, secretary M.J. Rubens

Attendance: S.H. Carpenter, J.-P. Collin, G. Frohnsdorff,
                 G. Frondistou-Yannas, T.C. Hansen, P.C. Kreijger,
                 H. Lambotte, P. Lindsell, B. Mather, C. Molin,
                 C. De Pauw, C. D. Pomeroy, D.M. Roy, M.J. Rubens,
                 R. Sierra, J.W.Weber, J.F. Young, S. Ziegeldorf.

7.1 INTRODUCTORY COMMENTS

Geoffrey Frohnsdorff

National Bureau of Standards

Washington, D.C. 20234

Our workshop on "Effects of Contamination on Reuse of Concrete" is primarily concerned with reuse of concrete as aggregate in new concrete. Emphasis will be on identifying gaps in knowledge and research needs. Valuable background information has already been provided by the lecture on "Contamination Problems in the Recycling of Concrete" by Professor Young[1]. To help establish the scope of the workshop, a contaminant will be defined as:

1. A material present in the concrete to be recycled which was not present in the concrete as originally cast.

2. Any material in concrete to be recycled which would be detrimental to the proposed reuse of the concrete.

Examples of possible detrimental effects of contaminants on new concrete are:

o Reduced durability
o Disruptive expansion
o Corrosion of reinforcement
o Less predictable performance of admixtures
o Reduced workability
o Increased water requirement
o Reduced rate of strength gain
o Increased shrinkage
o Increased creep
o Reduced bond to reinforcement
o Discoloration
o Health hazards

With the lecture by Young[1] as background, questions which the workshop might consider include:

o   What are the possible contaminants in recycled concrete?
o   Are there adequate methods for the detection and charac-
    terization of contaminants?
o   What are the permissible limits on various contaminants?
o   Can the likely contaminants be removed economically?
o   How do various possible contaminants affect reuse of
    concrete?
o   Can fines from recycled concrete be reused in mass concrete?
o   How are the forces of adhesion and cohesion in concrete
    affected by the possible contaminants?
o   Are there adequate codes and standards for recycled
    concrete?
o   How can the quality of recycled concrete be assured?
o   Are there institutional barriers of a technical nature
    which could be removed by research?
o   What are the gaps in knowledge and the research needs?

To the extent possible, the workshop should focus on materials science aspects of the gaps in knowledge and research needs. Examples of these are:

o   Phase compositions
o   Phase distributions
o   Nucleation and growth of phases
o   Phase equilibria
o   Pore distributions
o   Mechanical properties of phases
o   Interfacial forces
o   Internal stresses
o   Environmental effects

In considering the effects of contamination on reuse, it will be necessary to consider the form in which the concrete is to be recycled, whether as coarse or fine aggregate or, perhaps, as cement or mineral admixture. It may also be necessary to consider the type of concrete in which the recycled concrete is to be used (e.g., ready-mixed concrete, site-mixed concrete, factory production) and the end use (e.g., foundations, highways, factory-fabricated units).

Since the durability of concrete is almost always important, ways of evaluating durability must be considered. Guidance for development of durability tests is provided by ASTM Standard E 632[2] from which Figure 1 is taken. Important factors to consider are the end-use performance requirements of the recycled concrete, the expected type and range of degradation factors, and the possible degradation mechanisms. The analysis of factors likely to affect durability can be aided by considering all the possible interactions

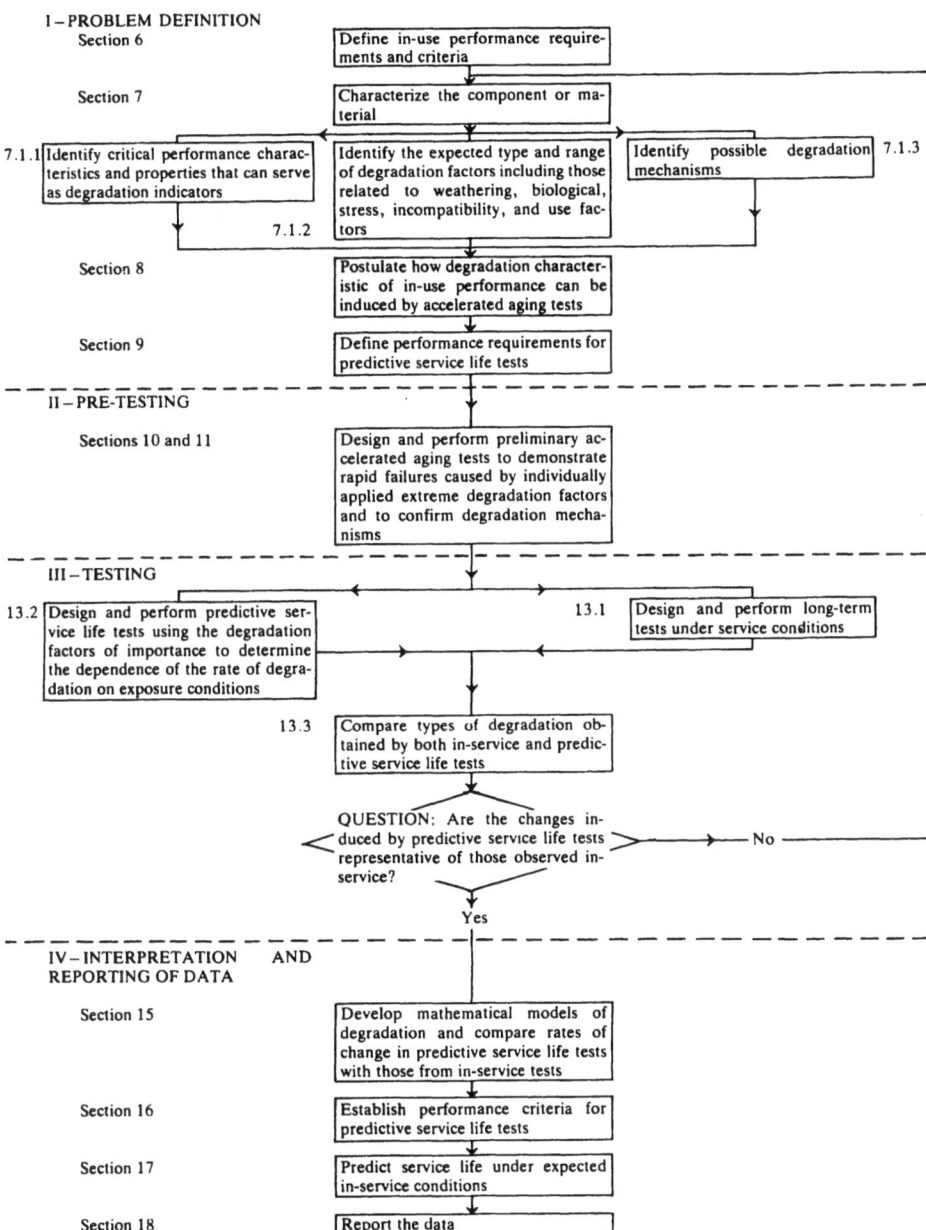

Figure 1.   Flow Chart for Development and Application of Durability
(Service Life) Tests as Given in ASTM Standard E 632.

Table 1.   Suggested Axes for a Compatibility Matrix

Recycled Concrete   New Concrete

   Aggregate

Inorganic   Inorganic

   Cement   Cement
   Coarse aggregate   Coarse aggregate
   Fine aggregate   Fine aggregate
   Mineral admixture   Mineral admixture
   Brick
   MgO
   CaO

   SO_4^{2-}

   Cl^-
   Alkalies

Organic   Organic

   Chemical admixture   Chemical admixture
   Plastics
   Paints
   Asphalt
   Wood

Metal   Metal

   Steel   Steel
   Aluminum

Other considerations concerning the recycled concrete must be the toxicity and radioactivity of its ingredients and the extent to which the compatibility of the concrete ingredients will be affected by conditions to which the new concrete will be subjected during curing and in service.

between the ingredients of the recycled concrete and the additional
ingredients to be included in the new concrete. To try to ensure
that no possibly detrimental reactions are overlooked, it may be
helpful to produce a "compatibility matrix" with one axis representing all known and likely ingredients of the recycled concrete
aggregate and the other representing the additional ingredients to
be included in the new concrete mixture. The intersections between
the entries on the two axes then represent the pairs of ingredients
for which questions must be raised about compatibility, in addition
to those which are normally addressed. Examples of entries which
might be included on the recycled concrete aggregate and new concrete axes of the matrix are given in Table 1.

7.2.1 MINUTES OF WORKSHOP 7

M. Geudelin

Direction de la Recherche

UTI, Paris

Having outlined the goal set for the workshop, <u>Frohnsdorff</u>
submitted several complementary points for discussion such as: need
for simple methods for characterizing contaminants in the field, definition of permissable limits of contaminants in recycled concrete,
effects of contaminants on forces of adhesion and cohesion, reuse
of fines in mass concrete, codes for recycled concrete in different
coutries and interaction with technology, quality assurance, institutional barriers in relation to codes and technical education, possible future trends in concrete technology.

Contaminants reduce durability, implying corrosion of reinforcement, less predictable performances of admixtures, reduction in
workability, rate of strength gain, bond to reinforcement and between
matrix and aggregate, increase in water requirements, shrinkage and
creep, not forgetting discoloration and health hazards.

<u>Frohnsdorff</u> then advocated examining the problem of contamina-
tion from the material science point of view to study the effect of
phase composition on the performance of finished concrete, distribution of phases and contaminants (scattered or concentrated), setting
and strength development of concrete, types of detrimental effect
according to modulus, repercussion on pore distribution, permeability,
therefore crack propagation, interfacial forces and bonding. More
knowledge is needed on the composition of contaminants also the size
and shape distribution of solid particles. Using flow sheets he illustrated the work for ASTM E 632 to develop accelerated tests for durability and prediction of service life, indicating that the first requirement is a definition of performance in use and criteria by identification the expected type and range of degradation factors and
mechanisms.

Further steps in this direction should consist of developing
matrixes surveying all the contaminants suspected, adding the ingredients of new concrete to identify possible deleterious interaction. This approach would supplement Kreijger's table 6 (see part
2. 1.1) to define permissible limits of contaminants content. After
Lindsell had remarked that site inspection or tests of extent of
damage was first necessary before setting limits, Kreijger traced
Dutch tests (see 7.3.1) of the effect of gypsum contamination on the
swelling of portland - and portland blastfurnace cement, explaining
the difference in behaviour and mechanisms and indicating too the
effect of the size of the gypsum grains. Lambotte gave a short presentation of his paper (see 7.3.2) on the influence of contaminants
on the quality and behaviour of recycled concrete, showing a fall
in workability as contamination content increases, also the effect
of gypsum on compressive strength. Regarding the rather favourable
results concerning the brick contamination on strength, Sierra
thought of the similarity with structures built by the Romans where
the binder was made of lime and crushed brick. This brings the pozzolanic effect of fire clay to the mixture. The clay remaining in
baked clay is beneficial but this is not the case for unfired clay.
We know that natural aggregate for concrete must be cleansed of all
essentially clayey soiling, if not it can cause lack of bond at
short or medium term between the cement paste and aggregates. All
concrete recovered from structures in contact with water, be they
marine or waterways structures'or partially underground foundation
in water table areas, are contaminated by the simple fact that water
circulates by permeability or capillarity in the particularly fine
clays (ultraclays), therefore harmful even at low contents. Such
factors must be taken into consideration in any study of contaminants
in recycled concrete.

Kreijger warned for use of brick rubble when SO3-content is
above 1% causing long term expansion referring to his table given
earlier (see 5.3.1), which however seems to be less important if
"no-fines"-concrete is made of such aggregates.

In discussion of the experiences reported by other participants
relating to the difference in effect of given amounts of gypsum contamination on concrete made using pure portland cement versus concrete
made using portland blastfurnace slag cement, Mather noted that work
he had reported in the Proceedings of the American Concrete Institute,
Vol 54, 1957, pages 205-232, failed to reveal such an effect. Using
all the commercial blended iron blast-furnace slag cements then in
production in the United States (which contained from 55 to 75 percent portland cement clinker) and one experimental cement which contained only 38 percent clinker, it was found that when tested using
the ASTM standard test (C 452) where gypsum is added to bring the
SO3 content to 7.0 percent by mass of the concrete it was found that
the blended cements performed proportionately to the C3A content of
the portland-cement constituent and no effect of the slag other than

as a diluent was detected. Work has been in progress since that
time to develop testing procedures that will properly test for the
sulfate resistance of blended cements where the slag or pozzolan
constituent has time to hydrate prior to entry of the sulfate solution into the concrete from the environment.

Discussion than tuned to finding quick convenient ways of detecting contaminants as was possible with indication strips for
chloride content(Hansen). Pomeroy announced the development of a
field kit for use by advisory engineers to detect chlorides, degree
of carbonation, Ph generally, also for taking samples of hardened
concrete (therefore possible with aggregate). All participants
expressed interest for this equipment.

The stage at which such tests are performed is important, De
Pauw emphasized. Tests prior to demolition would be benificial,
leading to more selective techniques, if planned for the recycling
stage, they would complicate the operations. Young added the problem of sampling and distinguishing between the harmful forms of
contaminants (for example different when present as ettringite or
gypsum).

De Pauw reviewed his written paper (see 5.3.2) whilst Roy
demonstrated work on reactivity and effects at the cement-aggregate
interface related to the presence of possible contaminants such as
glass. The interest of developing criteria for deciding whether
pulverized fuel ash could be used to deal with glass contained aggregates was underlined by Frohnsdorff. Young urged on return to
basic work on the "system" to determine for example, whether limits
set by certain tests are valid, also the degree of solubility of
contaminants which prompted Kreijger to draw attention for possible
danger of heavy metals in waste materials and possible radioactivity.

On the subject of the possible harmful effects of glasses as
a contaminant by the participation of the glass in a chemical reaction with the alkalies in the cement, Mather noted that a paper
reviewing the data on the reactions between cement and artificial
glass in concrete had been prepared by Mr. John W. Figg of Harry
Stanger Ltd., in England, for the Conference on Alkali-Aggregate
Reaction in Concrete to be held in Cape Town, South Africa, 30 March
- 3 April 1981. The paper includes 70 references.

General discussion turned to standards, considered reasonably
satisfactory in U.S. and applicable to recycled aggregates. When
Frohnsdorff expressed a wish for performance standards, Hansen stated
that the Danish texts were drafted in this principle but proved
difficult to use by the ready-mixed concrete industry. Kreijger
voiced interest for East German Standards (see Journal: Standardi-
sierung im Bauwesen) which give a worthwile background. Mather urged
participants to consults the ASTM STP 169 B "Significance of tests

and limits in specifications for concrete aggregate", although
Pomeroy asserted the need for research to know how much standards
can be relaxed to avoid a restrictive nature. In every case a performance standard is needed, prompted Mather, to evaluate newly
developed materials, following which a quick test to predict behaviour should be drafted.

On the subject of fines, in which an increased concentration
of contaminants can be expected in principle, Pomeroy considered
that more work was needed to define then chemical constituents.
To avoid trouble, it seemed preferable to eliminate reuse of fines
for the present (Lambotte).

Young pointed the economic impact of multiple tests, raising
the question of responsibility when deleterious materials remain after recycling. Participants considered that the authorities should
assume the cost of developing tests and the producers pay for equipment whereupon Mather defined the decisive role of governments which
should either prohibit exploitation of quarries or dumping of demolished concrete which would lead to two totally different situations.
In his summing up, Frohnsdorff stated that industry must be encouraged
to recycle since the problem with coarse recycled aggregate differs
little from that of natural aggregate. He considered that production
is well developed but not optimized, especially in the field of
contaminants removal. In spite of certain shortcomings , existing
standards for natural aggregate appear broad enough to cover recycled
products.

For beneficiation, more research is needed to define actual
contaminant content to avoid trying to remove detrimental materials
which in fact are not present. Assumptions on contaminants restrict
reuse and spoil chances of high strength. However caution is essential
when recycling special structures such as refractory concrete, also
aggregate obtained after high-temperature demolition. The more information available on provenance, the better the decisions.
Futher guidelines are needed on permissible levels of contaminants
in concrete. The chairman quoted the example of work by E. Dunstan
as revealed in his presentation of a paper to ACI Committee 201.
Dunstan's research is devoted to predicting the effects of sulphate
attack on concrete in different environments with various cements,
w/c-ratios etc. Frohnsdorff considered that this is yet another
approach for the future, as is the optimization of separation processes through a study of municipal solid waste disposal.

7.2.2 REPORT ON WORKSHOP 7

Geoffrey Frohnsdorff

National Bureau of Standards

Washington DC 20234

The workshop discussions were supplemented by a prepared presentation by Mr. Lambotte which is included in these proceedings[3].
The rest of the discussion is reported under the headings which
follow.

a. What are possible contaminants in recycled concrete?

Considering the use of recycled concrete as aggregate for new
concrete, the range of possible contaminants is large. However, it
is little different from the range of contaminants of natural aggregates, though the quantities initially present may be higher. Possible contaminants mentioned during the workshop session included:

Organic Matter:

o Wood and other cellulosic materials
o Plastics and rubber
o Bituminous materials
o Paints and coatings
o Mineral oil
o Vegetable and animal fats and oils

Inorganic Matter:

o Salts (e.g., sulfates, chlorides and carbonates)
o Glass
o Brick (clay brick and refactory brick)
o Clay

Metals:

o Steel
o Copper
o Aluminum
o Zinc

Other factors to be considered:

o Radioactivity
o Toxic substances

The range of contaminants likely to be present will vary with the source of the recycled concrete.For example, concrete from buildings is more likely to contaminated with brick other than concrete from highways, though highway concrete is more likely to be contaminated with chlorides from de-icing salts.

b. **Are there adequate methods for detection and characterization of contaminants in recycled concrete?**

Solid contaminants can usually be easily detected and identified by visual methods. Materials such as chlorides and sulfates in the pores of recycled concretes are more difficult to detect. Test papers are available for detecting chlorides in water but detection of sulfates is a bigger problem. The Cement and Concrete Association (United Kingdom) is developing a kit for identifying constituents of concrete in the field, but this is not yet available. Similarly, the U.S. Army of Corps Engineers has developed guidelines for use by its concrete inspectors which should be applicable to recycled concrete.

c. **What are the permissible limits of contaminants in recycled concrete?**

The permissible limits on any contaminant in recycled concrete will depend upon the application. For portland cements concrete of moderate strength which will not be used in severe environments, the limits indicated in Table 6 of Professor Kreijger's introduction[4] to this volume are ones which have been proposed in the literature.

A systematic study is needed of the effects of various levels of likely contaminants on the performance of concretes within the practical range. It must be borne in mind that the possibility of synergistic detrimental effects involving mixtures of contaminants might be encountered. It should also be recognized that the exposure conditions to which a contaminant has been subjected may affect its influence on the performance of the concrete. As a

possible example, there are indications that bituminous concrete
contamination sometimes, but not always, causes difficulties in
controlling air content of new concrete. It was suggested that
such an effect might be due to changes in surfactant concentrations
in bituminous concretes caused by certain exposure conditions.

d.   Can contaminants be removed economically?

Procedures commonly used for the beneficiation of natural
aggregates can remove many contaminants from crushed recycled
concrete. This applies particularly to low-density contaminants
and friable materials which collect in the fine fractions. These
include wood and other cellulosic materials, plastics, and gypsum
plaster. Contaminants which could be difficult to remove include
chlorides and sulfates within the pores of the concrete and
alkali reactive aggregates. While more information is needed before
a completely adequate answer to this question is given, it appears
that most detrimental contaminants could be removed economically
under many conditions.

e.   Effects of Contaminants on Reuse of Concrete

In general, the use of recycled concrete as coarse aggregate
in new concrete gives compressive strengths which are somewhat
lower than strengths for concrete made with the original natural
aggregate. This effect can be corrected for by changing the mixture
proportions in the new concrete. Additional losses of strength
may be associated with contaminants in the recycled concrete.
Gypsum plaster is generally considered to be the most probable
contaminant. If natural aggregate is replaced by an aggregate
which is 100 percent gypsum, the compressive strength of the
concrete may be only about 15 percent of that of the control
concrete with natural aggregate. A new study on the effects of
gypsum plaster and other possible contaminants is being carried out
by Lambotte and de Pauw[3]. A potentially important method for
predicting the effects of sulfates on the performance of concrete
has recently been outlined by Dunstan[6]. His scheme, which has not
yet been fully validated, would take into account the sulfate
concentration, the chemistry of the cement, the water/cement ratio,
and the environmental conditions. Using a 15 percent reduction in
compressive strength as the criterion, Takeshi[5] and his colleagues
showed the following to be equivalent levels of contamination:

7 percent gypsum plaster, 5 percent soil, 4 percent wood,
3 percent gypsum, 2 percent asphalt, and 0.2 percent vinyl
acetate paint. (These figures are by volume).

Another effect on reuse which must be considered is alkali-aggregate reaction. Alkali-aggregate reactions could result from glass or reactive natural aggregates in recycled concrete being used in a new concrete containing high levels of alkali, whether from the cement or contamination from the environment to which the recycled concrete was exposed.

There is some evidence that asphalt contamination from highways can have an overriding effect on air entrainment to the extent that intentionally added air entraining agents will produce little additional effect. It may be supposed that surface active constituents of recycled concrete may cause problems with the predictability of the performance of chemical admixtures.

The particle size distribution of coarse aggregate from recycled concrete should be similar to that of crushed rock aggregate and cause no unusual effects. Clay and other fine material on the surfaces of coarse aggregate particles could reduce the bonding to the matrix just as in natural aggregates.

f. Can fines from recycled concrete be reused in mass concrete?

In the crushing of portland cement concrete to generate aggregate for use in new concrete, about 10 percent of the product is classified as fines. The fines tend to contain matter from the more friable components of the concrete which include gypsum plaster and partially degraded concrete. For these reasons, until more is known about their compositions, caution should be exercised in using the fines in new concrete. Published results[5] also suggest that the loss of strength from use of recycled fine aggregate in the place of fine natural aggregate is greater than the replacement of coarse natural aggregate by coarse aggregate from recycled concrete. The possibility of processing the fines to make either a cementitious material or a mineral admixture should be explored.

g. How are the forces of adhesion and cohesion affected by contaminants in the recycled concrete?

Because of the large number of different interfaces in hardened concrete, it is difficult to give a general answer to this question. It must be supposed that the forces of adhesion between the matrix and the coarse aggregate from recycled concrete will be comparable to those in concrete made with natural aggregates, except where the surfaces have been contaminated with materials of lower surface energy such as paints and asphalt. Similarly, the adhesion to any plastic which contaminates the concrete is likely to be less than that to the usual concrete materials. Nevertheless, because the quantities of the low surface energy materials present in recycled concrete will normally be very low and have low surface areas, the effects on the forces of adhesion and cohesion are likely to be small.

h. What is the relevant theory?

The theory of recycled concrete is no different from that for new concrete made from natural aggregates. In neither case is the theory developed to the point where performance can be adequately predicted from knowledge of the ingredients and the mixture proportions. Presumably, the theory of composite materials should be applicable to the mechanical properties but, at present, it only leads to qualitative conclusions. In regard to mechanisms of degradation of recycled concrete, knowledge applicable to concrete made with natural aggregates is directly applicable and should be considered when deciding on the acceptability of recycled concretes containing chlorides, sulfates, other dissolved salts, reactive aggregates, and other materials known to be detrimental. In general, there is need to learn much more about the microstructures of hardened cements and concretes and their relationships to performance in service.

i. Are existing codes and standards adequate for recycled concrete?

For the most part, it appears that existing standards such as those of ASTM and the British Standards Institute can be applied to coarse aggregates from recycled concrete. Those standards which exist only in performance terms, such as the Danish standards, do not provide adequate guidance for aggregate selection. Where adequate standards exist, these are referenced in national concrete codes such as, for example, the American Concrete Institute code, and little or no change is required to allow the safe use of recycled concrete as coarse aggregate.

j. How can quality of recycled concrete aggregate be assured?

The quality of coarse aggregate from recycled concrete can be assured by requiring it to comply with appropriate standards for aggregates such as ASTM C 33. Additional levels of quality assurance could be provided by requiring the supplier to state the source of the recycled concrete and to report indications of possible contamination observed at the time of demolition. It is also possible that aggregate from recycled concrete could be blended in averaging piles to reduce batch-to-batch variations in properties. However, this is unlikely to be necessary unless recycled concrete is to be used in high strength concrete, or concrete to be exposed to severe environments.

k. <u>Institutional barriers requiring technical knowledge to overcome</u>

There was nearly no discussion of this point in the workshop.

l. <u>Gaps in knowledge and research needs</u>

It was generally agreed that the recycling of concrete as coarse aggregate did not pose large problems or give cause for much concern. However, gaps in knowledge which should be filled to make possible the most effective use of contaminated concrete include:

o   Effects of specific contaminants on strength and other aspects of concrete performance
o   Effects on concrete performance of interactions between groups of contaminants
o   Effects of distribution of contaminants on concrete performance
o   Effects of prior exposure conditions of contaminants on concrete performance

In general, the research needs for concretes containing contaminated recycled concrete are similar to those for concrete containing natural aggregates. They include development of equations to predict performance of concrete in terms of its ingredients and the distribution of ingredients, taking into account the effect of the expected service environment.

SUMMING UP

The workshop on "Effects of Contamination on the Reuse of Concrete" was concerned almost exclusively with the reuse of concrete as coarse aggregate. The consensus was that contaminated concrete can be reused as coarse aggregate provided that good standards for natural coarse aggregate are applied. More specific comments which summarized the views of the participants are:

1. As for natural aggregates, contamination of coarse aggregate from recycled concrete should seldom be a problem. However, the possibility that contamination might cause a problem should <u>never</u> be ignored.

2. Just as no two natural aggregates are exactly alike and few, if any, are free from unwanted ingredients, so it is with aggregate from recycled concrete. In the present state of knowledge, uncertainties about the performance of aggregates from recycled concrete may be somewhat greater than for natural aggregates, but the difference is likely to be reduced as experience is gained.

3. The technology for reducing the quantities of almost all potentially harmful contaminants in recycled concrete to safe levels is available. Ways of applying the technology in the most cost-effective way are not fully known.

4. Taking the ASTM standards of the United States as an example, standards for coarse aggregates are sufficiently broad to be able to encompass aggregates from recycled concrete satisfactorily. The situation regarding standards for fine aggregates is not so good.

5. Assumptions made about possible contaminants in a recycled concrete can affect the economics of processing and the risks of encountering deficient performance in a given application. It is therefore important to know as much as possible about the likely contaminants.

6. The types and quantities of contaminants in a recycled concrete depend on the source. For example, before processing, concrete from demolished buildings is likely to be more highly contaminated than concrete from highways. It is desirable, though not essential, that the range of possible contaminants from each source be known.

7. Processing of recycled concrete from conventional buildings and from highways to form coarse aggregate is generally capable of bringing the level of contamination down to a level which is acceptable for most applications. This applies to contaminants such as gypsum plaster, paint, wood, and brick.

8. Particular caution should be exercised when contamination may be refractory concrete, other refractory materials such as periclase brick, or fire-damaged concrete. This is because cements used in refractory concretes may not be compatible with portland cement, because hydration of periclase can cause disruptive expansion in concrete, and because fire-damaged concrete may contain hard-burned lime which may also cause disruptive expansion upon hydration.

9. It is desirable to know the source and history of recycled concrete, if its reuse in high-performance applications is being considered. Such applications might include high-strength concrete, concrete to be exposed to severe conditions of freezing and thawing, or concrete for use in contact with steel reinforcement under conditions of high humidity. Otherwise, it is probably not important to know the source of

the recycled concrete, provided the lightest weight and most friable materials have been removed and it meets standards similar to those for natural aggregates.

10. Some guidance as to the permissible levels of contaminants in recycled concrete is available in the literature. Additional information is needed and it is desirable that it should be provided in the form of easily used guidelines.

11. Ways of optimizing processes for removing contaminants from recycled concrete should be investigated. Experience gained in the separation of municipal solid wastes may be relevant. This information should lead to improvement in the practicality, reliability, and efficiency of processes for removing contaminants from recycled concretes.

12. The state of knowledge about the performance of concretes containing contaminated recycled concrete is about the same as for conventional concretes. In both cases, ability to predict the effects of various constituents on concrete performance is needed. For both conventional concrete and concrete containing recycled concrete, improved understanding is needed of the chemistry and micro-mechanics of concrete and their effects on durability performance.

13. Dunstan has recently proposed a comprehensive approach to predicting the effects of sulfates on the performance of concretes. If his approach can be validated for sulfates, it should be applicable to sulfate contamination. It might also suggest that a similar approach be sought for predicting the effects of other possible contaminants.

REFERENCES

1. J. F. Young, Contamination Problems in the Recycling of Concrete, these proceedings.
2. ASTM E 632, Recommended Practice for Development of Accelerated Short-Term Test for Prediction of the Service Life of Building Materials and Components, American Society for Testing and Materials, Philadelphia.
3. H. Lambotte and C. de Pauw, The Influence of Contaminants on the Quality and the Behavior of Recycled Concrete, these proceedings.
4. P. C. Kreijger, Introduction and Statement of the Problem, these proceedings.

5. M. Takeshi, K. Torao, N. Muneo, and K. Masafumi, Study on Reuse of Waste Concrete for Aggregate of Concrete, Japan – U.S. Seminar on Energy and Resources Conservation in Concrete Technology, p. 85, San Francisco, September 10-13, 1979.

6. E. Dunstan, A Spec Odyssey – Sulfate Resistant Concrete for the 80's, presented at the Verbeck Memorial Symposium, American Concrete Institute, Las Vegas, Nevada, March 1980; to be published in a Special Publication of the American Concrete Institute (in press).

7.3.1 THE EFFECT OF GYPSUM ON LENGTH CHANGE AND STRENGTH OF PORT-

LAND - AND PORTLAND BLASTFURNACE CEMENTSTONE SPECIMEN

Pieter C. Kreijger

University of Technology Eindhoven

Postbox 513, 5600 MB Eindhoven, the Netherlands

To get a rough idea about the effect of SO_3-contamination in recycled aggregates made from concrete debris or brick rubble, some measurements have been made on the effect of gypsum on length change and strength of portland - and portland blastfurnace cement-stone.

The gypsum was made from pure Ca SO_4 - 2 H_2O (Merck), heated (170°C) to Ca SO_4 - $\frac{1}{2}$ H_2O, mixed with water (water/gypsum = 1.0) and forming a gypsum plate which was broken in a ball mill to fractions: fine (diameter < 10μ, measured Blaine 2200cm ^{2}/g), middle fine (0.063 - 0.09 mm, calculated Blaine 250 cm ^{2}/g), middle coarse (0.5 - 1mm, calculated Blaine 30 cm^2(g) and coarse (4-8 mm, calculated Blaine 5 cm^2/g) The fraction were dried for 24 h. at 50°C and used as gypsum contamination. A $_F$ portland cement class B indicated further as P.C. (C_3A content 9.7%, SO_3 content 2.6%, Blaine 38 20 cm ^{2}/g and a portland blastfurnace cement class B indicated$_2$further as H.C. (slag content 55%, SO_3 content 2.8%, Blaine 4800 cm ^{2}/g) where used to make specimen 40 x 40 x 160 mm, with w/c-ratio 0.30. Series of specimen were made without gypsum addition and additions of the various types of gypsum so that the total SO_3-content was 4.25% and 6.0%. After curing the specimen for 1 day in moist air, the various series of specimen were placed in different conditions (direct in water saturated with Ca(OH)$_2$ and in air, in saturated Na$_2$ SO_4-solution after various times of water curing and drying for 2 days at 50°C, in air after various times of water curing. Length changes were measured regularly from 1 day age.

1. Length changes

Some results of length changes are given in fig. 1 from which

follows that the amount of gypsum contamination with grains
< 10 μm does not change shrinkage in air (22°C, 50% R.H.) but does
so in water (Ca(OH)$_2$ saturated, 20°C). A different behaviour is
found for P.C. and H.C. For P.C. as well increased SO$_3$ contamination,
as length of period in water or saturated Na$_2$ SO$_4$ - solution leads
to increased swelling. Although for H.C. increased SO$_3$ contamination
(grains < 10 μm) leads to strongly increased swelling, this swelling
remains constant after about 4-6 days (depending on the added SO$_3$ -
content and is found to be independent of a further stay either
in water or in saturated Na$_2$ SO$_4$ - solution.

For the highest SO$_3$ - contamination to H.C. (added 3.2%, total SO3 - content 6.0%) the effect of fineness of gypsum-addition
(expressed as Blaine value) is given in fig. 2. From this graph
it follows that between Blaine-values of 4 and 250 cm^2/g the increase
in swelling seems to be about proportional to the log- of the Blaine
value.

An explanation for the experienced behaviour of H.C. may be
based on the thesis of R. Bakker [1] who proved for H.C. with slag
content > 65% that Al$_2$ O$_4$$^=$ and SiO$_3^=$ ions of hydrating slag together
with Ca^{++}ions from hydrating cementclinker form a (nearly) imper-
meable precipitate after about 3-7 days. This should be the reason
for the good behaviour of most portland blastfurnace cements and
most pozzolanic cements against alkali-aggregate reactions and
sulphate attack (see fig. 3).

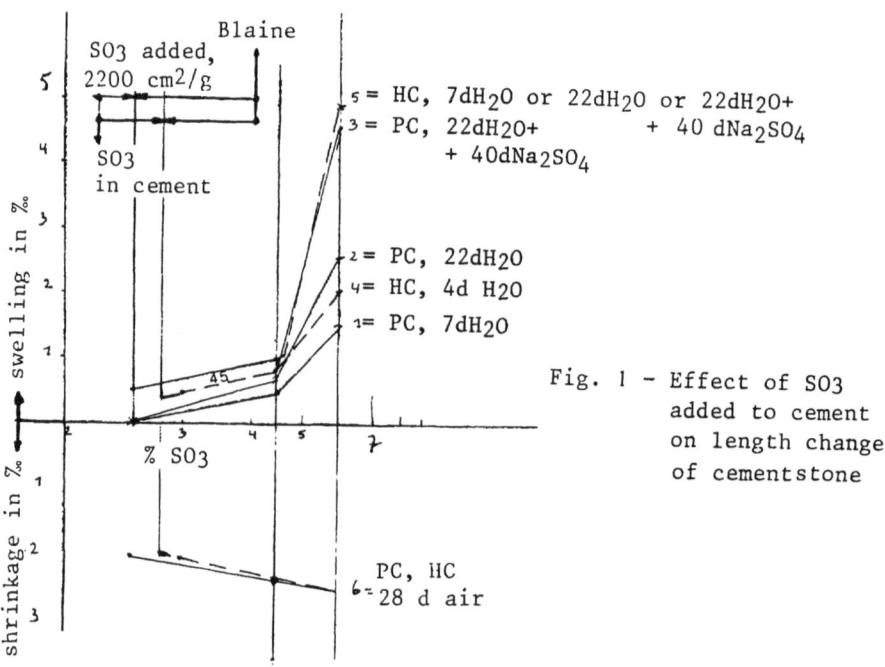

Fig. 1 - Effect of SO3
added to cement
on length change
of cementstone

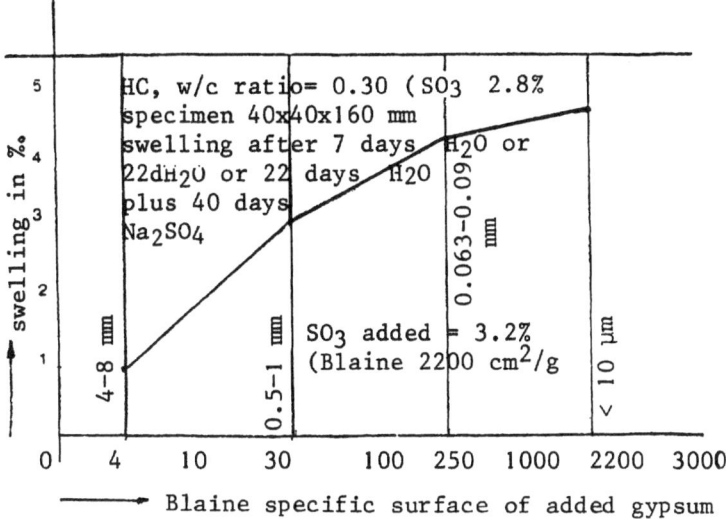

Fig. 2 - Effect of fineness of gypsum on swelling of portland
blastfurnace cement stone

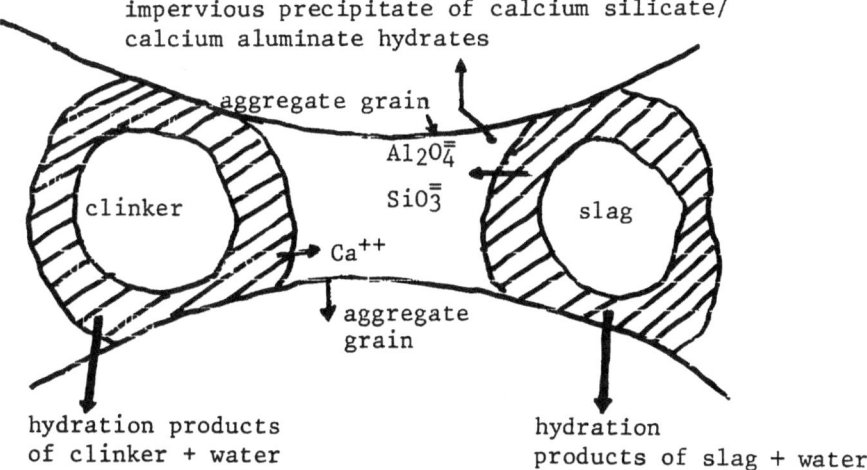

Fig. 3 - Mechanism of making portland blastfurnace cement with
slag content ⩾ 65% impervious to sulphates and to alkali-
aggregate reaction (1)

2. Strength

Specimen were tested after 7, 28 and 56 days first on bending strength (σ_{bt}) after which of the two halves compressive strength (σ_d) and splitting tensile strength (σ_t) were determined Results are given in table 1.

Table 1 - Effect of gypsum addition (specific surface 2200 cm^2/g on strength of cementstone (N/mm^2)

cement type (2200 cm²/g)	% SO₃ content added	curing in water (20°C)								
		σ d			σ bt			σ t		
		7d	28d	56d	7d	28d	56d	7d	28d	56d
P.C	2.6	73.4	90.0		0.9	7.6		5.2	5.4	
	2.6	82.4	102.6	100.6	9.6	6.1	4.8	6.3	5.8	5.4
	2.6	75.7	92.3	103.2	6.3	4.0	4.5	3.4	4.6	5.1
H.C	2.8	71.7	83.9		12.2	12.5		5.1	6.4	
	2.8	70.8	82.7	85.4	10.2	11.7	11.9	5.1	6.6	6.9
	2.8	49.7	58.8	90.8	3.9	8.7	8.4	4.7	5.1	6.7
		curing in air (22°C, 50% R.H.)								
P.C.	2.6	70.2	69.5		4.8	6.1		3.2	4.8	
	2.6	70.4	73.3	83.6	2.5	4.1	5.1	3.3	4.8	5.2
H.C.	2.8	56.3	56.6		2.7	3.0		2.3	2.9	
	2.8	48.2	52.0	47.1	4.1	3.9	2.3	3.5	4.1	3.2

Without SO₃-addition the well known effect is confirmed that H.C. is much more sensitive for drying out than P.C., while the ratio $\frac{\sigma_{bt}}{\sigma_d}$ is much higher for H.C. than for P.C. (2) for every type of curing. Addition of gypsum seems to effect σ_d of P.C. nearly not, of H.C. only at lower age. The larger addition of gypsum however seriously decreases σ_{bt} both of P.C. and of H.C. for curing under water, while for curing in air a decrease in σ_{bt} seems to be true only for P.C. Regarding σ_t, here the larger gypsum addition only affects P.C. at low age somewhat more seriously if curing under water has been used.

3. Conclusion

The effect of gypsum addition on length change and strength of
portland (P.C) and portland blastfurnace (H.C) cement specimen
was investigated. A different behaviour of both types of cement was
found and the cause indicated; H.C. being much less affected by
gypsum than P.C. while the size of the gypsum grains too has a great
influence. If the behaviour is tested by strength measurements,
bending strength will indicate a possible effect best.

Lit. 1 R. Bakker - Uber die Ursache des erhöhten Widerstandes von
                   Beton mit Hochofenzement gegen die Alkali-
                   Kieselsäurereaktion und den Sulfatangriff -
                   DissertationFakultät für Bergbau und Hüttenwesen
                   der Rheinisch-Westfälischen Technischen Hochschule
                   Aachen, 11 juni 1980 (144 p.)

2. P.C. Kreijger   et all - The shrinkage of hardening cement paste
                   and mortar -Cement and Concrete Research vol 5
                   pp 295-320, 1975

7.3.2. THE INFLUENCE OF CONTAMINANTS ON THE QUALITY

AND THE BEHAVIOUR OF RECYCLED CONCRETE

Prof. Ir. H. Lambotte   Ir. C. De Pauw

Magnel Laboratory   C.S.T.C.

University of GHENT

This research is set up by the C.S.T.C. (Centre Scientifique et Technique de la Construction) and subsidized by the I.R.S.I.A. (Institut pour l'Encouragement de la Recherche Scientifique dans l'Industrie et l'Agriculture). It is a part of a wider research on demolition, recycling and dismantling of concrete, that is set up in co-operation with Germany and the Netherlands. The tests are carried out at the Magnel Laboratory for Reinforced Concrete of the University of Ghent.

The investigation intends to determine the influence of contamination by gypsum, bricks and lime mortar on the quality and the behaviour of recycled concrete. It is expected that the results may be used to set limits to the percentages of contamination that are acceptable.

The test program was started in August 1980 and complete results will be available by the end of 1981.

This contribution describes the program and gives the results available at this date.

PROGRAM

Mixtures

Two different types of recycled concrete are being considered : type I with natural sand as fine aggregate and type II with recycled sand. For both types crushed concrete is used as coarse aggregate.

Type I

a. With portland cement (P40)
   Besides the reference (IPR), the following 11 mixtures are
   manufactured.

Contaminant	Percentage of contamination			
	0,2	1	5	10
fine gypsum	IPGf 0,2	IPGf 1	IPGf 5	
coarse gypsum		IPGg 1	IPGg 5	IPGg 10
brick		IPB 1	IPB 5	IPB 10
lime mortar		IPK 1	IPK 5	

The percentage of contamination is referred to the total weight
of fine and coarse aggregates. The contamination products are particles of grinded rubble. Fine gypsum contains particles up to
12 mm, coarse gypsum up to 20 mm.

b. With blast-furnace cement (HK40)
   Besides the reference (IHR), only two mixtures are manufactured :
   IHGf 1 : 1 % fine   gypsum
   IHGg 1 : 1 % coarse gypsum

Type II (only with Portland cement)
   Besides the reference (IIR), two mixtures are manufactured :
   II Gf 1 : 1 % fine   gypsum
   II Gg 1 : 1 % coarse gypsum

TESTS

For the type I concrete mixes with Portland cement a complete
series of tests is run as indicated in table 1, including splitting, creep, shrinkage and freezing tests. For the other mixes
these four tests are omitted. A survey of the tests is given in
table 1.

RESULTS

The results available until today are given in tables 2 and 3.

PRELIMINARY CONCLUSIONS

1. All the contaminants yet investigated affect the workability of the recycled concrete in the same way : with increasing percentage of contamination the workability decreases.

2. Contamination by gypsum decreases the compressive strength and other related characteristics of the recycled concrete. Contamination by bricks seems not to affect the compressive strength. It has to be stated that all contaminated mixes have the same cement content and the same W/C ratio as the reference mix and that the conclusions are depending on these conditions.

If the consistency instead of the water-cement ratio should be kept constant, the necessary addition of water would lead to a decrease of strength in the case of contamination by gypsum.

TABLE 1 - SURVEY OF THE TESTS

Age in days	Test		
0	consistency : jolting table slump bulk density		
1			shrinkage(1)
7	non-destructive compressive strength		
28	non-destructive compressive strength flexural strength splitting tensile strength(1) static modulus of elasticity	creep(1)	
	water absorption 15 freezing-thawing cycles(*)(1)		
90	non-destructive compressive strength		
182	non-destructive compressive strength		
364	non destructive		
547	non-destructive		
728	non-destructive static modulus of elasticity		

(1)   only for the mixes of type I with Portland cement

(*)   at the age of 90 days these test specimens are tested
       in a non-destructive as well as in a destructive way.

TABLE 2

| MIX | CONSISTENCY | | BULK DENSITY | | COMPRESSIVE STRENGTH Cube 200 mm | | FLEXURAL STRENGTH 28 days | SPLITTING STRENGTH 28 days | WATER ABSORPTION (1 dm³) |
	jolting table	slump cm	Fresh concrete kg/m³	7 days kg/m³	7 days N/mm²	28 days N/mm²	N/mm²	N/mm²	% by weight
IPR	1,89	10,5	2315	2310	38,1	48,5	5,07	3,42	7,1
IPGf0,2	1,66	6,5	2315	2295	39,2	51,7	4,65	3,70	7,0
1	1,57	5,5	2280	2285	43,2	54,3	4,48	3,72	8,3
5	1,12	0,5	2300	2270	36,9	45,4	4,70	3,45	8,3
IPGg 1	1,67	6,5	2315	2295	45,8	54,3	5,25	3,97	7,7
5	1,22	1,5	2290	2275	37,4	44,8	5,13	3,47	8,1
(*) 10	1,06	1,0	2250	2225	31,9	37,8	4,17	3,08	9,9
IPB 1	1,62	6,5	2330	2315	45,5	54,3	5,28	4,22	7,1
5	1,35	4,0	2290	2300	43,8	53,9	5,55	4,23	
10	1,20	1,0	2290	2285	46,5				

(*) 20 litre water is added to this mix

TABLE 3

MIX	NON-DESTRUCTIVE TEST (prism 100 x 100 x 400 mm)						E_s 28 days KN/mm^2	COMPRESSIVE STRENGTH prism 200x200x500 mm 28 days N/mm^2
	E_{dL} 7 days KN/mm^2	E_{dL} 28 days KN/mm^2	G_{dT} 7 days KN/mm^2	G_{dT} 28 days KN/mm^2	V_L 7 days m/sec.	V_L 28 days m/sec.		
IPR	34,42	36,21	14,53	15,27	4050	4170	32,85	41,7
IPGf 0,2	35,45	37,54	14,81	15,72	4100	4200	33,10	45,6
IPGf 1	35,02	36,88	14,78	15,39	4100	4210	32,80	46,5
IPGf 5	34,07	35,01	14,05	14,69	4060	4090	29,50	42,0
IPGg 1	37,65	39,32	15,92	16,41	4180	4220	34,55	50,7
IPGg 5	32,59	33,90	13,65	14,34	3950	4080	29,85	40,7
IPGg 10	30,98	31,66	12,84	13,36	3960	4010	26,75	32,4
IPB 1	35,19	38,54	14,98	16,00	4160	4170	35,15	53,3
IPB 5	35,76		15,27		4190			
IPB 10	34,64		14,57		4030			

E_{dL} longitudinal dynamic modulus of elasticity

G_{dT} dynamic modulus of rigidity (torsional mode)

V_L longitudinal velocity of sound

E_s static modulus of elasticity

8. WORKSHOP 8 FUTURE DEMOLITION-FRIENDLY MATERIALS

8.1 INTRODUCTORY COMMENTS

Della M. Roy

Materials Research Laboratory
The Pennsylvania State University
University Park, PA. L6802, USA

When considering the problems relating to building materials
recycling, and factors which determine to what extent it is possible
to build with "demolition-friendly" materials, there are certain
analogies to the situations facing early Roman architects and
builders. The annual figures projecting magnitudes of expected
demolition (and hence potentially recyclable) concrete structures
seem astronomical. One might be led to ask if greater attention
should not be given to preservation. However, once societal decisions
are made that demolition is desirable, then technology must be ready
to provide alternatives. It is even likely that the knowledge
applicable in one area will be applicable in the other as well.
Further, for both early and present builders, it appears to be a
given and accepted premise that recycling is a social necessity,
or at least a good.

Thus, the factors of concern are both technical and non-technical,
and the two categories are independent. A few of the factors of
concern are herewith listed, which either are known to be important
or which are likely to influence choices of building materials,
their methods of fabrication or their demolition methods. It is
hoped to discuss some of these, as well as other factors in the
Workshop. A brief comment on some of them is also given here:

Economic factors

Aesthetics/community values

Disposal/transport factors and costs

Aggregate type and availability

Nature of cement matrix/aggregate bond

Effect of new factors influencing matrix behavior

Extent of materials damage upon demolition

Differences in problems concerning materials recovered
   from matrix and aggregate.

Thus, a number of specific questions might be posed:

1. How reconcilable are the two apparently contradictory qualifi-
cations: the requirement of strength and durability of materials,
and the desirability of superior performance; vs the potential for
easy recycling?

2. What properties must a binder for concrete have in the event that
an easy demolition is desired after a fixed lifetime, or if social requirements place a shorter limit on lifetime, at an early time; and
what effect is consequent in the concrete?

3. What characteristics will a concrete have which possesses the
best combination of performance during its lifetime, with the
possibility for recycling, i.e. optimization tradeoffs? Can concrete
be truly tailor-made for such circumstances?

4. Would we make cement matrices any differently than in current-
day practice in view of probable recycling? What are the disadvantages of cement as a binder?

5. Is it possible to recover unhydrated cement?   Larger grains in
a cement typically serve as partially unhydrated "micro-aggregate"
and thus have potential binding value remaining. Much of this
presumably can be re-used; but in such an event, would substantial
portions be re-utilizable in conventional concrete, or in other
applications such as masonry cement? How efficient can fractiona-
tion be in order to recover the desired fractions?

6. What effects do certain relatively new types of additives forming part of the cement paste matrix have upon the properties:
either potentially favorable or deleterious effects? What are the
effects of relatively reactive components added to the paste? How
does one predict the long-term performance of relatively "new"
materials such as superplasticizers?

7.   Should portland cement paste as a matrix be either supplemented
or supplanted by other binders such as sodium silicates?

8.   What information does the nature of the aggregate - interfacial
region with the paste yield regarding the possibility for reconciling
the conflicting factors of adequate structural performance vs.
easy demolition and recycling?

9.   What variations are possible in selection and beneficiation of
both coarse and fine aggregate? How important are the specific
chemical and mineralogical composition of the aggregates, vs.
microstructural, textural and mechanical features?

10.   What effects do specific reasons for demolition, or methods
used in demolition have upon the re-usability of aggregates? Is it
possible to compensate for phase changes brought about in aggregate
minerals, or effects of thermal stresses which are consequent upon
fires in structures which may necessitate demolition?

11.   What types of tailoring of concrete mixtures can be done in
order to optimize the characteristics of the matrix-aggregate bond?
Do factors such as zeta potentials of matrix and aggregate
surfaces have a major influence upon the ease of demolition and
suitability for re-use? If so, could suitable "tailoring" be
achieved in such characteristics?

8.2 MINUTES OF WORKSHOP 8

M. Geudelin

Directions de le Recherche

UTI, Paris

Professor <u>Roy</u> had prepared some questions to introduce her introductory lecture directed at examining the possibility of building structure in such a way that demolition, when necessary, can be accomplished with relative ease, whilst producing materials suitable for recycling. After a historical recapitulation drawn from Roman practice, she pointed to the possibility of replacing demolition by dismantling. Since this is not always feasible, there remained the problem of defining the properties of a binder to facilitate demolition coupled with the characteristics of concrete for best performance during lifetime. Could the matrix be modified and as such what are the disadvantages of cement as a binder? This pointed to examining ways to recover unhydrated cement from the mortar and the effects of relatively reactive components. Other binders such as sodium silicates could be investigated. Perhaps the size of aggregate could be of influence. Ensuing discussion should also deal with compensating phase changes and ways to optimize bond through study of zêta potential and surface conditions.

<u>Kreijger</u> taking up the performance during lifetime, formulated an approach whereby the designers, through knowledge of rates of processes of deterioration coupled to coefficients of variation of material properties, could foresee when durability approached limits for maintenance, repairing or demolition (see 8.3.1).
<u>Mather</u>, recalling ACI work, emphasized the macro-approach to structures to be designed with built-in devices (joints,, etc.) to make them self- destructive at a time chosen by the designer. He considered this safer then tampering with concrete composition. This was also the opinion of <u>Lindsell</u> using the example of post-tensioned concrete beams whereby the irreversible processes (grouting, tensioning and cropping) prevented the members from being reused. <u>Mather</u> stressed

the advantage of non-bonded cables (for checking corrosion and ten-
sioning) whereas Lindsell called for research on futural alternative grouts using phase change coupled to use of compressed air for
removal, or new anchorage design.

Frohnsdorff argued that the concept of self-destructive materials implied a change in the aesthetic values for society. The word
"friendly" had been chosen carefully, asserted Kreijger, to imply
no interference with people or environment by dangerous processes.
Would it help to know exactly why materials bound so that the process would be reversed to desintegration?

Hansen suspected contradiction in a situation where easy reuse
was sought from a material designed for strength and durability .

Roy referred back to the subject of bond between matrix and
aggregate. The surface potential on rocks is negative but hydrating
cement paste quickly develop a positive surface due to the calcium
ions adsorbed from calcium hydroxide going into solution. With superplasticizers there is an increased negative zêta potential resulting in dispersion in solution but these conditions change with time
and are not always disclosed by manufacturers. Mather asserted that
superplasticizer action is finished after 30 minutes. To strip aggregates clean, it is preferable to use aggregate particles with minimum adsorbtion capacity and maximum surface smoothness and round-
ness to obtain maximum effect from difference in thermal coefficient
of expansion. If the system is heated, it will desintegrate.
With siliceous aggregate, paste can be removed with acid.

Discussion than moved back to demolition methods presented by
Molin, with a view to the future of expansive liquids, sodium silicates
percussion drill hammers, explosives, water jetting, pulsing twinned
with built-in holes (Kreijger), dummy joints, instant oxidation
(Mather), microwaves (Roy), plasma torch, micro biology-bacteria,
special aggregates and erosion principles (avitation).

Frohnsdorff reported on Professor Idorn's work on abolishing
aggregate, also Rebinder's lecture to the P.C.A. underlining the
use for improved use of concrete. The Sovjet appraoch involves developing rapid hardening cement systems with plasticizers, water
reducing agents and fine-ground sands, avoiding reinforcement.
Mikhailov had lectured at the N.B.S. on self-stressed concrete -
100,000 ton of which were used for Olympic structures. Kreijger
in this connection recalled the possibilities of the use of swelling
cement coupled to the method of curing.

A short discussion took place on possible uses of lower grade
aggregate (Lindsell), also "fines" (Kreijger): in theory the smaller
the particles the greater the surface forces. If concrete is broken
up into particles similar to fly ash, Mather asserted that it could

be competitive as mineral admixture, but the real problem is one of
research and markets.

The session ended by reviewing demolishing techniques which in
the nearly future could have the best chance to approach the subject
og the workshop and a general consensus seems to exist in using conventional breaking coupled to a system of built-in holes and/or
joints, infact the design of demountable structures, and more research
was advocated regarding the structural behaviour of demountable joint-
connections, as was going on now in Belgium and the Netherlands
(Kreijger).

8.3.1 SOME NOTES REGARDING THE PERFORMANCE OF WASTE MATERIALS AND

CONCRETE IN BUILDING COMPONENTS

P.C. Kreijger

University of Technology Eindhoven

Postbox 513, 5600 MB Eindhoven, the Netherlands

One of the most important properties of building components
is the durability, defined as loss of performance as function of time.
Apart from this the functional properties should have an intrinsic
value which suits the purpose.

1. Durability in general

A building component→ joining of materials / materials parts

functions (economic→ action (adhere,   of stony materials,
constructing, protec-   bricklaying,   glass, steel, aluminium
ting, aesthetic)   glueing etc.)   wood, polymers, butimen,
                                           paints etc.

durability=loss of
function = F(time)   loss in acting together + use and by attack by
determined by ———▶ = decrease of adhesion/ mechanical, physical
                        cohesion in the   chemical, biological
                                             influences in the
                        PRACTICAL SITUATION PRACTICAL SITUATION

main functions are: created by climate, situation, living- and
- bearing (or not)   use conditions, type of cleaning, maintenance
- protecting (or not) connected with type and function of the buil-
  against moisture,   ding, placing in, -on,- under-,upon the buil-
  temperature, frost, ding and indoors/outdoors.
  fire, sound-indoors
  or outdoors
- aesthetical (visual
importance (or not)- unhindered transport

More/less sensitive for loss of function →	Apart from normal wear in use there are:
	detrimental influences which might be specific for certain materials/material parts and/or certain joinings of them. Examples:

Incompatibility of materials:	Mechanical: overloading, fatigue
- contact corrosion	phys./chem: weather, temperature, salts, pollutants, erosion, frost, special climates, radiation etc.
- materials with different coefficients of expansion, shrinkage, E-modulas etc.	biological: bacteria, moulds, mosses, algae, insects, animals.

loss of function is indicated externally by →

attack which starts at surface and gradually penetrates into the material:
- without worth mentioning decrease in thickness: loss of gloss, colour, transparency
- with a uniform decrease in thickness/section over the surface
- non-uniform decrease in thickness over the surface: cracks, perforations

loss of function is not indicated externally by →

2. deformation: warping, bubbling, folding etc. change of properties characterized by decreasing cohesion through out the whole material: fatigue, hydrogen brittleness, decrease in strength by high temperatures, alkali-aggregate reactions, frost attack, fracture.

LIFETIME is indicated or determined by →

The practical situation(s) during lifetime coupled to detrimental influences which might occur for the combination of materials/material parts in question - which in joining have made them to a building component

2. Waste materials used for building components

In many cases waste materials (concrete, debris, brick rubble. colliery waste, incinerator residues, sludges fosfogypsum, sulfur, slags etc.) are tried for use in (un/reinforced building components that is to say the junction of small building products each consisting of junction of different materials, mostly:
a binder (mostly cement) + waste materials + water.
The compatibility is determined by:
- Regarding contaminants: aluminium - and zinchydroxide, phosphates (regarding cement)

- gasforming contaminants: metallic Al, Zn, Mg, Pb.
- organic rests   : humus, phulvo acids
- uncombustible rests   : loss of ignition
- glass   : regarding alkali-silicate reactions and
                           alkali-carbonate reactions (mainly for
                           temperatures between 10°C and 60°C)
- staining contaminants   : iron, vanadium (change of bi--▶ tri)
- efflorescence causing
  contaminants   : chlorides, sulphates, carbonates, nitrates
- corroding contaminants : chlorides, sulphites
- health effecting con-
  taminants   : heavy metals, radioactivity
- the amount of particles < 4 mm and < 63 µm
- regarding cement as binder, important here are fineness, C3A-con-
  tent, slag content and alkali-content.

The importance of the above is connected with the functional requirements of the formed building element, so in general:reinforced (or
not), bearing (or not), protecting (or not), used indoors or outdoors,aesthetic (or not) while such performances in its turn are
determined by the intrinsic properties of the products (1) like
strength, stiffness,shrinkage, coefficient of expansion, moisture
adsorbtion, isolation aspects and visual aspects.
In table 1 a simple matrix is given between functions and properties
to judge in first instance a possible use of waste materials in
building. Partly these properties also say something about possible
solutions for junctions or joints between the products (so material
+ action, see ad. 1). Special reference should be made to the development and application of durability (service life) tests as
given in ASTM standard E 632 (see also 7.1)

3. Rough estimate of concrete durability

Use of concrete generally is judged by its (compressive) strength
and each is familiar with strength curves as function of time like
are sketched in fig. 1, (6) although many constructing engineers
seem to look only to the "magic" 28 days-strength.
Even looking only to the compressive strength of cement, one has
to accept a coefficient of variation of about 15%, concluding to
a characterictic compressive strength (5% failures accepted) of the
value: average strength (μ) - 1.64x0.15 μ= 0.75μ. Accepting a safety factor of 1.7 - 1.8 the limiting value of μ consequently is
0.75 μ/1.8 $\simeq$ 0.42 μ and in which long term effects are enclosed, so
after, say, 100 years a limiting value of 0.8 x 0.42μ = 0.34μ can be
accepted or about a third of the average 28-days compressive strength.

Now the safety factor is related to, as is stated in many standards "structures unfit for use" which can be related to strength
and stiffness, to appearance or to corrosion of reinforcement.

Table 1 – Matrix between performance and intrinsic properties

intrinsic material properties	reinforced	not reinforced bearing	in-/outdoors	heat insulating	sound insulating	aesthetic	contaminants set retarding	gasforming	organic rests in-combustible ests	glass	staining	salts
1 specific mass				O	O							
2 porosity		O		O	O	O	O					
3 effect of moisture equil.moist.cont.	O		O	O			O			O	O	O
wateradsorbtion	O	O	O	O		O	O				O	O
solubility in water	O	O	O	O	O	O			O			O
water permeability coeff	O	O	O					O				
capillary suction	O		O	O			O	O			O	O
shrinkage / swelling		O	O		O	O	O					O
4 effect of temperature coeff. of expansion	O	O	O			O	O					
spec. heat capacity				O								
melting point / glass trans. temp.		O	O									
thermal conductivity				O				O				
5 strength and mod. of elasticity		O	O			O		O		O	O	O
σd,σb,σt,σ0.2 viscosity	O	O	O				O	O	O	O		O
creep	O	O					O					
hardness		O	O					O	O	O		
impact strength	O	O	O									
6 effect of sound coeff. of adsorbtion					O		O					
coeff. of reflection					O		O					
coeff. of transmission					O		O					
7 sparks,dirt electric resistivity			O			O		O				O
8 frost resistance			O					O	O			O
9 fire fire resistance	O	O	O	O	O			O				O
fire extention		O	O									
fire flash over		O	O									
10 attack corrosion	O							O				O
chemical attack	O	O	O	O	O	O		O	O	O	O	O
biological attack	O	O	O	O	O	O	O	O				
11 durability	O	O	O	O	O	O	O	O	O	O	O	O
12 possible detrimental eff.	O	O	O	O	O	O	O	O	O	O	O	O
13 coeff. of friction			O			O	O					
contaminants first glance, better looking to properties 1-13 set retarding	O	O	O	O	O	O						
gas forming	O	O		O	O							
organic rests inconbustible			O	O								
glass						O						
staining /efflorescence						O						
salts/corroding	O	O	O			O						

Regarding climates with (much) precipitation and frost cycles there is a good chance that after five years the concrete can be weathered seriously on the outside. Also in such climates it is found that rusting of reinforcement is the most frequently occurring type of damage. A rough estimate of the time such damage will take place can be based on the process of carbonation which process often is formulated as a diffusion process (2,3,4) and as such can be calculated easily, although more recently correction factors have been added (5). For application to practice 10 years ago the author constructed (6) a graph which is given in fig. 2 and may be used for a rough estimate of the incubation period after which rusting of reinforcement will take place. Starting with the cover of the reinforcement (= carbonation depth) which can be taken from drawings or measured non-destructive, and putting in constant R for the effect of type of cement (P.C.=1, portland blastfurnace cement= 2-2.6), T for the type of aggregate (sand and gravel= 1, lightweight aggregate = 2), K for the climate (time of concrete surface wetted (indoors=1, outdoors mostly 0.25-0.75), the w/c-ratio of the concrete in question leads to the estimate of the incubation period during which the whole concrete cover of reinforcement has been carbonated.

This and other attacking processes like wear, erosion etc. can also be dealt with stochastically and many such attempts are in evalution since (7) appeared.

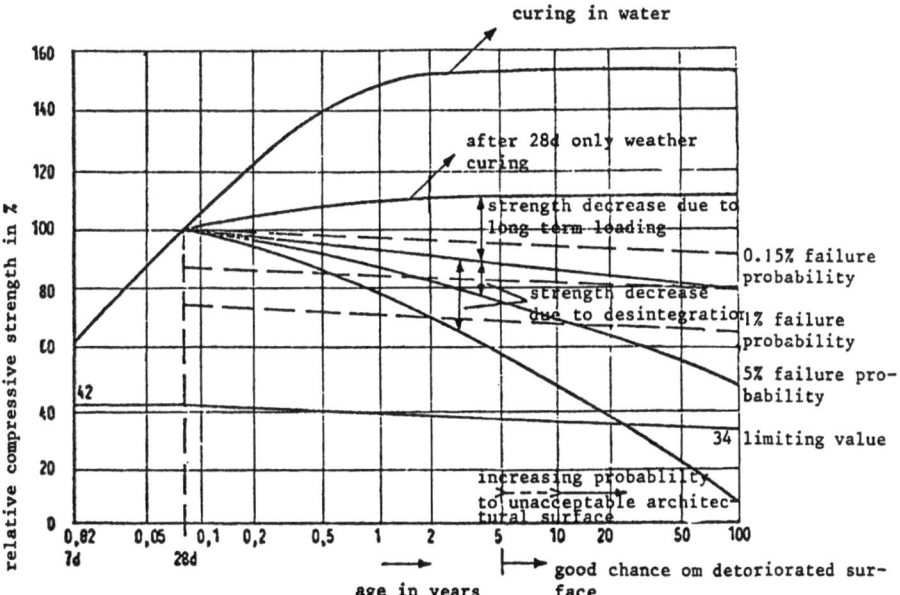

Fig. 1 Analysis of compressive strength results

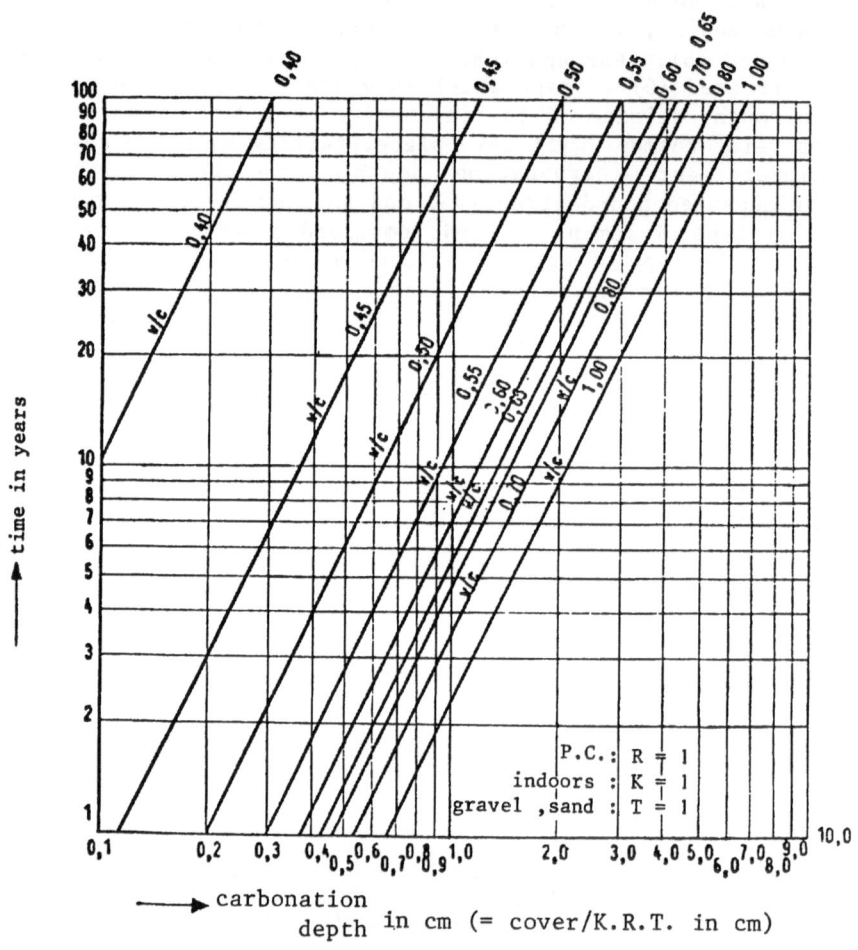

Fig. 2 Carbonation depths as function of w/c ratio and time or
         carbonation time as function of w/c ratio and concrete
         cover.

Finally it may be remarked that the use of a constant safety factor in construction calculations is rather one-sized since it does not connect the type of damage to economic consequences. It is quite different if for example a glass pane collapses or a whole structure. Therefore it seems better to take into account the expected costs of damage and the probability(p) of such a damage. The costs of damage may be related to (n) the foundation costs K which gives the risk of a damage $R = n.K.p$. It has been shown (8) that $(K+R))$ will be a minimum if $R \simeq 0.1 \times K$ or $n.p \simeq 0.1$. So if one can make a reasonable estimate of n - for example for a glass pane $n \simeq 1-2$, for floors and roofs $n \simeq 2-10$, the failure probability can be calculated. While in our today standards the safety factor seems to be based on $p \simeq 10^{-5}$, the foregoing leads to different values for the safety factor. There seems to be a general consensus that regarding roofs and floors, $p \simeq 10^{-2}$, while regarding the occurrence of fire $p \simeq 5.10^{-3}$ (9) and that of an explosion $p \simeq 2.10^{-4}$
Such an approach necessitates the designer to look for risks of what might happen in future and the cost consequences and may lead to structures more fit for the purpose then is the case today.

Literature

1. CIB Master list for structuring documents relating to buildings, building elements, components, materials and services - Report no. 18
2. Outline of the studies in Japan regarding to the neutralization of alkali or carbonation of concrete - T. Nishi. RILEM Symposium Durability of Concrete, Prague 1961, Final report II p. 485 - 489.
3. Karbonatisierung von Schwerbeton - A. Meyer, H.J. Wierig, K. Hausmann, D.A.S. Heft 182 (1967)
4. Einfluss von CO_2 und feuchtigkeit auf die Beschaffenheit von Beton als Korrosionschutz für Stahleinlagen, D.A.S. Heft 182 (1967)
5. RILEM Symposium Carbonation - Wexham Springs, April, 1976
6. PDOB Cursus Duurzaamheid van Bouwmaterialen Sept. 1971 - P.C. Kreijger (Post - University Course - Durability of Building materials)
7. Model of failure - L.B. Gertsbakh, K.h Kordonskiy, Springer Verlag Berlin, Heidelberg, New York 1969
8. Hoe veilig moet een constructie eigenlijk zijn? F.K. Ligtenberg T.N.O.- Nieuws 26, March 1971, p. 271-276 (How safe should a structure be?)
9. Optimum fire resistance - ACWM vrouwenvilder - Heron 22, 1977 no. 3, 18p

PART 4   CONCLUDING REMARKS

PART 4 CONCLUDING REMARKS

Pieter C.Kreijger
University of Technology, Eindhoven
Postbox 513, 5600 MB Eindhoven
The Netherlands

The essential ingredients of this symposium can in fact be considered to co- and adhere as can be seen from the following scheme:

ADHESION PROBLEMS
in the
RECYCLING OF CONCRETE

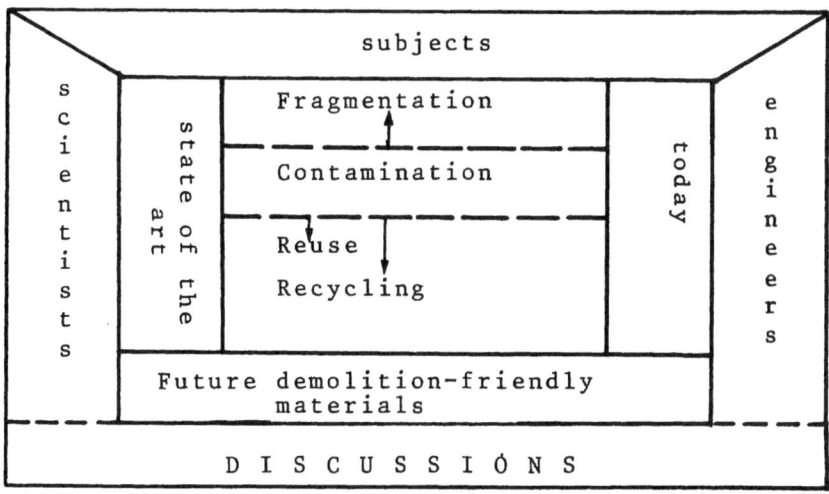

The subjects were elucidated in lectures of a high level, while during the discussions, scientists and engineers interwove theoretical considerations and practical arguments in an endeavour to build a bridge between today's state-of-the-art and the future needs.

One of the benefits has been the exellent understanding between scientists and engineers which resulted in a clear notion of each other's difficulties, developing collective approaches to the various problems and so being really complementary to one another. For example, the recycling and reuse of concrete today starts with demolition and is followed by fragmentation: workshop 1, 4, 5 and 8 in this sequence leading to research needs, to be quoted here briefly: The first process is related to the structure and consequently to the damage caused, whereas the second process is a material problem related to energy. The problem to get crushed particles for reuse or recycling, therefore, has the purpose of relating damage to energy.

The <u>fundamental approach</u> should develop and integrate valid fracture mechanics parameters for concrete and the best applicable damage theory, while in both surface physics and chemistry of debonding and fracture should be considered.

The <u>engineering approach</u> should set up acceptable empirical measurements of damage and energy and find empirical relations between them for the various processes used in practice so that distributions of energy and damage will be known in different fracture modes.

<u>Both approaches</u> should take into account a number of specified <u>parameters</u> and such studies combine the two approaches to obtain the optimal energy-damage ratio, depending on the method of breaking and the product to obtain. Moreover we should take advantage of studying progress in other diciplines, like rock mechanics, mining and tunelling.

It is hoped that by defining seperate experimental problems to be dealt with by seperate sets of people, it would be possible to propose guidelines for the rational selection of the appropriate procedures in demolition and fragmentation.

Regarding the problem of contamination (workshop 7) it should be realized that the permissible limits on any contaminant -organic matter, inorganic matter or metals - will depend upon the application of the crushed particles. For example, for some types of reuse contamination needs not be a problem at all when considering the use of recycled concrete as an aggregate for new concrete. In fact, the problem is slightly different from the range of possible contaminations in natural aggregates, though the quantities initially present may be higher and/or different (radioactivity, toxic substances).

So the general consensus was that research needs for concrete containing contaminated recycled concrete as similar to those for concrete containing natural aggregate.

They include as is said in the report of workshop 7, the development of equations to predict performance in terms of its ingredients
and the distribution of ingredients, taking into account the effect
of the expected service environment. Improved understanding is needed
of the chemistry and micro-mechanics of concrete and their effects
on durability performance.

Since ingredients may include contaminants, more specific
knowledge is necessary about effects of specific types and quantities of contaminants and their interactions as to strength and
other aspects of long-term concrete performance. Because the contaminants depend on the source where the crushed originally comes
from, it is desirable that the range of possible contaminants from
each source be known. Special attention in this connection is
called for refractory materials, fire-damaged concrete and concrete
attacked by freeze-thawing and/or alkali-aggregate reactions.

It is also desirable, to know the source and history of the
aggregates if its use in high-performance applications is being considered for example including high-strength concrete or concrete to
be exposed to severe conditions. Otherwise it would probably not be
important to know the source, provided the lightest weight and most
friable materials have been moved, meeting standards similar to
those for material aggregates.

Ways of optimizing processes for removing contaminants from
crushed aggregates should be investigated. Today's procedures deal
mainly with low density contaminants and friable materials which
collect in the fine fractions. They include wood, plastics and
gypsum which can be removed economically under many conditions
while chlorides, sulphates, asphalt and paints within the pores
of the crushed concrete and alkali reactive aggregates might be
difficult to remove. Generally speaking, most caution should be
exercised in using fine fractions of recovered aggregates and the
possibility of processing the fines for other uses should be considered first. Here too research should be considered in order to
make either a cementitious material or a mineral admixture.

Although the use of coarse fractions of recovered aggregates
in recycled concrete does not seem to give much trouble, there
is a need to learn more about some properties of the new concrete
apart from the research needs already mentioned, as is stated in
the report of workshop 2: for example strength variability, toughness properties, mechanics of failure and all long term-performance
properties. To make such research results comparable, first better
used while possibly a classification for such aggregates could be
developed from the research results and might be useful for new
concrete standards as well.

In the area of production technology it was strongly felt
that a pilot plant that includes sorting facilities is needed
as a first step to promote reuse of recovered concrete. Quality control and assurance procedures should be designed during
the early steps of "new" aggregate, even if in principle all
concrete should be regarded as a quarry of sedimentary rock, implying that that requirements for recycled and natural aggregates
should be identical.

In workshop 3, the use of fine and very fine fractions of
fragmented concrete and brick rubble and the hydrated cement
herein was discussed again in the light of those of other used
than as aggregated for concrete. Possibilities could be investigated for agricultural use (neutralizing acid soil or as
a source of silica, for example in highly leached lateritic soils),
solidifying waste sludges to retain it in place or to transport it to dumps but especially to avoid transfer of pollution
by checking all the harmful and toxic elements concentrated
there, chemical scrubbers for removal of SO_2, low quality stabilization, upgrading to a kind of masonry cement. In connection
with ultra fines, study of kilndust technology was advocated.

On the other extreme side of the size of aggregates there is
the application as ballast on mattresses or as rip-rap and artificial reefs. So already during the stage of demolition the possible
uses should be considered independent and dependent on type and
size of aggregate, taking into account transport distances.

Most of the recycled concrete aggregate in the U.S. is coming
form pavements and uses again for pavements in several applications
varying from low quality soil stabilization, filter drainage layers
and subbase or base materials to concrete surface dressings with
cement or bitumina as binders. In the latter case there is of course
the problem of contaminants that has to be examined, especially the
effect of de-icing chemicals but also the effect of asphalt contamination on air entrainment of the new concrete and the effect
of "D"-cracked concrete being used as aggregate for new concrete.

For non-pavement uses other than those already mentioned on
could think of use of recovered concrete aggregate for new structural concrete (main application) but also for roofing granules (if
meeting the specifications) and for heat capacity in thermal reservoirs. There seems to be a possibility of reusing concrete slabs
as buildings stone for masonry construction after prenotching them
prior to demolition, leading to aesthetic surfaces composed of
fractured or exposed aggregate depending on the strength of the
concrete. It was not expected that these exotic applications
would use as much demolished concrete however.

There was the general opinion that chances of reusing steel in construction were minimal, the only real use was considered to be as an ingredient in steel manufacture.

Regarding the connection between today and the future with a view to demolition-friendly materials (workshop 6) the first contribution came from workshop 4 and concerns the demolition of prestressed structures or constructions containing prestressed elements: from such buildings long-term records of design calculations, as-built drawings of all construction details included structural function, erection sequence, location of tendons, anchorage design, chemical details of grouting, concrete mix, strength and demolition procedures should be kept by engineering and local authorities. Case histories should be organized at national level by professional bodies. Tension release in this type of demolition work is a very acute problem and needs more research, as the design of easily removable anchorage devices.

Alternative demolition methods like work on controlled explosives (shaped charges), cutting techniques (water jetting) and development of alternative grouts comprising for example the possibility of phase changes and reducing the grout to powder easily removable by compressed air, were proposed.

The best chance to approach the realization of the subject of workshop 6, so was the general consensus, seems to consist in making use of conventional breaking methods for building components designed with built-in devices like holes and/or joints to make them self-destructive at an arbitrarily chosen time. This in fact has its limit in the design of demountable structures, research on which already being done in Belgium and the Netherlands.

Regarding the material side, so related to easy stripping of aggregates, a lot of speculative ideas were launched which deserve a broader analysis.
Other methods like the better use of adsorbtion sensitive fracture are considered worthwhile and research already going on in the Netherlands and the USA should be stimulated to continue, not at least because from research better suited theories on the cohesion and adhesion forces in concrete will be developed which on their turn will contribute to a better recycling of concrete.

The foregoing might serve as a summary of some of the outcomes of the ARI as given in parts 2 and 3 of these Proceedings.

It is believed by the author that a closer study of the texts will prove that this first Symposium on the recycling of concrete was a valuable one, not at least by collecting relevant data and ideas from various disciplines.

LIST OF PARTICIPANTS

1. I <u>Belgium</u>   1. Prof.Ir. H. Lambotte – Ghent State University,
Laboratorium Magnel voor gewapend beton
Grote Steenweg Noord 12
B 9710 Gent (Zwijnaarde)
Tel.: 091 – 225755
Telex: 12754 RUGENT

2.   2. Ir. C. De Pauw, project leader Wetenschappelijk
en Technisch Centrum voor het Bouwbedrijf
Lombardstraat 41
1000 Brussel
Tel.: 02 – 6538801
Telex: 407695057

3. II <u>Canada</u>   1. Prof.Dr. S. Mindess, University of British
Columbia – Department of Civil Engineering
2324 Main Mall
Vancouver B.C. V6T1 W5

4.   2. Dr. V.S. Ramachandran, Head Building Materials
Section, Division of Building Research,
National Research Council of Canada
Ottawa, Ontario KIAOR 6
Tel.: (613) 9931596

5. III <u>Denmark</u> 1. Prof.Dr. T.C. Hansen, Technical University
of Denmark
Building 118,2800 Langby
Tel.: (01) 883511
Telex: 37529

6. IV <u>France</u> 1. Dr. R. Brepson, Laboratoire de Glaciologie
et Géophysique de l'Environnement
Centre National de la Recherche Scientifique
2 Rue Très Cloitres
38031 Grenoble Tel.: (76) 420527

7.   2. Mr. Ph. Briquet, Laboratoires des Ponts et
                           Chaussées
                           Département Bétons et Métaux
                           58 Boulevard Lefebre
                           75732 Paris Cedex 15
                           Tel.: (1) 5323179
                           Telex: 200361

8.   3. Mr. J.P. Collin, Engineer Materials Department.
                           CATED,
                        9 Rue la Pérouse, 75784 Paris Cedex 16
                           Tel.: (1) 720 10 20, Telex: 611975

9.   4. Mr. P. Cormon, Head Engineer Materials
                              Department, CATED,
                              9 Rue la Pérouse, 75784 Paris Cedex 16
                              Tel.: (1) 720 10 20 Telex: 611975

10.   5. Dr. R. Sierra, Laboratoires Central des
                              Ponts et Chaussées
                              Centre de Nantes
                              BP 19 - 44340 Bouguenais
                              Tel.: (40) 651488
                              Telex: 710805

11. V W.- Germany   1. Dipl.-Ing.B. Armbruster, Kneucker & Co. GmbH
                              Abbruch vom Industrieanlagen, Brücken
                              und Gebäuden
                              6800 Mannheim 1
                              Friesenheimerstrasse 17b
                              Tel.: (0621) 312084

12.   2. Prof.Dr.-Ing. H.K. Hilsdorf
                              Institut für Baustofftechnologie
                              University of Karlsruhe
                              Kaiserstrasse 12
                              P.O. Box 6380
                              7500 Karlsruhe 1
                              Tel.: (0721) 608–3890
                              Telex: 07826521

13.   3. Dipl.-Ing. U. Neck, Forschungsinstitut
                              der Zementindustrie Düsseldorf
                              Tannenstrasse 2
                              D 4000 Düsseldorf 30
                              Tel.: (0211) 45781
                              Telex: 08584867

14.

4. Dr.-Ing. J.H. Weber, Institut für Bau-
forschung der Rheinisch - Westfälischen
Technischen Hochschule Aachen
Schinkelstrasse 3
D 5100 Aachen
Tel.: (0241) 805111
Telex: (08) 32 704

15.

5. Dr. S. Ziegeldorf, Institut für Baustoff-
technologie
University of Karlsruhe
Kaiserstrasse 12
P.O. Box 6380, 7500 Karlsruhe 1
Tel.: (0721) 608 - 3890
Telex: 07826521

16. VI Israel

1. Prof.Dr. O. Ishai, Faculty of Mechanical
Engineering
Technion, Israel Institute of Technology
Haifa
Tel.: 04-292717
Telex: 46650

17. VII Luxembourg

1. Dr. E.R.G. Hoffmann, Laboratoires des
Ponts et Chaussées
Rue Albert 1er
7 - 11 - Luxembourg

18. VIII Netherlands

1. Prof.Ir. P.C. Kreijger, Technological
University of Eindhoven
Department of Building
Architecture and Planning
Group Science of Materials
Postbox 513
5600 MB Eindhoven
Tel.: (040) 472292
Telex: 51163

19.

2. Ir. A.T.F. Neerhoff, Technological
University of Eindhoven, Department of
Building, Architecture and Planning
Group Science of Materials
Postbox 513   5600 MB Eindhoven
Tel.: (040) 473610
Telex: 51163

20. IX <u>Sweden</u>   1. Dr. C. Molin, Swedish Cement and Concrete
            Research Institute
            Institute of Technology
            Fack S 10044
            Stockholm 70, Tel.: 08-233570

21. X <u>United Kingdom</u> 1. Mr. P. Lindsell, University of Surrey,
            Department of Civil Engineering
            Guildford Surrey G U 2 5 X H
            Tel.: (0483) 71281
            Telex: 859331

22.   2. Dr. C.D. Pomeroy, Cement and Concrete
            Association
            Research and Development Division
            Wexham Springs
            Slough SL 3 6 PL
            Tel.: (02816) 2727
            Telex: 848352

23.   3. Prof. Dr. D. Tabor, University of
            Cambridge, Department of Physics
            Cavendish Laboratory
            Madingley Road
            Cambridge CB 3 OHE
            Tel.: (0223) 66477
            Telex: 81292

24. XI <u>U.S.A.</u>   1. Prof.Dr. S.H. Carpenter, University of
            Illinois at Urbana-Champaign
            Department of Civil Engineering
            Urbana, Illinois 61801
            Tel.: (217) 333 - 3812

25.   2. Dr. G. Frohnsdorff, Leader Building
            Composites Group
            Structures and Materials Division
            Center for Building Technology, NEL,
            National Bureau of Standards
            Washington D.C. 20234, Telex: 898493

26.   3. Dr. S. Frondistou-Yannas, President
            Management and Technology Associates Inc.
            149 Baldpate Hill Road
            Newton, Massachusetts 02159
            Tel.: (617) 965 5259

27.

4. Mr. B. Mather, U.S. Army Engineer
Waterways Experiment Station
Chief Structures Laboratory
P.O. Box 631
Vicksburg,Mississippi 39180
Tel.: (601) 634 3264

28.

5. Dr.J.J. Mills, Martin Marietta Laboratories
1450 South Rolling Road
Baltimore
Maryland 21227
Tel.: (301) 247 0700
Telex: 710-236 9076

29.

6. Prof.Dr. Della M. Roy, The Pennsylvania
State University
Materials Research Laboratory
University Park
Pennsylvania 16802
Tel.: (814) 865
Telex: 510 670 3532

30.

7. Prof.Dr. S.P. Shah, University of
Illinois at Chicago Circle
College of Engineering
Department of Materials Engineering
P.O. Box 4348
Chicago, Illinois 60680
Tel.: 996 3428

31.

8. Prof.Dr. J.F. Young, University of
Illinois, Urbana-Champaign
Department of Civil Engineering, Urbana
Illinois 61801
Tel.: (217) 333 - 3812

417